World Scientific Series in Current Energy Issues Volume 2

Solar Energy

World Scientific Series in Current Energy Issues

Series Editor: Gerard M Crawley *(University of South Carolina & Marcus Enterprise LLC, USA)*

Published

Vol. 1 Fossil Fuels: Current Status and Future Directions
edited by Gerard M Crawley

Vol. 2 Solar Energy
edited by Gerard M Crawley

Forthcoming

Energy from the Nucleus: The Science and Engineering of Fission and Fusion
edited by Gerard M Crawley

World Scientific Series in Current Energy Issues — Volume 2

Solar Energy

Editor

Gerard M Crawley

Marcus Enterprise LLC, USA

&

Professor and Dean Emeritus

University of South Carolina, USA

World Scientific

NEW JERSEY • LONDON • SINGAPORE • BEIJING • SHANGHAI • HONG KONG • TAIPEI • CHENNAI • TOKYO

Published by

World Scientific Publishing Co. Pte. Ltd.
5 Toh Tuck Link, Singapore 596224
USA office: 27 Warren Street, Suite 401-402, Hackensack, NJ 07601
UK office: 57 Shelton Street, Covent Garden, London WC2H 9HE

Library of Congress Cataloging-in-Publication Data
Names: Crawley, Gerard M., editor.
Title: Solar energy / [compiled by] Gerard M Crawley (University of South Carolina, USA & Marcus Enterprise LLC, USA).
Other titles: Solar energy (World Scientific (Firm))
Description: [Hackensack] New Jersey : World Scientific, 2016. | Series: World Scientific series in current energy issues ; vol. 2 | Includes bibliographical references.
Identifiers: LCCN 2015031779 | ISBN 9789814689496 (alk. paper)
Subjects: LCSH: Solar energy.
Classification: LCC TJ810 .S488175 2016 | DDC 621.47--dc23
LC record available at http://lccn.loc.gov/2015031779

British Library Cataloguing-in-Publication Data
A catalogue record for this book is available from the British Library.

In-house Editor: Dipasri Sardar

Typeset by Stallion Press
Email: enquiries@stallionpress.com

Printed in Singapore by Mainland Press Pte Ltd.

Contents

Foreword to the World Scientific Series on Current Energy Issues

Gerard M. Crawley

Sometime between four hundred thousand and a million years ago, an early humanoid species developed the mastery of fire and changed the course of our planet. Still, as recently as a few hundred years ago, the energy sources available to the human race remained surprisingly limited. In fact, until the early 19th century, the main energy sources for humanity were biomass (from crops and trees), their domesticated animals and their own efforts.

Even after many millennia, the average per capita energy use in 1830 only reached about 20 Gigajoules (GJ) per year. By 2010, however, this number had increased dramatically to 80 GJ per year.[1] One reason for this notable shift in energy use is that the number of possible energy sources increased substantially during this period, starting with coal in about the 1850s and then successively adding oil and natural gas. By the middle of the 20th century, hydropower and nuclear fission were added to the mix. As we move into the 21st century, there has been a steady increase in other forms of energy such as wind and solar, although presently they represent a relatively small fraction of world energy use.

Despite the rise of a variety of energy sources, per capita energy use is not uniform around the world. There are enormous differences from country to country, pointing to a large disparity in wealth and opportunity (see Table 1). For example, in the United States the per capita energy use

Table 1: Primary Energy Use Per Capita in Million Btu (MMBtu).[2]

Country	2007 (MMBtu)	2011 (MMBtu)	Percentage change
Canada	416.1	393.7	−5.4
United States	336.9	312.8	−7.2
Brazil	52.7	60.2	14.2
France	175.7	165.9	−5.6
Germany	167.8	165.4	−1.4
Russia	204.0	213.4	4.6
Nigeria	6.1	5.0	−18.0
Egypt	36.4	41.6	14.3
China	57.1	77.5	35.7
India	17.0	19.7	15.9
World	72.2	74.9	3.7

per year in 2011 was 312.8 million Btu[a] (MMBtu) and in Germany, 165.4 MMBtu. In China, however, per capita energy use was only 77.5 MMBtu, despite its impressive economic and technological gains. India, weighs in even lower at 19.7 MMBTU per person.[2] The general trends over the last decade suggest that countries with developed economies generally show modest increases or even small decreases in energy use, but that developing economies, particularly China and India, are experiencing rapidly increasing energy consumption per capita.

These changes, both in the kind of resource used and the growth of energy use in countries with developing economies, will have enormous effects in the near future, both economically and politically, as greater numbers of people compete for limited energy resources at a viable price. A growing demand for energy will have an impact on the distribution of other limited resources such as food and fresh water as well. All this leads to the conclusion that energy will be a pressing issue for the future of humanity.

Another important consideration is that all energy sources have disadvantages as well as advantages, risks as well as opportunities, both in the production of the resource and in its distribution and ultimate use. Coal, the oldest of the "new" energy sources, is still used extensively to produce electricity, despite its potential environmental and safety concerns in mining both underground and open cut mining. Burning coal releases sulphur and nitrogen oxides which in turn can leads to acid rain and a cascade of detrimental consequences. Coal production requires careful regulation and

[a]Note 1 GJ = 0.947 MMBtu

oversight to allow it to be used safely and without damaging the environment. Even a resource like wind energy using large wind turbines has its critics because of the potential for bird kill and noise pollution. Some critics also find large wind turbines an unsightly addition to the landscape, particularly when the wind farms are erected in pristine environments. Energy from nuclear fission, originally believed to be "too cheap to meter"[3] has not had the growth predicted because of the problem with long term storage of the waste from nuclear reactors and because of the public perception regarding the danger of catastrophic accidents such as happened at Chernobyl in 1986 and at Fukushima in 2011.

Even more recently, the measured amount of carbon dioxide, a greenhouse gas, in the global atmosphere has steadily increased and is now greater than 400 parts per million (ppm).[4] This has raised concern in the scientific community and has led the majority of climate scientists to conclude[5] that this increase in CO_2 will produce an increase in global temperatures. We will see a rise in ocean temperature, acidity and sea level, all of which will have a profound impact on human life and ecosystems around the world. Relying primarily on fossil fuels far into the future may therefore prove precarious, since burning coal, oil, and natural gas will necessarily increase CO_2 levels. Certainly for the long term future, adopting a variety of alternative energy sources which do not produce CO_2 seems to be our best strategy.

The volumes in the World Scientific Series on Current Energy Issues explore different energy resources and issues related to the use of energy. The volumes are intended to be comprehensive, accurate, current, and include an international perspective. The authors of the various chapters are experts in their respective fields and provide reliable information that can be useful to scientists and engineers, but also to policy makers and the general public interested in learning about the essential concepts related to energy. The volumes will deal with the technical aspects of energy questions but will also include relevant discussion about economic and policy matters. The goal of the series is not polemical but rather is intended to provide information that will allow the reader to reach conclusions based on sound, scientific data.

The role of energy in our future is critical and will become increasingly urgent as world population increases and the global demand for energy turns ever upwards. Questions such as which energy sources to develop, how to store energy, and how to manage the environmental impact of energy use will take center stage in our future. The distribution and cost of energy will have powerful political and economic consequences and must also be addressed. How the world deals with these questions will make a crucial difference to the future of the earth and its inhabitants. Careful consideration of our energy use today will have lasting effects for tomorrow. We intend that the World Scientific Series on Current Energy Issues will make a valuable contribution to this discussion.

References

1. Our Finite World (March 2012). World energy consumption since 1820 in charts. Available at: http://ourfiniteworld.com/2012/03/12/world-energy-consumption-since-1820-in-charts/. Accessed February 2015.
2. US Energy Information Administration. Independent statistics and analysis. Available at: http://www.eia.gov/cfapps/ipdbproject/iedindex3.cfm?tid=44&pid=45&aid=2&cid=regions&syid=2005&eyid=2011&unit=MBTUPP. Accessed March 2015.
3. The quote is from a speech by Lewis Strauss, then Chairman of the United States Atomic Energy Commission, in 1954. There is some debate as to whether Strauss actually meant energy from nuclear fission or not.
4. NOAA Earth System Research Laboratory. Trends in atmospheric carbon dioxide. Available at: http://www.esrl.noaa.gov/gmd/ccgg/trends/. Accessed March 2015.
5. IPCC, Intergovernmental Panel on Climate Change, Fifth Assessment report (2014). Available at: http://www.ipcc.ch/. Accessed March 2015.

Chapter 1

Introduction to Solar Energy

R. Corkish,*,† W. Lipiński‡ and R. J. Patterson*
* *University of New South Wales, Sydney 2052, Australia*
† *Australian Centre for Advanced Photovoltaics, Sydney 2052, Australia*
‡ *The Australian National University, Canberra 2601, Australia*

This chapter introduces the sun as our planet's principle and sustainable energy source, noting its main relevant characteristics, and the history as well as pros and cons of its energetic use. It gives a high-level overview of the various solar energy capture and conversion technologies discussed in the book and summarizes the economics and policy aspects.

1 What is Solar Energy?

The sun drives 99.98% of the world's energy supply,[1] including thermal, photovoltaic, photochemical, photobiological and hybrid solar, hydro, wind, wave, and biomass energy conversion. It originally grew the biomass that we now access as fossil fuels. Other sources include tidal, geothermal and nuclear.[1] The sun's energy comes from fusion reactions in its core. These reactions have been "burning" for 4.5 billion years and are expected to continue for another 6.5 billion years. The total power radiated out into space by the sun is about 3.86×10^{26} W. Since the sun is approximately 1.5×10^{11} m from the earth, and because the earth is about 6.3×10^{6} m in radius, it intercepts only 0.000000045% of this power.[2] This still amounts to a massive 1.75×10^{17} W.

Most of this radiation is in the visible and infrared part of the electromagnetic spectrum, with less than 1% emitted in the radio, UV and X-ray spectral bands.

The sun's electromagnetic radiation approximates that of a black body with a temperature around 5778 K, with its peak in the yellow range of the visible spectrum. This is sometimes "rounded up" to 6000 K for simplicity.[3]

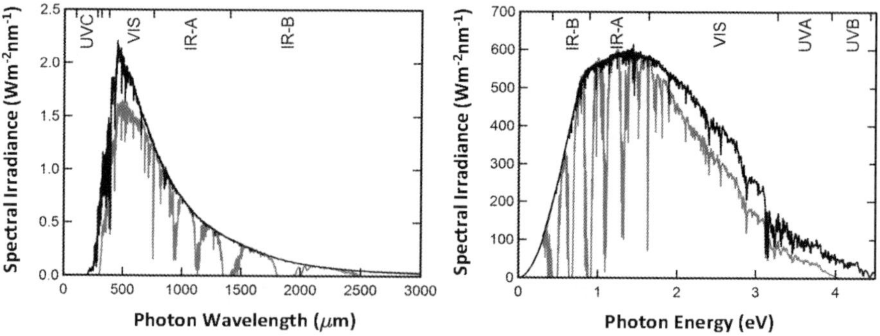

Fig. 1. The AM0 spectrum of solar radiation, as would be observed at the top of the earth's atmosphere. The spectral irradiance is shown as a function of photon wavelength (left) and photon energy (right). The spectral regions are indicated. Image credit: Prof. Pietro Altermatt.[5]

Figure 1 shows the spectrum of radiation from ultraviolet to infrared, as seen outside the earth's surface. This standard "AM0" spectrum[4] is that used to model and predict and qualify solar cells for use in space. The "AM" in the name refers to the "air mass", the thickness of terrestrial atmosphere through which the radiation has passed before it is observed, in this case zero.

The solar radiation that reaches the earth is reduced in intensity and the spectrum is changed by absorption and scattering as it passes through the atmosphere, and by reflection from the surface.[6] Scattering means that radiation reaches a receiver not only directly from the visible solar disk but also from the rest of the sky. Standard spectra and models have been derived for use in simulations and estimations.[7] For example, tables presented in ASTM Standard G173[8] are commonly used to represent terrestrial solar radiation in two forms (see Fig. 2):

1. AM1.5D: Direct Normal Incidence radiation is that received on earth directly from the solar disk, as would be received by a concentrating collector. The total integrated irradiances for the standard direct spectrum is $900.1\,\mathrm{Wm}^{-2}$;
2. AM1.5G: Hemispherical on 37° Tilted Surface radiation adds the direct radiation and the scattered diffuse radiation to yield the hemispherical or global radiation, impinging on a sun-facing 37°-tilted surface, useful for modeling flat plate (non-concentrating) collectors. The total integrated irradiances for the standard hemispherical tilted spectrum is $1000.4\,\mathrm{Wm}^{-2}$.

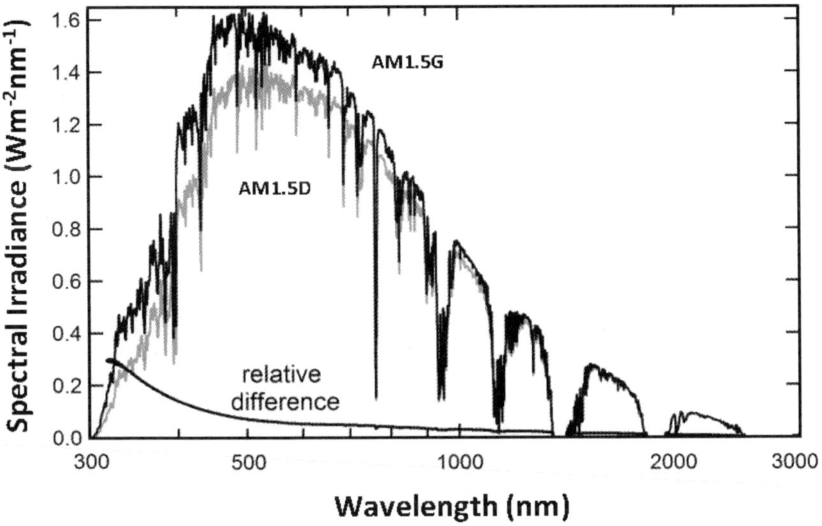

Fig. 2. AM1.5 global (AM1.5G) and direct normal (AM1.5D) standard terrestrial solar and their relative difference as a function of wavelength. Image credit: Prof. Pietro Altermatt.[11]

Note that there are many narrow and broad notches[9] in these spectra, resulting from absorption by particular atmospheric gas and vapor molecules. The atmosphere is opaque to ultraviolet radiation with wavelengths less than approximately 300 nm due to absorption by molecules such as H_2, O_2, and N_2. In the near-UV spectral range the main absorber is ozone, which also impacts on the visible range (380–780 nm), as do NO_2 and water vapor. Water vapor also introduces several broad absorption bands in the infrared range and carbon dioxide and oxygen also have an effect.[10]

The long-term average of the total solar irradiance at the average distance of the earth's orbit, one Astronomical Unit of $1.4959787066 \times 10^{11}$ m, is called the solar constant, S. Its reference value[12] is $1366.1\,\mathrm{Wm}^{-2}$.

Actually, the "solar constant" is not constant but varies by ±3% due to the earth's elliptical orbit. There is an additional variation of about ±0.1% in the solar constant which is due to a variation in the total luminosity of the sun itself over the 11-year solar cycle. Researchers have tried to model this variation over the last 400 years by correlation with recorded sunspot numbers, suggesting that the sun may have varied in its power output by up to 1%.[2]

When diurnal and seasonal variations are taken into account, approximately 342 W is available for every square meter of earth's surface.[13] This equates to an annual energy input[14] of 5.46×10^{24} J. About 29% is reflected back to space by clouds, atmospheric particles or ground surfaces. About 23% of intercepted solar energy is absorbed in the atmosphere and 48% passes through to be absorbed by the surface.[15]

The geographical distribution of solar radiation reaching the earth's surface is studied and monitored extensively and satellite-derived estimates are available for the whole globe.[16] Detailed mapping resources for USA and Australia, for example, are provided by the National Renewable Energy Laboratory[17] and the Bureau of Meteorology.[18] Free online maps for the world and many of its regions are provided by SolarGIS.[19]

As might be expected, there tends to be less solar radiation towards the poles and cloudiness around the equator reduces the radiation reaching the ground there, leading to the regions of highest insolation being found in the high tropical and low temperate latitudes.

Measurements of solar radiation are made at many sites globally[20] with calibrated pyrheliometers,[21] the most sophisticated of which use shadow bands to measure separately the diffuse and direct components. Historical data and long-term averages are available through various databases, including the World Radiation Data Centre[22] and the World Radiation Monitoring Center (WRMC), which maintain the central archive of the Baseline Surface Radiation Network (BSRN).[23] The directly measured data is of limited use to engineers to design solar energy collection systems and it undergoes considerable processing to generate useful formats such as typical meteorological year (TMY) or monthly average daily global radiation on a horizontal surface.[24]

Additionally, solar radiation is estimated from satellite-based instruments.[25,26] For example, the NASA Atmospheric Science Data Center[16] provides long-term estimates of surface solar energy flux for 1° longitude by 1° latitude grid covering the entire globe. The data is readily accessible via the freely available *RETScreen* Clean Energy Management Software system.[27]

2 Advantages and Disadvantages Associated with Solar Energy Use

Solar energy has several major advantages when compared with other sources. The resource is distributed, though unequally, to every location on the globe. The resource is abundant, to the extent that many countries

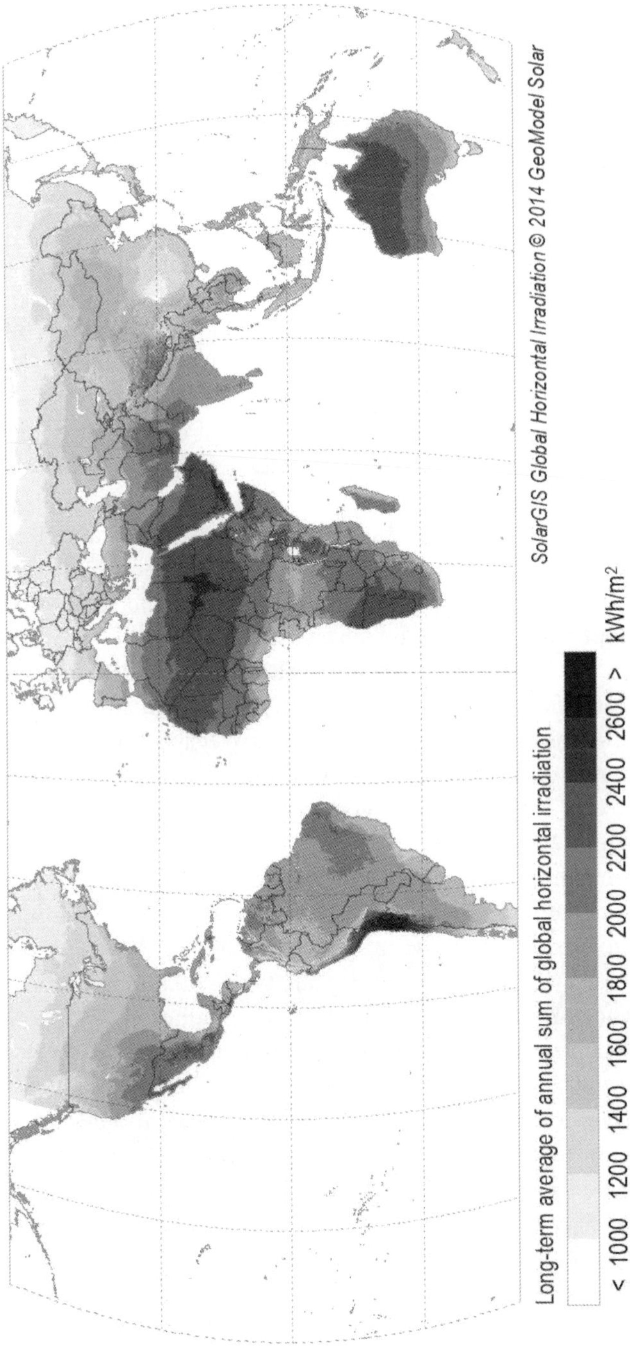

Fig. 3. Global solar radiation over the land surfaces of the world. Reprinted with permission from SolarGIS Global Horizontal Irradiation © 2014 GeoModel Solar.

have far more than they need to supply their energy needs from solar alone. It is effectively renewable on a human timescale, since the sun is expected to maintain similar production of its essential radiation at about the current rate for billions of years before eventually cooling to become a red giant.[28]

Arguably, untraded solar energy already dominates the global energy supply as it grows our forests and crops that provide basic energy services to a large fraction of the world's population, warms our passive solar buildings, evaporates seawater to produce our industrial salt supply and even dries our crops, clothes and fuels outdoors.

Collection and conversion to various useful energy forms is generally quiet and clean, with little or no local pollution from operation, including greenhouse gas pollution. Solar energy generally offers very low risk to public or operator safety and therefore, the location of solar energy plants can be flexible. Its use at large industrial scales is believed to be environmentally benign.

Where it is accessible through relatively clear skies, sunlight may be concentrated up to a theoretical maximum of 46,300 times.[29] This corresponds to catching all the light leaving a unit area of the solar surface and generating its optical image on the surface of a terrestrial absorber, which would require a vast elliptical reflector with the sun and the earth at the foci. Note that while this book is concerned with solar *energy*, ultimately it is *exergy*, a measure of the useful work it is possible to extract from an energy flow, that is of greater interest.[29] Higher concentration of sunlight permits lower exergy destruction in conversion stages so it allows more efficient conversion of solar energy to useful forms such as mechanical work, electricity, chemical fuels or high-temperature process heat. Conversion of non- or low-concentrated solar radiation has less strict requirements for dynamic alignment of the collectors towards the sun, as compared with concentrating solar applications, and such types of photovoltaic or solar thermal receivers are appropriate for urban installation. Many forms of solar energy conversion are small in scale and modular, so there are low barriers to incremental introduction.

On the other hand, solar energy also has disadvantages. It is much more diffuse than, for example, fossil or nuclear fuels, so large surface areas are required to collect large quantities of energy. At any given location on earth, the intensity and spectrum of sunlight varies in both predictable and less predictable ways, introducing intermittency of supply. The tilt of the earth's axis means seasonal changes in solar radiation availability, with changes in the extent of daylight and the apparent elevation of the sun in the sky. The elevation varies the thickness of "air mass" through which the sunlight passes to reach the earth's surface, affecting its intensity and spectrum. Less predictably, clouds and atmospheric aerosols, such as dust and smoke, reduce intensity and modify the spectrum. Technical responses to intermittency[30] include using solar energy as a "fuel saver" in fossil fueled networks, hybridization of systems with multiple solar or other renewable or non-renewable sources, coupled energy storage, the topic of another volume in this series, and the proposed collection of solar energy from orbiting power stations that then transmit to earth.[31] Some forms of solar energy collection and conversion, such as biomass and the domestic solar water heater, are relatively easily coupled to storage but it can be expensive and difficult for other forms.

Another disadvantage is geographic specificity. Economic introduction of solar energy to existing energy distribution systems needs to conform to existing infrastructure that was built around competing energy sources. For example, the optimal location of large extra-urban solar power stations requires access to existing electricity grid infrastructure with capacity to accept the solar supply, collocated with available, un-conflicted and low-cost land and high solar radiation.

Finally, there is a large array of solar energy conversion technologies, some simple and some technically complex, that, while together competing with established fossil and nuclear energy, also compete with each other. Researchers in all areas offer promise of better, cheaper technologies. Some investors are bewildered by the range choice and refrain altogether, awaiting clarity to avoid "betting on the wrong horse" and for the promised future technologies to become available. However, great attention is now being paid to this and good quality advice is available.[32,33] Some forms of solar energy conversion have traditionally been much more expensive than fossil fuel use, so long as economic externalities are excluded, but, at least for non-concentrating photovoltaics, this is now changing.[34] Funds are increasing flowing out of the incumbent technologies and into solar.[35] Large-scale adoption of solar energy will necessarily require large amounts of de-risked capital, requiring the solar energy industries to pay great attention to systems quality and performance prediction and monitoring.

3 History of Solar Energy Use

Solar energy has long supported humanity, with at least two forms, passive solar energy and biomass fuel use. Thus solar energy has been our partner throughout the progress of mankind. The growth of agriculture in the sunny "cradle of civilization" played a critical role in the development of civilization.[36] People have used the sun for drying crops, bricks, etc. since prehistoric times. The first known crop drying installation has been found in France and dates from around 8000 BC.[37] There is evidence from around the world of dryer development in many civilizations and this relatively simple solar technology continues to change lives and economies for the better, even today, in remote locations all over the planet.[38]

The US Department of Energy timeline provides a series of important historical milestones for solar energy.[39] Butti and Perlin[40] describe that history, beginning with ancient classical Greek and Roman over-consumption of biomass and including the passive solar dwelling and city design. In the case of the Roman Empire, the architect Vitruvius recommended different passive solar building designs for different latitudes, outlining principles that are still applied today. Solar access rights for buildings were included in the Justinian Code of law in the sixth century AD. Both ancient Greek and Chinese cultures developed concentrating solar reflectors to generate high temperature ignition for religious, civil and military purposes. "Burning mirrors" have since then been designed and used by many cultures through the centuries. Glazed heat traps in buildings were developed by the Romans and the idea was revived much later in Europe as the conservatory or greenhouse for horticulture of plants outside their natural ranges or out of season.

The commercial availability of the Climax Solar Water Heater at the end of the 19th century in the USA initiated the mass availability of affordable solar domestic heating of water that has continued to drive the development of flat plate and evacuated tube heaters ever since. The harnessing of the sun for mechanical power began at least as early as the 1st century AD with solar water syphons built in Alexandria. The invention of the first solar steam engine has been attributed to Augustin Mouchot in France in 1866. He went on to develop solar cooking ovens and solar thermoelectric generators. The early 20th century saw an explosion of applications for solar engines for water pumping and other remote energy applications in the American west and elsewhere.[40]

Three main forms of concentrator have been developed to generate either high temperatures in solar thermal collectors or high conversion efficiencies in photovoltaic collectors: parabolic troughs that focus light onto a line, parabolic dishes that focus light onto a point and arrays of heliostats focusing onto a central receiver mounted on a tower. Concentrating solar power has been developed significantly since the oil shocks of the 1970s, principally in the US, Spain, Australia, and Israel. "Solar One", a 10 MW central-receiver demonstration project which opened in the US in 1982, was the first of several large solar concentrators constructed in the modern phase of growth to establish feasibility. It generated steam to drive a turbine for electricity generation. Solar One was expanded and upgraded to Solar Two in 1995, including molten salt thermal energy storage.

There are several good histories documenting the beginnings of photovoltaics, among them that by Crossley *et al.*[41] The French scientist Edmond Becquerel discovered the photovoltaic effect in an experimental photoelectrochemical setup in 1839.[42] At that time it was not possible to distinguish between chemical and photoelectric effects and the explanation of these experiments was originally in terms of chemistry. It was not until 1914 that Goldmann and Brodsky[43] made a photoelectric interpretation. In the 1870s, William Gryllis Adams and R.E. Day investigated "*whether it would be possible to start a current in the selenium merely by the action of light*". The result was positive, "*clearly proving that by the action of light alone we could start and maintain an electrical current in the selenium*". They did not, however, understand the processes at work in their devices, explaining the voltage as being due to extra light-induced crystallization in the material. Charles Fritts[44] foresaw great potential for solar power from selenium photovoltaics. There were at least four American manufacturers of selenium photovoltaic cells by 1949. Copper–cuprous oxide cells were also under investigation since 1917[45] and there was intense rivalry between groups in Germany and USA through the 1920s, when copper-based cells were commercialized. The photovoltaic effect was found in germanium in 1944 in USA but all of these materials were eclipsed by the success of silicon as a photovoltaic material. Russel Ohl of Bell Laboratories filed patents in 1941 that were granted in 1946 and 1948 for the p–n junction photovoltaic effect in silicon and markets gradually grew for terrestrial and space applications. Silicon underpinned the development of serious and significant application of photovoltaics but cells based on alternative materials, cadmium–telluride, copper–indium–gallium–diselenide and III-V semiconductors (i.e. compounds of elements from Groups III and V of the Periodic Table), have also been developed and commercialized in the late 20th century.

4 Modern Applications of Solar Energy

Methods to collect solar energy and convert it to useful forms range from the simple and traditional to modern and highly sophisticated. Outputs include low grade heat, high temperature industrial process heat, hydrogen, synthesis gas, synthetic hydrocarbons and other chemical energy carriers such as ammonia and metals, and intermittent or dispatchable electricity. These technologies are all at different developmental stages and associated cost of energy. We introduce a range of them in this section before they are treated in detail in the following chapters.

4.1 *Photovoltaics for Large-Scale Electricity Production*

Photovoltaics have undergone continual development in the decades since 1950s until silicon and cadmium telluride photovoltaics now present a credible challenge to fossil fuels and are one of the most promising methods to continue to provide electrical services to society in a carbon-challenged world. Government support programs for photovoltaics integration into electricity grids, most importantly and effectively in Japan and then Germany in the late 20th century, initiated the recent boom in demand and consequent mass production. Research associated with those programs solved the major safety concern about potential creation of live "islands" in otherwise closed grids and power quality and utility scale photovoltaics installations sprang up in many countries to compete with the incumbent wholesale electricity generators (see Fig. 4). The output of these systems is not associated with particular consumers. These systems are normally 24 ground-mounted and independent of nearby activities. The International Energy Agency's Photovoltaic Power Systems Programme (IEA PVPS)[46] tracks the different categories of installations in its 24 member countries. Such systems have been particularly popular in China and USA, where their installed new capacity exceeds that for decentralized grid connected systems. The main market for utility scale systems has shifted from Europe to Asia in recent years.

Perhaps the most significant challenge to future growth of this market segment is the growing resistance of electricity utilities and influential incumbent generators. Costs of large systems are generally lower than for small distributed systems but they also compete for a lower price product, bulk, wholesale electricity. Concentrating photovoltaics technologies have continued to struggle for a significant market share for a range of well-known reasons.[47] See Chapter 3 for an in-depth discussion of large-scale photovoltaics.

Fig. 4. Part of a 9 MW utility scale photovoltaics installation at Stone Mills, Ontario Canada.
Photo credit: R. Corkish.

4.2 *Photovoltaics for Small Scale on Homes and Commercial Buildings*

Resistance from incumbent grid managers and electricity retailers is even more evident for smaller scale distributed grid-connected systems. These systems, commonly on building roofs, can connect on the customer side of consumption meters and avoid the purchase of electricity from utilities and threaten their traditional business models (Fig. 5). Costs of these systems are generally higher than for utility scale systems due to their smaller scale but they compete at the retail price level. In many markets, such systems now need to be designed for self-consumption of their output, avoiding export to the grid, to make them economical. This is generally easier and more valuable for commercial than for domestic loads since they tend to be predominantly daytime demand. These are the types of systems that were facilitated by the Japanese and German support schemes and their followers in other countries and the IEA PVPS records show that the grid-connected market worldwide has evolved from dominance of distributed systems to roughly half utility scale now. The situation is patchy across the world though, with conditions in some countries favoring small systems.

Fig. 5. House with grid-connected photovoltaics and domestic solar water heater. Photo credit: R. Corkish.

New market segments are opening rapidly with encouragement through innovative schemes such a leasing of rooftop systems, especially in the USA. The household rooftop market is perhaps larger than would be justified on strict financial grounds because some householders buy or lease solar power systems driven by other motivations, particularly its benign environmental image. Note that such decisions are made easier by the rapidly declining prices of these systems in recent years.

4.3 *Photovoltaics for Small-Scale Off-grid Applications*

Prior to the rapid increase in the number of grid-connected systems in the late 20th century, the sustaining markets for the photovoltaics manufacturing industry were in those locations where all energy services are expensive. These locations are usually remote from grid electricity and other services, such as islands and rural areas. The main applications for the smaller systems, up to 1 kW, were and are water pumps, lighting and remote homes while those for larger installations were and are for commercial or industrial applications such as telecommunications repeaters, pastoral or mining power systems or cathodic protection of pipelines. In off-grid systems, photovoltaic modules are commonly connected through a charge controller to a battery and backup generation is also often incorporated. Water pumping systems commonly operate without batteries or backup. These markets have not gone away but now attract less attention because of the great increase in the mass grid-connected demand, which now dwarfs the earlier markets.[46]

4.4 *Concentrating Solar Thermal Electricity*

The concentration of the incident solar flux, for either photovoltaics or concentrating solar thermal (CST) technologies, is limited to regions with consistently high, direct, solar irradiation. This implies a separation from many of the areas where people have chosen to congregate in the world. It also demands mechanical tracking of collectors to keep them pointed at the sun, with consequent reliability concerns. CST technologies have had a much more positive history than has concentrating photovoltaics and several installations have been built in Spain and the USA.[48] Other countries including China, India, and Australia have also built CST plants mainly on a test basis and the world has approximately 3800 MW capacity in operation with more than double that in the pipeline. The economics are currently not as attractive as those of flat plate photovoltaics but CST has the distinct advantage of easy coupling to relatively inexpensive storage, as sensible or latent heat or in thermochemical conversion (see next section). Integrated storage offers "dispatchable" energy, available on demand.

One of the main uses for CST is the generation of high temperatures to drive turbines for electricity generation, in both stand-alone solar power plants and as a fuel-saving adjunct to existing fossil-fueled power stations. Others are to drive industrial processes directly, facilitate endothermic chemical reactions (see Chapter 9), and provide cooling services. Higher concentrations are required for higher temperatures and conversion efficiencies.

Linear parabolic trough reflectors or Fresnel lenses, dishes and power towers are all used.[49] Linear focusing collectors with single-axis solar tracking, in north-south rows, generate medium temperatures, around 400°C and organic Rankine cycle turbines convert the energy to a rotating mechanical form to spin a generator. Paraboloidal dishes and centralized power towers need dual axis tracking and commonly have higher concentration ratios and, hence, higher temperatures. Some dish collectors incorporate a Stirling cycle engine at the focus. In a central tower solar thermal power plant, an array of heliostats reflect light onto a receiver at the top of a tower. The working fluid transports the heat to a Rankine or Brayton cycle engine on the ground.

Solar thermal generation of electricity currently remains expensive relative to photovoltaics and to some fossil fuel technologies so it would seem that installations in the near term are likely to be restricted to clear-sky situations in which a premium for storage is justified. However, it remains to be seen whether costs will come down significantly with increasing experience and larger-scale production in China and India.

4.5 *Solar Thermochemical Processes*

Options to reduce the impact of intermittency on concentrating solar thermal plants are to store sensible or latent heat or to use it to drive endothermic thermochemical processes.[50] The product chemicals can be transported over long distances or stored and delivered to reformer reactors in which heat is released in exothermic reactions. Alternatively, solar heat may be applied to change the chemical composition of a fossil fuel to increase its calorific content or endothermic industrial chemical conversion processes may be directly utilized. These processes are likely to be applied only at very large scale, and are currently at a pre-commercial stage of development.[51] This storage option has the advantage over sensible and latent heat storage since that storage is at ambient temperature. In the case of solar-produced fuels, significant storage capacity may be afforded by the volume of existing reticulation systems at low cost.[51] A special class of processes are those for production of commodities such as lime, cement, ammonia, metals, which are not obtained for energy storage purposes. However, substitution of conventional fossil-based process heat with solar heat in the highly-endothermic reactions allows for drastic reduction in fossil energy use, and the associated greenhouse gas emissions. There are a large number of outstanding technical issues to be solved for this complex technology, including reversibility, material discovery, stability and durability, reliability and efficiency of solar-driven reactors, and aspects associated with high-temperatures and transients. These technologies are discussed in Chapter 9.

4.6 *Solar Water Heating*

Solar water heating, both domestic and commercial/industrial, is so common and mainstream in many regions of today's world that it is sometimes overlooked as a significant renewable energy technology.[52] Like non-concentrating photovoltaics, it is a technology that is easily incorporated into urban infrastructure and displaces imported energy services. Two main collector technologies are already in mass production: flat plate and evacuated tube. In the former, the solar radiation heats a dark metal plate and heat is transferred to either a heat transfer fluid or directly to the water itself. Except for very low temperature applications such as swimming pool heating, an air gap and glazing above the plate allows higher temperatures to be generated by the greenhouse effect (visible light passes through the glazing to the plate while upwardly directed infrared radiation from the plate is blocked by the glass). Performance may be further improved by the use of spectrally "selective absorber" coatings on the plate that further limit radiative losses. Circulation of the fluid in contact with the plate can be driven by the thermosyphon effect or by a pump if the storage tank cannot be mounted directly above the collector, as it is in the example in Fig. 5.

The other main technology uses a double walled glass evacuated tube in which the gap between the walls is evacuated to block heat loss by conduction and convection.[53] A spectrally selective coating is applied to the inner wall or to a long, narrow absorber plate which allows a circulating working fluid to be heated to higher temperatures than with flat plates. Research into improved collectors includes the potential use of nanofluids, fluids with suspensions of nanoparticles, for direct absorption of sunlight by the working fluid.

A third technology in widespread use is arguably defined as a solar water heating technology. Domestic scale heat pump water heaters entered the mass market with ground mounted water storage tanks and roof-mounted and sun-exposed panels to evaporate the working refrigerant. However, customer resistance to the high installation cost, including the need for licensed refrigeration tradesmen to install and certify the necessary refrigerant pipework, led to the packaging of the evaporator panel wrapped around the tank, trading reduced performance for greater market acceptance. See Chapter 8 for further information on solar water heating.

4.7 *Passive Heating of Buildings and Solar Architecture*

As has already been outlined in Sec. 3 of this Introduction, the deliberate and designed-in use of solar radiation has an extremely long history. The idea had a strong resurgence of interest and application following the two oil price shocks in the 1970s and passive solar architecture was codified and specialized for climate zones to a much greater extent than had previously been done.[54] The basic principles of Glass, thermal Mass and Insulation (GMI) are simple and clear and essentially the same as in ancient times but the necessary compromises in their application to different climate regions and for seasonal and diurnal variations requires careful and subtle engineering. Passive solar building design became less popular during the 1990s with lower fossil fuel costs but the old ideas together with new developments are experiencing resurgence. Various rating schemes[55] exist around the world and there are numerous regional schemes for mandating or, at least encouraging, more efficient building design and operation. This is discussed further in Chapter 8.

4.8 *Evaporative Cooling*

Like domestic solar water heating, evaporative cooling is a solar technology that has achieved mass market acceptance and is sometimes overlooked as a contribution of solar energy to human comfort and convenience. Evaporative cooling is a ubiquitous air conditioning choice in many of the less humid climate zones of the world. It works by using solar-heated ambient warmth to extract heat from a fan-forced flow of air by the latent heat of evaporation of water. Even more simply, passive forms rely on prevailing wind to blow air across ponds. It is the technology that supports the cooling towers prevalent in commercial/industrial air conditioning and industrial cooling and in fossil fueled and nuclear generation of electricity so it may be seen that evaporative cooling makes a huge contribution to human energy services but is generally excluded from renewable energy statistics.

The state of the art of evaporative cooling is described in Chapter 8 of this volume.

4.9 *Biomass and Biofuels*

Modern application of biomass and biofuels encompasses a wide range of technologies from efficient cook stoves, mass-scale pelletization of wood and waste materials, the capture and combustion for electricity generation of methane from decaying municipal waste (landfill gas), to the highly technical production of liquid fuels from agricultural and forestry crops or wastes, organic municipal waste, and cultivated algae for the subsequent extraction of oil.

The use of liquid biomass fuels has entered the mass market in some regions of the world, most notably in Brazil with sugar-derived ethanol powering a large part of the car transport there. Competition for arable land and water resources between energy production and food production is highly controversial. Also controversial is the ongoing debate about whether the optimal energy carrier for transportation is a liquid or gaseous biofuel or solar generated electricity.[56] Of particular interest here is the prospects for future air travel that is independent or at least less dependent on fossil fuels. The options in a carbon-constrained world, given the vast fertile land areas that would be required to produce sufficient biofuels, seem to be limited. In a very different arena, biofuels from wastes or from low-value crops such as copra[57] may help underpin energy self sufficiency in some remote but fertile parts of the world, such as some Pacific Island Nations but, here too, competition with food production will be relevant.

4.10 *Artificial Photosynthesis*

The solar-driven photocatalytic production of hydrogen or hydrocarbon fuels by reduction of H_2O and CO_2 into hydrocarbon fuels, is an extremely attractive path for energy collection. These artificial processes simulate natural photosynthesis in green plants and are therefore termed artificial photosynthesis. This pairs CO_2 fixation with water oxidation in a carbon-neutral cycle, thereby diverting or removing problematic CO_2 from the atmosphere and generating a storable and transportable fuel.[58] However, no commercially viable system based on earth-abundant elements has yet emerged, despite an intense research effort over decades and experimentation with a very large range of materials and structures.[58,59] Conversion efficiencies are still low. Advances in all the areas of catalysis, semiconductor physics and engineering, surface chemistry, photocatalysis, and nanotechnology are needed. Nevertheless, there remain promising avenues and the desirability of "green" fuel production, absolutely necessary for some of the services, including air transport, on which modern life depends and the additional benefit of CO_2 sequestration mean that research in this field will continue to be critically important. Chapter 6 in this volume explores this field in depth, outlining the challenges and the most likely paths forward.

5 Economics of Solar Energy Use

We write at a time when some solar energy technologies, most spectacularly photovoltaics, are fulfilling their long-term promise to reduce costs and prices to the point of credible direct competition with fossil and nuclear

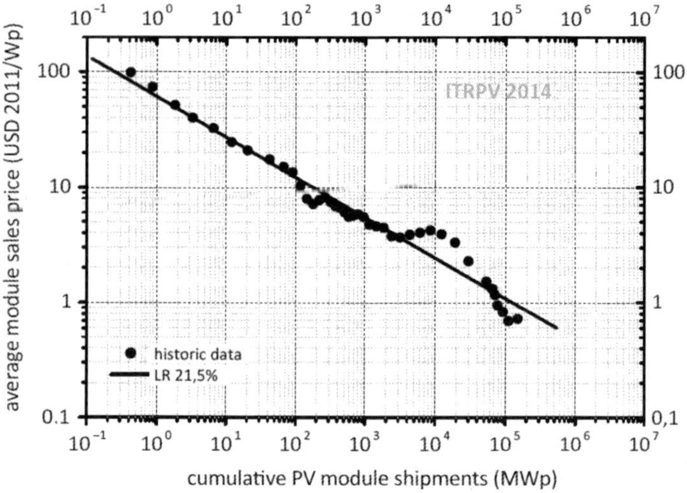

Fig. 6. Historic price reductions and the so-called "PV experience curve" for PV cells.[61]

fuelled electricity generation. For example, prices in the USA of residential and commercial PV systems fell, on average, 6–7% per annum from 1998–2013 and then rapidly, by 12–14% per year from 2012–2013. An additional 3–12% decrease is expected in 2014.[48,60] The ranges account for different market dynamics and cost structures for systems of different sizes. Much of the historical cost decrease has resulted from cost reductions in cell and module production, following a beneficial "experience curve" (Fig. 6),[61] which may flatten in the near term as the industry recovers from overcapacity resulting from the global financial crisis. However, drops in balance of system components and installation methods and soft costs[62] (including the non-hardware costs such as sales and marketing, financing, contracting, permitting, grid connection inspection, installation and O&M) are likely to allow the system price to continue falling. USA prices for installed photovoltaic systems have been reported to be more than twice the price in Germany, for instance,[60] suggesting the potential for significant reductions in the USA.

To compare the economics of different electricity generation technologies it is necessary to estimate the cost of the energy produced. Levelized cost of energy (LCOE), is defined as the long-run marginal cost of electricity generation, based on capital, fixed and variable, operations and maintenance, and fuel costs. Many studies have been carried out to compare and rank energy generation technologies, particularly in the form of electricity, in various regions.[32,63,64] Many of these studies tend to ignore the cost of delivering the energy to consumers so the advantage for rooftop photovoltaics (and solar hot water) of siting directly at the load is commonly neglected. LCOE values of electricity generated from solar radiation also vary with intensity of sunlight, so they vary regionally. The best regarded photovoltaics industry roadmap[61] projects a 41% decrease in LCOE between 2013 and 2024, from a base, for large systems, of 0.056 USD/kWh (sunny regions) and 0.111 USD/kWh for places with half as much solar radiation. An Australian study[64] estimated (updated estimates in 2013 for 2012) a large range of LCOE for fixed-tilt photovoltaics from about 0.16 to 0.27 USD/kWh and 0.03 to 0.13 USD/kWh under different sets of assumptions. A report from a consultancy, Lazard, estimated for the US ranges of 0.18–0.265 USD/kWh for rooftop residential installations, 0.126–0.177 USD/KWh for commercial and industrial systems and 0.072–0.086 USD/KWh for utility scale.[32]

Costs for concentrated solar thermal generated electricity are less well defined since the market is still small and many projects are "proof of concept" or experimental. This technology shares with concentrated solar photovoltaics many barriers to reaching commercial markets, clearly outlined more than a decade ago including restriction to clear-sky regions that, commonly are remote from human habitation (although the energy transmission costs are at least partly balanced by low land cost), the lack of markets for small-scale initial installations to build cash flow, inconsistent government policies, negative perceptions, competition from flat-plate photovoltaics and direct competition from fossil fuel power plants.[47] Nevertheless, estimates for LCOE from solar thermal electricity are still reasonably attractive and projected to fall. The Australian study mentioned above found that LCOE for solar thermal electricity without storage was in the range 0.21–0.35 USD/kWh and slightly higher for systems with storage in 2012 (updated in 2013)[63] and projected a very wide range on 0.05–0.26 USD/kWh in 2050. However, the same study found that coupling solar thermal to existing fossil-fueled power plants as fuel savers offers much more competitive prices.[65] Another estimate, for the US, estimated LCOE for concentrated thermal electricity with storage at 0.118–0.13 USD/kWh currently.

Low temperature water heating has contributed massively to CO_2 savings, particularly in China and Europe.[66] The economics are complex, depending on solar radiation, water usage patterns, local energy prices, government support, type of solar water heater, etc. A comprehensive study of domestic solar water heater performance in the contiguous states of the USA[67] found a wide range of break-even prices, varying by a factor of five. The large variation is mainly due to varying incentives and electricity prices. The study predicts energy savings of 1600–2600 kWh per year for typical households, with annual electric bill savings of about 100 USD to more than 300 USD.

Solar energy technologies producing electricity must compete either with systems supplying the grid, or, for systems at the location of the consumer, with electricity produced by fossil fuels or nuclear fission. The reports mentioned in Sec. 5.1[32,63] are examples of studies that compare a wide range of energy generation technologies, assuming competition at the wholesale level, for the conditions pertaining to a particular country. Others have used a global approach.[48] Since they tend to use different sets of assumptions, care must be taken in comparing their results but internal consistency appears to have been sought in each case.

The World Energy Council[48] concluded that LCOE for electricity from silicon-based photovoltaics varies widely across the world, with lower costs in China and India relative to Western Europe, the US and Japan. This study used LCOE data for actual projects collected quarterly from across the globe. It noted that the cost of finance varies by technology and location and this has a significant effect on LCOE, as do resource availability and siting. These wide, extremely wide for some technologies, ranges (Fig. 7) make it challenging to draw conclusions for the whole world since there is a great deal of overlap in the ranges of LCOE for competing technologies. However, Fig. 7 also includes weighted averages, denoted "central". We can see that, on average and for the set of assumptions adopted, photovoltaics for bulk power generation are not yet cheaper than coal or combined cycle gas turbine, large hydro or nuclear electricity generation but it can be in some cases. However, prices for photovoltaics are dropping rapidly compared to the current incumbent generators[34,60] and photovoltaics competes at retail, not wholesale, rates in rooftop systems. The balance is likely to be further tipped against the fossil fuel technologies by increasing divestment for financial and ethical motivations.[35,68] Other solar and renewable energy technologies, currently cheaper, will ally with photovoltaics to minimize intermittency and reduce carbon emissions: Examples are small hydro, anaerobic biomass digestion and biomass combustion (see Chapter 5 of this volume), landfill gas, geothermal and onshore wind. All the concentrating solar thermal technologies tend to be more expensive at present.

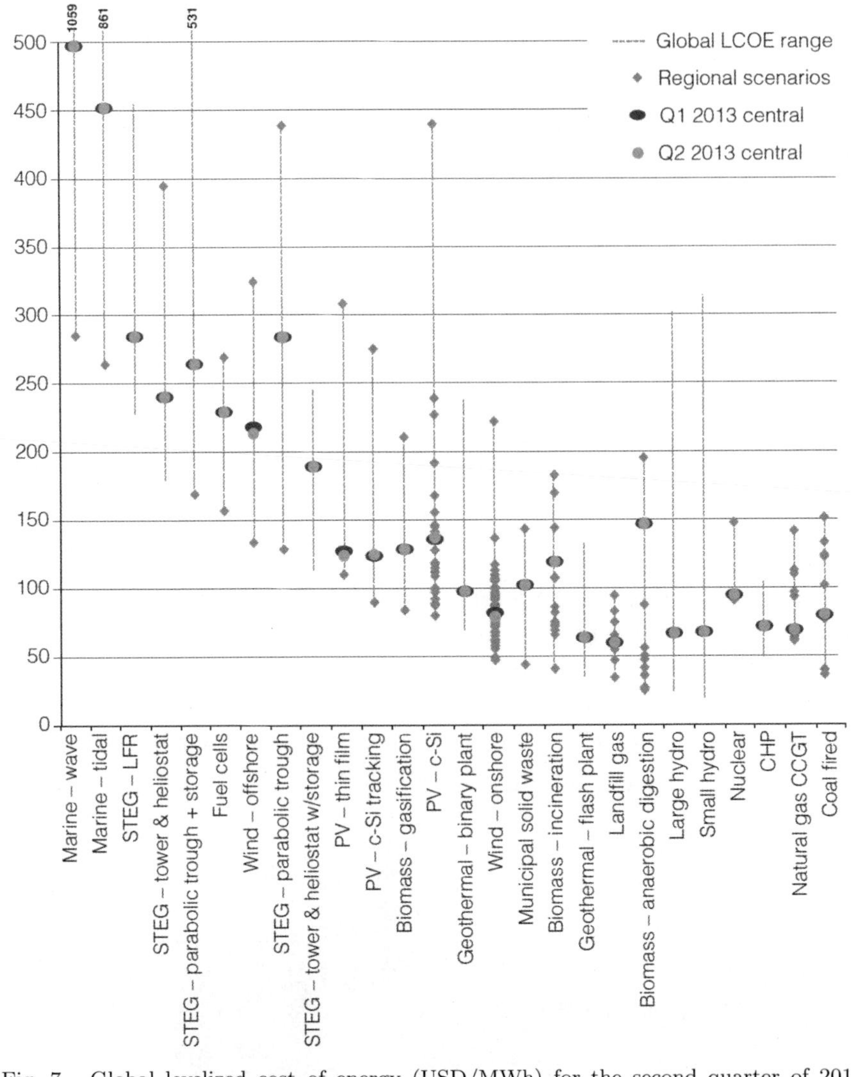

Fig. 7. Global levelized cost of energy (USD/MWh) for the second quarter of 2013 (Used by permission of the World Energy Council, London, www.worldenergy.org).

6 Social and Policy Aspects

Solar energy advancement in recent years has been underpinned by various forms of support from governments and, occasionally, utilities. The policies have tended to be both variable between countries and in a continuous state of uncertainty and change. Motivations for policies encouraging solar energy use have included environmental concerns and the support of fledgling industries likely to be important in the future and therefore to generate local jobs. In the case of grid connected solar energy, utilities have the duty to ensure electrical power quality and may see benefits or be mandated to facilitate the connection of distributed solar electricity generators. Related regulations can be used to control, encourage or inhibit the growing penetration of solar technologies. Standards and accreditation schemes generally enhance quality and performance but at some compliance cost.

IEA PVPS tracks photovoltaics policy changes across its member nations each year and summarizes the situation in its trends report (Fig. 8). The schemes used to support photovoltaics are similar to those for solar energy generally.[46,69] The most widespread support scheme is the feed-in tariff (FiT),[70] in which electricity produced from a photovoltaic system and supplied to the grid is paid for at a predefined price and guaranteed during a stated period. The most successful applications of FiT systems have been in China, Germany, Italy (until 2013) and Japan. The cost can be supported from tax income or by a levy on all or most electricity consumers in the network. Setting correct levels for FiT schemes or limiting the overall financial value or otherwise regulating the market, especially in the context of rapidly falling prices for photovoltaics, is critical. Too-generous FiT rates caused unsustainable market booms in Spain in 2008, in the Czech Republic in 2010, in Italy in 2011, and in many other jurisdictions.

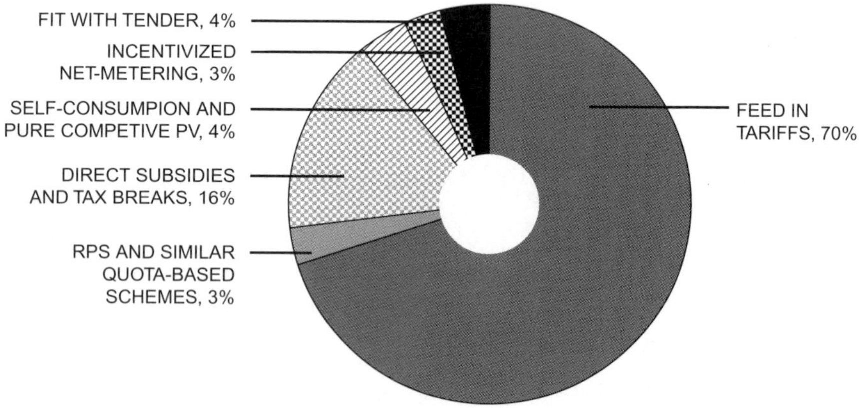

Fig. 8. Market incentives and enablers in 2013 for photovoltaics, in terms of drivers of market share. (Used with permission of International Energy Agency Photovoltaic Power Systems Programme (IEA PVPS)).

Direct capital subsidies aimed at reducing the upfront investment represented around 16% of the incentives in 2013. These subsidies are derived from taxation and may support grid or off-grid applications. They have been used, among others, in Australia, Belgium, Sweden, Japan, the USA, Italy, and China. Tax credits are a special form of direct subsidy. Tax credits have been used in Canada, the USA, Belgium (until 2010), Switzerland, France, Japan, the Netherlands, and other countries.

Renewable Portfolio Standard (RPS) and related approaches mandate a share of electricity to be produced by specific sources that utilities must use, either by producing it themselves or by buying certificates. These certificates allow renewable electricity producers to get a market-based remuneration for their output to the grid.

Sustainable building regulations are an important motivator for uptake of photovoltaics, solar thermal water heating and passive solar construction. These solar technologies may be included in a suite of options for reducing the energy footprint of a new building or could be specifically mandated for new buildings.

The declining cost of distributed generation has allowed it to compete directly with retail electricity from the grid. Several jurisdictions have permitted schemes allowing local consumption (self-consumption or net-metering) of locally produced electricity. These allow reduction of imported electricity and the related cost for the system owner, on site or even, occasionally, between distant sites. Solar generated electricity can be consumed by the PV system owner, reducing the electricity bill and any excess electricity can then be exported to the grid. This system was used, for example, in Germany until 2012. Many regions encouraged self-consumption by funding a bonus above the retail electricity price. Several methods are used to price the excess electricity sent to the grid: injected electricity is not paid; excess electricity is paid at the market price, with or without a bonus; a FiT remunerates the excess electricity at a pre-defined price, either lower or higher than the retail price of electricity; and net-metering, sometimes with either incentives or taxes. Export to the grid is expressly forbidden in some cases, with requirements for expensive equipment to block it.

Since 2012 opposition from utilities and grid operators grew against net-metering and FiT schemes. In situations where both improved energy efficiency and increased solar energy penetration have reduced demand for fossil-fueled electricity to the extent that incumbent generators fear that their assets may be stranded, all forms of support for solar energy integration are being questioned.[71] In many cases, electricity companies have been allowed to levy fees on connection of photovoltaic systems since they need to have grid capacity available both to accept the photovoltaic generation and to provide back-up. The estimation of a fair value for such charges is highly controversial.[72] Several regulators in Europe and elsewhere are expected to introduce capacity-based tariffs rather than energy-based tariffs for grid costs. This could discourage both solar energy production and energy efficiency and delay solar energy's competitiveness in some countries. Support in many areas has recently been reduced or withdrawn altogether, sometimes even retrospectively. These regulatory changes have motivated both system designs to minimize generation at times with likely low load[73] as well as the use local battery storage.[74]

Carbon taxes, introduced in various forms in several jurisdictions, have tended to have a rather limited impact on solar energy production since these incentives have operated directly at the wholesale energy market level, where lower cost options are frequently available (see Fig. 7). While the foregoing discussion has focused on grid-connected solar electricity generation, Zhang[75] has assessed the impact of World Bank policies on the encouragement of photovoltaics uptake in developing countries since the 1990s. The World Bank contributed USD790 million over the period 1992–2009 to the promotion of small-scale solar home systems in 34 developing countries. The results have been mixed and many barriers have been identified. The study found that using output-based producer subsidies and relying on microfinance for consumers, have been effective in encouraging the uptake of solar home systems. The World Bank experience suggests that power grid development can be leapfrogged by solar home systems for households with access to good microfinance services.

The rapid fall in prices for some forms of renewable energy generation are beneficial but have raised some serious challenges for policy makers.[76] These include:

- Rapidly falling costs have made it difficult to set public sector support at optimal levels;
- Some support schemes have proven to be relatively expensive burdens for consumers and/or tax-payers;
- Unanticipated competition between distributed and incumbent generators;
- Growing levels of intermittent generation are providing technical challenges for grid operators.

The International Renewable Energy Agency has developed a set of policy directions to address these issues in different contexts.

7 Summary/Conclusions

Solar energy is on the verge of a massive boom. Together with wind energy, it directly challenges the incumbent dominant forms of traded energy, fossil and nuclear. This chapter outlines the rapidly improving economics of solar energy, particularly, flat-plate photovoltaics, in an inexorable march towards dominant market share as global concern grows about the impact of atmospheric carbon on climate change. We are already seeing the beginnings of divestment of fossil fuelled energy by influential investors, as fears grow about stranded assets. The chapter also addresses the policy environment that has encouraged the development of solar energy in recent decades. However, some of these policies are under threat as energy companies, especially in the areas of electricity transmission and distribution, realize the challenge posed by solar to their traditional business models and learn to adjust those models to include distributed generation.

This volume describes the recent advances and current status of a wide range of approaches to capturing and exploiting solar energy to serve humanity's needs. It is a timely summary of the technologies becoming mature and breaking into the current markets as well as those still in the R&D realm. In spite of many challenges ahead, solar energy is likely to dominate traded human energy soon, as it has always dominated untraded energy.

References

1. K. Hubbert, *Sci. Amer.* **225** (1971), p. 61.
2. J. Kennewell and A. McDonald, *The Solar Constant* (2015). Available at: http://www.ips.gov.au/Educational/2/1/12. Accessed 15 August 2015.
3. P. Altermatt, Comparison with Blackbody Radiation — Plancks Law, Altermatt Lecture: The Solar Spectrum (2012). Available at: www.pvlighthouse.com.au/resources/courses/altermatt/The%20Solar%20Spectrum/Comparison%20with%20blackbody%20radiation%20-%20Plancks%20law.aspx. Accessed 29 September 2014.
4. ASTM, 2000 ASTM Standard Extraterrestrial Spectrum Reference E-490-00, Solar Spectra: Air Mass Zero, Renewable Resource Data Center (2000). Available at: http://rredc.nrel.gov/solar/spectra/am0/. Accessed 29 September 2014.
5. P. Altermatt, The extraterrestrial (AM0) solar spectrum, Altermatt Lecture: The Solar Spectrum (2012). Available at: www.pvlighthouse.com.au/

resources/courses/altermatt/The%20Solar%20Spectrum/The%20extraterrestrial%20%28AM0%29%20solar%20spectrum.aspx. Accessed 29 September 2014.

6. P. Altermatt, The influences of Earth's atmosphere on the solar spectrum, Altermatt Lecture: The Solar Spectrum (2012). Available at: www.pv-lighthouse.com.au/resources/courses/altermatt/The%20Solar%20Spectrum/The%20influences%20of%20Earths%20atmosphere%20on%20the%20solar%20spectrum.aspx. Accessed 29 September 2014.
7. RReDC Solar spectra, Renewable Resource Data Center (RReDC) (2000). Available at: http://rredc.nrel.gov/solar/spectra/. Accessed 29 September 2014.
8. ASTM, Standard tables for reference solar spectral irradiances: direct normal and hemispherical on 37° tilted surface, ASTM International (2012). Available at: http://www.astm.org/Standards/G173.htm. Accessed 29 September 2014.
9. P. Altermatt, Two mechanisms for atmospheric absorption of sunlight, Altermatt Lecture: The Solar Spectrum (2012). Available at: www.pvlighthouse.com.au/resources/courses/altermatt/The%20Solar%20Spectrum/Two%20mechanisms%20for%20atmospheric%20absorption%20of%20sunlight.aspx. Accessed 29 September 2014.
10. P. Altermatt, Atmospheric absorption — an overview, Altermatt Lecture: The Solar Spectrum (2012). Available at: www.pvlighthouse.com.au/resources/courses/altermatt/The%20Solar%20Spectrum/Atmospheric%20absorption%20-%20an%20overview.aspx. Accessed 29 September 2014.
11. P. Altermatt, The global standard spectrum (AM1.5g), Altermatt Lecture: The Solar Spectrum (2012). Available at: www.pvlighthouse.com.au/resources/courses/altermatt/The%20Solar%20Spectrum/The%20global%20standard%20spectrum%20%28AM1-5g%29.aspx. Accessed 29 September 2014.
12. ASTM (2014). *Standard Solar Constant and Zero Air Mass Solar Spectral Irradiance Tables* (ASTM International, 2014).
13. NASA, The balance of power in the earth–sun system, NASA Fact Sheet FS-2005-9-074-GSFC (2005). Available at: http://www.nasa.gov/pdf/135642main_balance_trifold21.pdf. Accessed 28 September 2014.
14. W. Shepherd and D. W. Shepherd, *Energy Studies*, 2nd edn. (Imperial College Press, London, 2003).
15. NASA, Earth's energy budget (2014). Available at: http://earthobservatory.nasa.gov/Features/EnergyBalance/page4.php. Accessed 28 September 2014.
16. NASA, Surface meteorology and Solar Energy. A renewable energy resource web site (release 6.0) (2014). Available at: https://eosweb.larc.nasa.gov/sse/. Accessed 28 September 2014.
17. NREL, Solar Maps (2013). Available at: http://www.nrel.gov/gis/solar.html. Accessed 28 September 2014.
18. BOM Daily solar exposure for Australia (2015). Available at: www.bom.gov.au/jsp/awap/solar/index.jsp Accessed 15 August 2015.
19. SolarGIS, Free download of solar radiation maps: global horizontal irradiation (GHI), SolarGIS (2013). Available at: http://solargis.info/doc/free-solar-radiation-maps-GHI. Accessed 29 September 2014.

20. WRDC, The world radiometric network (1964–1993), WRDC Online Archive (2014). Available at: http://wrdc-mgo.nrel.gov/html/mapap.html. Accessed 29 September 2014.
21. M. Iqbal, *An Introduction to Solar Radiation* (Academic, Toronto, 1983).
22. WRDC, WRDC online archive (2014). Available at: http://wrdc-mgo.nrel.gov/. Accessed 29 September 2014.
23. WRMC, Baseline surface radiation network (2014). Available at: www.bsrn.awi.de/. Accessed 11 November 2014.
24. Anon, Solar radiation data for Australian sites, EcoGeneration (2012). Available at: http://ecogeneration.com.au/news/solar_radiation_data_for_australian_sites/076419/. Accessed 29 September 2014.
25. R. Perez, T. Cebecauer and M. Šúri, in *Solar Energy Forecasting and Resource Assessment*, ed. J. Kleissl (Academic, Amsterdam, 2013), p. 21.
26. M. Šúri and T. Cebecauer, in *ASES SOLAR 2014 Conference* (San Francisco, 2014), p. 1.
27. RETScreen, RETScreen Software Suite Home page (2014). Available at: http://www.retscreen.net/ang/home.php. Accessed 29 September 2014.
28. ASP, Biography of a star: Our suns birth, life, and death, astrosociety (2014). Available at: http://www.astrosociety.org/edu/publications/tnl/39/sun2.html. Accessed 29 September 2014.
29. A. De Vos, *Endoreversible Thermodynamics of Solar Energy Conversion* (Oxford University Press, Oxford, 1992).
30. B. Sorensen, *Energy Intermittency* (CRC Press, Boca Raton, 2014).
31. J. C. Mankins, in *The First International Assessment of Space Solar Power: Opportunities, Issues and Potential Ways Forward* (International Academy of Astronautics, Stockholm, 2011).
32. Lazard, Lazard's Levelized Cost of Energy Analysis — Version 8.0 (2014).
33. Bloomberg New Energy Finance, *2014 Sustainable Energy in America Factbook* (2014). Available at: http://bnef.com/InsightDownload/9271/pdf/. Accessed 23 September 2014.
34. M. W. Parker *et al.*, Bernstein Energy and Power Blast: If Solar Wins Who Loses? (2014). Available at: http://reneweconomy.com.au/wp-content/uploads/2014/04/Bernstein-solar.pdf. Accessed 29 September 2014.
35. Rockefeller Brothers Fund, Fund announces plans to divest from fossil fuels (2014). Available at: http://www.rbf.org/post/fund-announces-plans-divest-fossil-fuels. Accessed 28 September 2014.
36. J. M. Diamond, *Guns, Germs, and Steel: The Fates of Human Societies* (W. W. Norton, New York, 1999).
37. V. Belessiotisand E. Delyannis, *Sol Energy* **85** (2011), p. 1665.
38. Anon., South Pacific Nuts to conduct free extensive training for Vanuatu farmers, *Vanuatu Daily Post* (2014), p. 25.
39. EERE, *The History of Solar* (2005). Available at: https://www1.eere.energy.gov/solar/pdfs/solar_timeline.pdf. Accessed 20 September 2014.
40. K. Butti and J. Perlin, *A Golden Thread. 2500 Years of Solar Architecture and Technology* (Van Nostrand Reinhold, New York, 1980).
41. P. A. Crossley *et al.*, *Review and Evaluation of Past Solar-Cell Development Efforts* (RCA Astro-Electronics, Washington, 1968).

42. E. Becquerel, Compte rendus hebdomadaires des séances de l'Académie des sciences **9** (1839), p. 145.
43. A. Goldmann and J. Brodsky, *Ann. Phys.* **44** (1914), p. 849.
44. C. E. Fritts, *Proc. Am. Assoc. Adv. Sci.* **XXXIII** (1884), p. 97.
45. E. H. Kennard and E. O. Dietrich, *Phys. Rev.* **9** (1917), p. 58.
46. IEA-PVPS, *Trends 2014 in Photovoltaic Applications*, International Energy Agency Photovoltaic Power Systems Programme (2014). Available at: http://www.iea-pvps.org/fileadmin/dam/public/report/ statistics/IEA_PVPS_Trends_2014_in_PV_Applications_-_lr.pdf. Accessed 29 September 2014.
47. R. M. Swanson, *Prog. Photovolt.: Res. Appl.* **8** (2000), p. 93.
48. World Energy Council, *World Energy Perspective: Cost of Energy Technologies* (World Energy Council, 2013). Available at: http://www.worldenergy.org/publications/2013/world-energy-perspective-cost-of-energy-technologies/. Accessed 11 September 2014.
49. V. S. Reddy, S. C. Kaushik, K. R. Ranjan and S. K. Tyagi, *Renew. Sust. Energy Rev.* **27** (2013), p. 258.
50. P. Pardo, A. Deydier, Z. Anxionnaz-Minyielle, S. Rouge, M. Cabassud and P. Cognet, *Renew. Sust. Energy Rev.* **32** (2014), p. 591.
51. C. Agrafiotis, H. von Storch, M. Roeb and C. Sattler, *Renew. Sust. Energy Rev.* **29** (2014), p. 656.
52. S. Sadhishkumar and T. Balusamy, *Renew. Sust. Energy Rev.* **37** (2014), p. 191.
53. R. Shukla, K. Sumathy, P. Erickson and J. Gong, *Renew. Sust. Energy Rev.* **19** (2013), p. 173.
54. M. Parnell, and G. Cole. *Australian Solar Houses* (Second Back Row/Solar Scope, Leura, 1983).
55. L. Perez-Lombard, J. Ortiz, R. Gonzalez and I. R. Maestre, *Energy and Building* **41** (2009), p. 272.
56. C. Podewils, *Photon Int.* **2007-04** (2007), pp. 106–113.
57. S. Fazio and L. Barbanti, *Renew. Energy* **69** (2014), p. 233.
58. W. G. Tu, Y. Zhou and Z. G. Zou, *Adv. Mater.* **26** (2014), p. 4607.
59. Z. J. Han and R. Eisenberg, *Acc. Chem. Res.* **47** (2014), p. 2537.
60. D. Feldman *et al.*, *Photovoltaic System Pricing Trends. Historical, Recent, and Near-Term Projections* (National Renewable Energy Laboratory, 2014).
61. H. Forstner, *et al.*, *International Technology Roadmap for Photovoltaic (ITRPV)* (2014).
62. DOE. "Soft Costs", sunshot program (2014). Available at: http://energy.gov/eere/sunshot/soft-costs. Accessed 29 September 2014.
63. *Australian Energy Technology Assessment (AETA) 2013 Model*, Bureau of Resources and Energy Economics (2013). Available at: http://www.bree.gov.au/publications/australian-energy-technology-assessments.
64. *Australian Energy Technology Assessment (AETA) 2014 Model*, Bureau of Resources and Energy Economics (2014). Available at: http://www.bree.gov.au / publications / australian-energy-technology-assessments. Accessed 29 September 2014.

65. *Solar Augmentation Project at Liddell Power Station*, EcoGeneration Jan./Feb. (2013). Available at: http://ecogeneration.com.au/news/solar_augmentation_project_at_liddell_power_station/079478/. Accessed 29 September 2014.
66. M. R. Islam, K. Sumathy and S. U. Khan, *Renew. Sust. Energy Rev.* **17** (2013), p. 1.
67. H. Cassard, P. Denholm and S. Ong, *Renew. Sust. Energy Rev.* **15** (2011), p. 3789.
68. F. Jotzo, *Outrage at ANU divestment shows the power of its idea*, The Conversation (2014). Available at: http://theconversation.com/outrage-at-anu-divestment-shows-the-power-of-its-idea-32736. Accessed 13 October 2014.
69. M. Miller and S. Cox, *Overview of Variable Renewable Energy Regulatory Issues* (National Renewable Energy Laboratory, 2014). Available at: http://www.nrel.gov/docs/fy14osti/61350.pdf. Accessed 13 October 2014.
70. M. Mendoça, *Feed-In Tariffs. Accelerating the Deployment of Renewable Energy* (Earthscan, London, 2007).
71. J. Massola, *Ian Macfarlane defends proposed cut to renewable energy target*, Sydney Morning Herald (26 October 2014). Available at: http://www.smh.com.au/federal-politics/political-news/ian-macfarlane-defends-proposed-cut-to-renewable-energy-target-20141026-11bypd.html. Accessed 26 October 2014.
72. APVI, *Impacts of PV, AC, and Other Technologies and Tariffs on Consumer Costs*, Australian Photovoltaics Institute (2014). Available at: http://apvi.org.au/impacts-of-pv-ac-and-other-technologies-and-tariffs-on-consumer-costs/. Accessed 26 October 2014.
73. A. Orioli and A. Di Gangi, *Renew. Sust. Energy Rev.* **21** (2013), p. 13.
74. G. Mulder, F. De Ridder and D. Six, *Sol. Energy* **84** (2010), p. 1284.
75. F. Zhang, *Renew. Sust. Energy Rev.* **38** (2014), p. 811.
76. S. Vinci, D. Nagpal, R. Ferroukhi, E. Zindler and A. Czajkowska, *Adapting Renewable Energy Policies to Dynamic Market Conditions*, International Renewable Energy Agency (IRENA, 2014). Available at: http://www.irena.org/DocumentDownloads/Publications/policy_adaptation.pdf. Accessed 26 October 2014.

Chapter 2

Fundamentals of Photovoltaic Cells and Systems

Ignacio Rey-Stolle
Instituto de Energía Solar, Universidad Politécnica de Madrid
(Solar Energy Institute, Technical University of Madrid)
ETSI de Telecomunicación, Avda. Complutense 30, 28040 Madrid (Spain)

The present chapter summarizes the current state and perspectives of photovoltaic solar energy. The fundamentals of solar radiation and photovoltaic (solar) cells are covered; the basic technology of photovoltaic modules and systems is briefly discussed; and a general outlook is provided for the uses, market and environmental impact of photovoltaic solar energy.

1 Introduction

The Earth receives annually around $1.5 \cdot 10^{18}$ kWh of solar energy, which is by far the most abundant energy resource available for mankind so far. If adequately harnessed, only a minuscule fraction of this energy (~0.01%) would suffice to supply the world's primary energy demand, which in 2012 was about $1.55 \cdot 10^{14}$ kWh.[1] The primary energy is processed by the energetic system into different types of readily usable energy, among which electricity is considered the key technology for the next decades. Accordingly, the direct generation of electricity — the preferred consumable form of energy — from solar radiation — the richest resource — is a topic of the highest relevance and is the essence of Photovoltaics (PV). From the discovery of the PV effect in 1839 by French physicist Alexandre-Edmond Becquerel to the first successful application of PV panels to power the Vanguard I satellite launched in 1958 more than a century went by.[2] Since those pioneering works, many steps forward have been made and the PV industry has evolved from the Watt-ranged applications of the early days to the GW systems planned today. In the first one and a half decades of the 21st century, with more than 140 GW installed worldwide, PV technology has demonstrated the maturity to become a major source of power for the world. That robust and continuous growth is expected to continue in the decades ahead in order to turn PV into one of the key players in the pool of technologies involved in generating electricity for the 21st century.

2 Solar Radiation

2.1 *Fundamentals*

Solar radiation is a general term that refers to the electromagnetic energy flux emitted by the sun's surface (i.e. the photosphere). This emission of electromagnetic waves does not take place at a single wavelength but spans a continuum of different wavelengths from X-rays to deep infrared photons. Accordingly, the **solar spectrum** is the distribution of the electromagnetic power emitted by the sun (per unit wavelength and unit area) as a function of wavelength. The solar spectrum reaching the earth outside the atmosphere may be well approximated by that of a blackbody at 5778 K,[3] as shown in Fig. 1.

The integral of the solar spectrum in Fig. 1 yields the solar power per unit area reaching the outer surface of Earth's atmosphere. This magnitude is known as the **solar constant** (B_0) and its most accepted average value is 1367 W/m^2. Throughout the year (and from year to year) the solar constant varies slightly (~7%) as a result of the change in the sun–earth distance

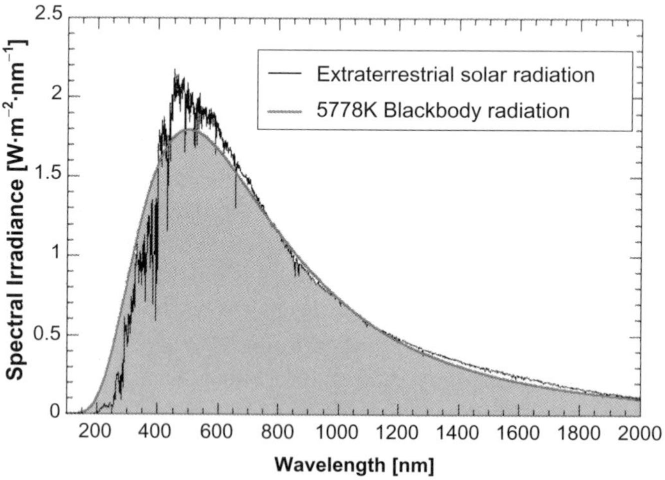

Fig. 1. Extraterrestrial solar spectrum compared to the spectrum emitted by a blackbody at 5778 K.

along the Earth's orbit and variations in the sun's activity (solar cycles). The actual value of B_0 can be simply calculated just by multiplying the solar constant by a correction factor (ε_0) taking into account the eccentricity of the Earth's orbit as a function of the ordinal day in the year (d_n; for 1st January $d_n = 1$; while for 31st December $d_n = 365$):

$$\varepsilon_0 = 1 + 0.033 \cdot \cos(2\pi \cdot d_n/365). \tag{1}$$

2.2 *Basic PV Terminology and Notation for Solar Radiation*

The **irradiance** at a given surface is the total energy per unit area per second (i.e. power per unit area) reaching that surface in the form of solar radiation. Units used for irradiance are W/m^2 and is typically noted using the letter G. Of particular interest — due to the wealth of experimental data available — is the **Global Horizontal Irradiance** (*GHI*) which is the amount of irradiance falling on a surface horizontal to the surface of the Earth.

The **irradiation** at a given location is the total energy per unit area reaching that location in the form of solar radiation in a certain period. Typical units used for irradiation are Wh/m^2 or J/m^2 and is noted using the symbol G_p, where the subscript p denotes the period over which the irradiation is calculated as

$$G_p = \int_p G \cdot dt. \tag{2}$$

Typical periods for calculating irradiation are hours (G_h), days (G_d) or months (G_m). It is also common to use averages of daily or hourly irradiation taken over different periods. In this case no uniform notation is found in the literature but in general subscripts are added to indicate the period for averaging (G_{dm} = average daily irradiation taken over a month).

2.3 *Components of the Solar Radiation*

When the sun's radiation traverses the atmosphere it is partially reflected by the clouds, partially absorbed and partially scattered by particles and gases. Thereby the radiation reaching the Earth's surface is largely attenuated, filtered, and randomized in terms of direction. Accordingly, the irradiance reaching a surface on Earth may be expressed in terms of three main components, namely, beam or direct irradiance (B), diffuse irradiance (D)

and albedo or reflected irradiance (R):

$$G = B + D + R. \tag{3}$$

The **beam** component refers to the irradiance coming directly from the solar disc; whilst the **diffuse** component refers to the irradiance coming from the rest of the sky. The **albedo** or reflected component accounts for radiation that reaches the target surface after reflection from the ground, buildings, snowy hills or any other reflecting surface. Figure 2(a) visually depicts these three components.

Three variants of these components are especially relevant in PV as a result of the wealth of experimental data coming from meteorological databases. The **Direct Normal Irradiance** or **DNI** refers to the beam irradiance impinging on a surface perpendicular to the rays (i.e. a surface tracking the sun). The **Diffuse Horizontal Irradiance** or ***DHI*** represents the diffuse irradiance reaching a horizontal surface. Analogously, the ***GHI*** represents the amount of global radiation reaching a horizontal surface. A common way to calculate *GHI* is as the sum of the *DHI* and the fraction of *DNI* impinging on a horizontal surface:

$$GHI = DHI + DNI \cdot \cos(\theta_{ZS}), \tag{4}$$

where θ_{ZS} is the solar zenith angle, which is a function of latitude, hour of the day, and day of the year. Figure 2(b) illustrates the concept of solar zenith angle.

On clear days, when attenuation is the main effect of the atmosphere, the global irradiance on a horizontal plane on the Earth surface (i.e. *GHI*) can be approximated with a very simple empirical expression:

$$GHI = B_0 \cdot \varepsilon_0 \cdot 0.74^{AM \times 0.678}, \tag{5}$$

where B_0 and ε_0 are the solar constant and eccentricity factor as defined in Sec. 2.1; whilst $\boldsymbol{AM}$ is the **air mass**, which is defined as the ratio of the length of the beam irradiance path through the atmosphere to the vertical length of the atmosphere. Accordingly, in PV terminology **AM0** refers to the extraterrestrial irradiance; **AM1.5** to an irradiance traversing an atmosphere length 1.5 times its vertical length, and analogous definitions can be given for other AM values. AM varies through the day and season (as apparent sun movement does) and depends on location (latitude). Simple geometrical considerations lead to[3]:

$$AM = 1/\cos\theta_{ZS}, \tag{6}$$

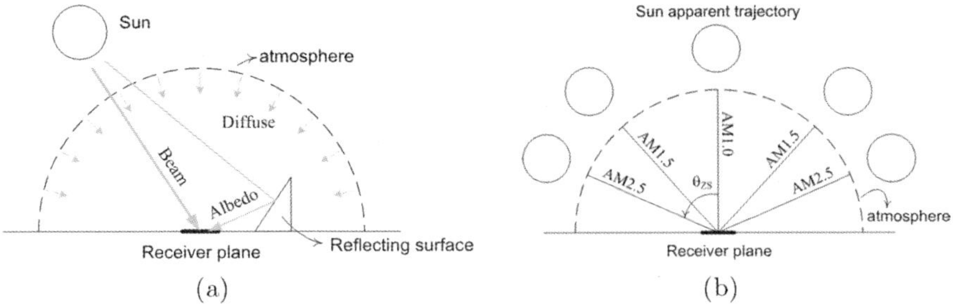

Fig. 2. (a) Components of irradiance; (b) Optical paths corresponding to different AM attenuations.

where θ_{ZS} is the solar zenith angle, which is itself a function of latitude, hour of the day, and day of the year. Figure 2(b) illustrates the concept of AM.

Following Eq. (3), irradiation can also be expressed in terms of the different components of irradiance integrated over a period (p) of time:

$$G_p = B_p + D_p = \int_p B \cdot dt + \int_p D \cdot dt, \tag{7}$$

where the albedo component has been neglected since it is site-dependent and thus it is typically measured.

2.4 *World Distribution of Solar Radiation*

Among energy resources, solar radiation is the one more evenly distributed on a world scale. Virtually any location with latitude below 55°, is suitable for installing PV systems. Figure 3 visually summarizes the world availability of the solar resource.

2.5 *Solar Radiation Collected by PV Systems*

In PV engineering the key question around solar radiation is how much irradiance will reach the surface of a PV system installed at a given location, this is, the so-called **in-plane irradiance**. Following Eq. (3) to calculate the in-plane irradiance, it is necessary to model or to measure the beam and diffuse irradiances (and albedo when applicable) on the target surface for the desired location during the course of the day and throughout the seasons in a year.

A starting point for this calculation is the so-called **typical meteorological year (TMY)**. A TMY is a collection of selected solar irradiation

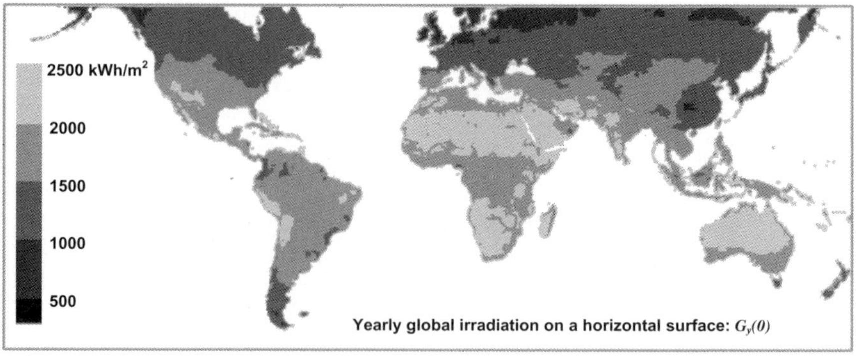

Fig. 3. Yearly global irradiation [kWh/m^2] on a horizontal surface.

data for a year and for a specific location, generated from a database much longer than a year in duration. There is no standard establishing what a TMY should contain but typically it includes hourly values of GHI, DNI, wind speed and ambient temperature, among other meteorological variables. The TMY is not constructed by just simply averaging hourly values in the database but by the concatenation of actual representative months. In other words, the data for the month in the database that has the average radiation most closely equal to the monthly average over the whole measurement period is chosen as the TMY data for that month. The TMY is then constructed by applying this method to all months.

Once the horizontal irradiance values are known for the location of interest, many approaches exist for calculating the in-plane irradiance in PV systems[3–5] that provide the average irradiance and irradiation of arbitrarily oriented surfaces for diverse periods of time (hours, days, months, year).

Another approach, which is seeing increasing use in PV engineering, is to obtain these values from reference databases which integrate complex solar radiation models with extensive data from ground stations or satellites. Examples of these databases are PVGIS[6] for Europe and Africa; and the NREL US dynamic solar atlas.[7]

3 Solar Cells

3.1 *Definition*

A **solar cell** or **PV cell** is a device that directly transforms solar radiation into electrical energy by means of the PV effect without any thermal cycles, mechanical cycles or chemical reactions.[8–10] Classical ways for implementing solar cells are p/n semiconductor junctions[11] in which light produces the photogeneration of charge carriers (electrons and holes) that are efficiently separated by the built-in voltage in the junction and flow to the outer circuit providing electrical power.[9,10]

3.2 *Basic Solar Cell Equations and Equivalent Circuits*

3.2.1 *Simple equivalent circuit for a solar cell and I–V characteristic*

The simplest circuit model associated with a p/n junction solar cell is the one shown in Fig. 4. This circuit consists in the parallel connection of (1) a current source modeling the photogenerated current (I_L); (2) a diode modeling the dark current through the p/n junction (I_D); (3) a parallel resistor (R_P) modeling possible shunts across the junction; and these three elements are connected in series with an additional resistor (R_S) modeling ohmic losses in elements adjacent to the junction (contacts, substrate, ...).

According to this equivalent circuit, the I–V curve of a solar cell has the following expression:

$$I = I_L - I_0 \left[\exp\left(\frac{V + I \cdot R_S}{n \cdot V_t} \right) - 1 \right] - \frac{V + I \cdot R_S}{R_P}, \qquad (8)$$

where I_L is the photogenerated current in Amperes; I_0 is the diode dark current in Amperes; n is the diode ideality factor; R_S is the series resistance in Ohms; R_P is the parallel or shunt resistance in Ohms; and V_t is the thermal Voltage in Volts given by the well-known expression:

$$V_t = \frac{kT}{q} \Rightarrow V_t \cong 0.026\,\text{V} \quad \text{for T} = 300\,\text{K}, \qquad (9)$$

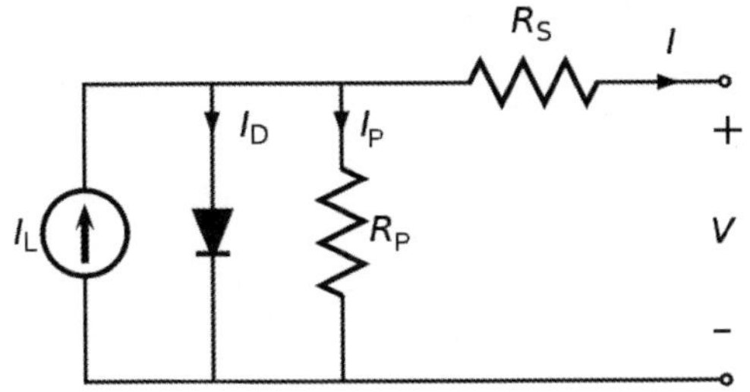

Fig. 4. Equivalent circuit of a solar cell with a single diode.

where k is the Boltzmann constant ($k = 1.3806 \cdot 10^{-23}$ J/K); q is the elementary charge ($q = 1.6022 \cdot 10^{-19}$ C); and T is the absolute temperature in Kelvin.

3.2.2 *General equivalent circuit for a solar cell and I–V characteristic*

For some solar cells a single conduction mechanism across the p/n junction is not enough to model the dark losses.[8–10] In this case, a circuit with multiple diodes in parallel is considered, where each diode accounts for a different mechanism (recombination in neutral regions, recombination at the space charge region, recombination at the junction perimeter, tunneling, ...). In such case, the new expression for the I–V curve is:

$$I = I_L - \sum_{i=1}^{n} I_{0i}\left[\exp\left(\frac{V + I \cdot R_S}{n_i \cdot V_t}\right) - 1\right] - \frac{V + I \cdot R_S}{R_P}, \qquad (10)$$

where $I_{0\mathrm{i}}$ is the diode dark current for conduction mechanism i in Amperes; and n_i is the diode ideality factor for the ith conduction mechanism.

3.3 *The I–V Curve of a Solar Cell*

3.3.1 *General look and key parameters*

The I–V curve of a solar cell under illumination is schematically depicted in Fig. 5. This curve has three characteristic points which are: (1) the **short circuit current** (**I_{SC}**) which is the current produced by the solar cell at zero voltage; (2) the **open circuit voltage** (**V_{OC}**) which is the voltage produced by the solar cell at zero current; and (3) the **maximum power point** or **MPP** (V_m, I_m) which is the point at which the power delivered by the solar cell is maximum (i.e. the product $V \cdot I$ reaches its maximum).

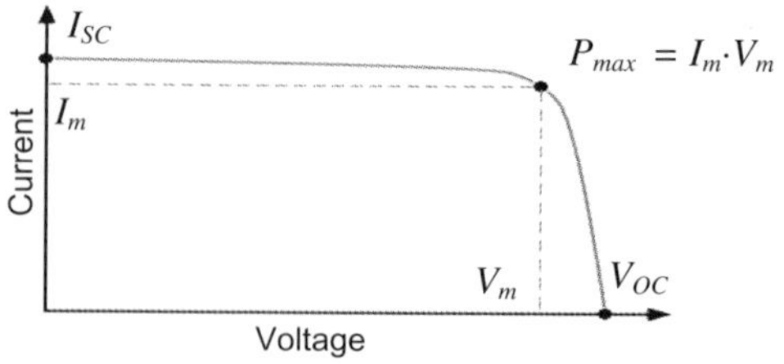

Fig. 5. Solar cell I–V curve.

Based on the curve in Fig. 5, it is possible to define a key figure of merit for a PV cell known as the **fill factor** (***FF***) where:

$$FF = \frac{I_m \cdot V_m}{I_{SC} \cdot V_{OC}}. \tag{11}$$

The efficiency of a solar cell is given by the ratio of the maximum power delivered to the load over the optical power received on the cell surface:

$$\eta = \frac{I_m \cdot V_m}{G \cdot A} = \frac{FF \cdot I_{SC} \cdot V_{OC}}{G \cdot A}, \tag{12}$$

where G is the irradiance on the solar cell in W/m^2; and A is the area of the solar cell in m^2.

Table 1 summarizes the key parameters of a solar cell including symbols and units.

3.3.2 *Effect of variations in series and parallel resistance on the I–V curve*

The effect of decreasing the magnitude of the parallel resistance or increasing the magnitude of the series resistance can be seen in Figs. 6(a) and 6(b), respectively. When the parallel resistance decreases, the solar cell becomes effectively shunted, and V_{OC} and FF of the cell decrease and the slope of the I–V curve changes markedly in the vicinity of I_{SC}, which remains unchanged. When the series resistance increases, the internal load in the solar cell increases, and thus the I_{SC} and FF of the cell decrease and the slope of the I–V curve changes markedly in the vicinity of V_{OC}, which remains unchanged.

Table 1: Key parameters of a solar cell with symbol and units.

Parameter	Symbol	Units
Short circuit current	I_{SC}	A
Open circuit voltage	V_{OC}	V
Current at maximum power point	I_m	A
Voltage at maximum power point	V_m	V
Fill factor	FF	%
Efficiency	η	%

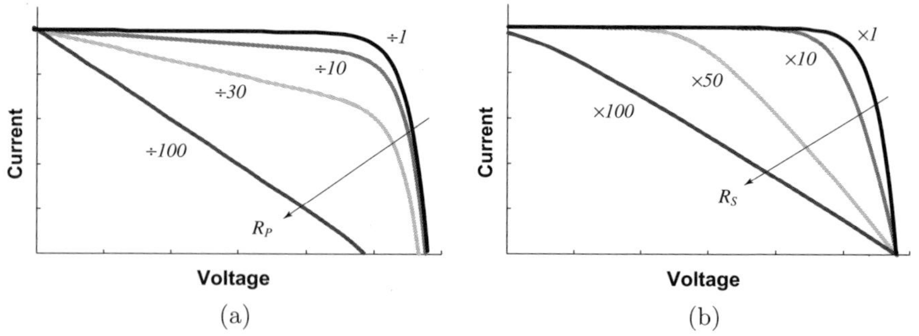

Fig. 6. (a) Effect of decreasing the parallel resistance and (b) increasing the series resistance on a solar cell I–V curve.

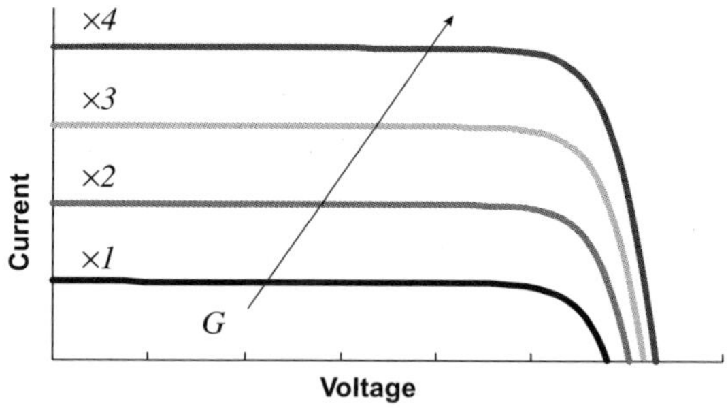

Fig. 7. Effect of increasing the irradiance on a solar cell I–V curve.

3.3.3 *Effect of variations in irradiance on the I–V curve*

The effect of changing the irradiance impinging on a PV cell can be visualized as a vertical shift of the I–V curve (Fig. 7).

Let G_{STC} be a reference irradiance and G any other irradiance value. We can then define the **concentration factor** (X) (sometimes referred to simply as **concentration**) as the ratio between G and G_{STC}:

$$X = G/G_{STC}. \tag{13}$$

The evolution of I_{SC} and V_{OC} of a solar cell with the concentration factor (i.e. with irradiance) are rather simple:

$$I_{SC}(G) = I_{SC}(G_{STC}) \cdot X, \tag{14a}$$

$$V_{OC}(G) = V_{OC}(G_{STC}) + V_t \cdot \ln X. \tag{14b}$$

On the other hand, the variation of FF (and thus maximum power point and efficiency) with irradiance is more complicated. At moderate irradiance levels, when the effects of series resistance are negligible, the fill factor increases with irradiance, while at high irradiance levels series resistance losses counterbalance this effect and the fill factor starts to decrease with concentration.

3.3.4 *Effect of variations in temperature*

When the operating temperature of a solar cell increases (Fig. 8) I_{SC} increases slightly, V_{OC} decreases more significantly and P_{max} also decreases. All these variations are quantified using three linear coefficients associated with I_{SC}, V_{OC}, and P_{max}.

For silicon solar cells the I_{SC} temperature coefficient (α) is around $\alpha = 0.1\%/°C$; the V_{OC} temperature coefficient (β) is around $\beta = -2\,mV/°C$; and the P_{max} temperature coefficient (γ) is around $\gamma = -0.5\%/°C$.

3.3.5 *Standard test conditions for solar cells*

In order to be able to compare the performance of different solar cells (and PV modules) in a consistent manner, efficiencies are reported using **standard test conditions** (STC). These conditions are a device temperature of 25°C and an irradiance set to the value and spectral distribution as defined in **reference spectra**, included in ASTM standard G173-03.[12] The most frequently spectra used in PV are **AM0** for space solar cells **AM1.5 global** for conventional non-concentrating solar cells and **AM1.5 direct** for concentrator solar cells.

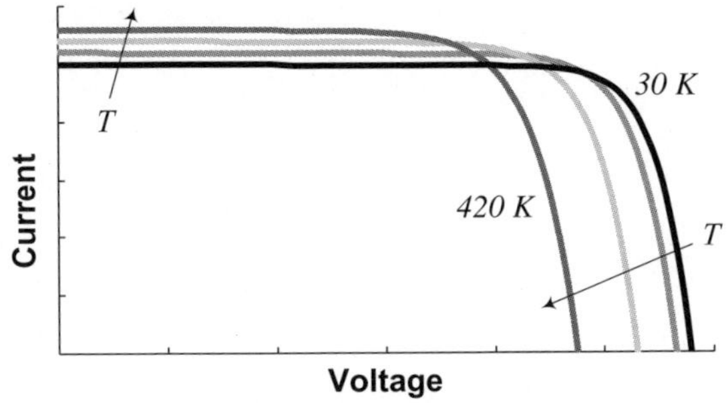

Fig. 8. Effect of changing the operating temperature on a solar cell I–V curve.

Table 2: Overview laboratory and commercial results for different solar cell technologies.

Technology	Silicon		Thin film				CPV
Material	Mono-c	Multi-c	a-Si:H	a-Si/ μc:Si	CdTe	CIGS	III–V
Record cell efficiency	25.6%	20.4%	10.4%	13.2%	21.0%	21.7	45.6%
Commercial cell efficiency	16–22%	14–18%	4–8%	7–9%	10–12%	7–12%	38–42%
Commercial module efficiency	14–20%	11–15%					27–32%
Module area needed per kWp*	~7 m^2	~8 m^2	~15 m^2	~12 m^2	~10 m^2	~10 m^2	

*For a definition of this unit (W_p = Watt peak) see Sec. 4.3.1.

3.4 *Overview of Solar Cell Technologies*

The core of any solar cell is a light absorbing material, where the photons are absorbed and generate charge carriers (free electrons and holes) via the PV effect.[8–10] In the history of PV, many materials have been investigated — most of them crystalline semiconductors — in the quest for an optimum low-cost high-performance solution. Today, the portfolio of materials used in PV is dominated by crystalline silicon, either mono-crystalline or multi-crystalline, though the so-called thin film materials, which include hydrogenated amorphous silicon (a-Si:H), cadmium telluride (CdTe) and copper indium gallium diselenide (CIGS), are starting to gather momentum.[13–15] III–V crystalline semiconductors are today limited to space applications though they might be a key player in concentrator photovoltaics (CPV).[16] Other PV materials, which are today limited to laboratory research or marginal production, include dye-sensitized,[17] organic polymers[18] and perovskites.[19,20] Table 2 gives an overview of laboratory and commercial results for different solar cell technologies.[21–23]

4 PV Modules

4.1 *Fundamentals*

4.1.1 *Concept and mission*

A **PV module** or **solar panel** is an interconnected assembly of solar cells packaged in a robust, weather-proof casing. Solar cells are brittle devices and provide voltage and current levels inadequate for almost any electrical load. Thereby, the role of a solar panel is (1) to provide adequate levels for electrical current and voltage; (2) to preserve solar cells from ambient degradation and mechanical damage; (3) to maintain electrical insulation; and (4) to facilitate the convenient handling of solar cells and their mounting outdoors.

4.1.2 *Construction*

Figure 9 schematically shows the construction of PV modules made of wafer-based silicon solar cells. These are constructed as a laminate where the assembly of solar cells is embedded into a transparent encapsulant (EVA); which is then sandwiched between a highly transparent (low iron) tempered glass and a tedlar backsheet. The laminate is then framed with anodized aluminum inserting a sealant to prevent moisture penetration. Electrical terminals are made accessible in a plastic weather-proof (IP65) connection box fixed at the rear of the module.

Figure 10 shows a schematic representation of a solar panel based on thin film technology. In this case, the module is not based on the connection of discrete solar cells previously fabricated. On the contrary, in

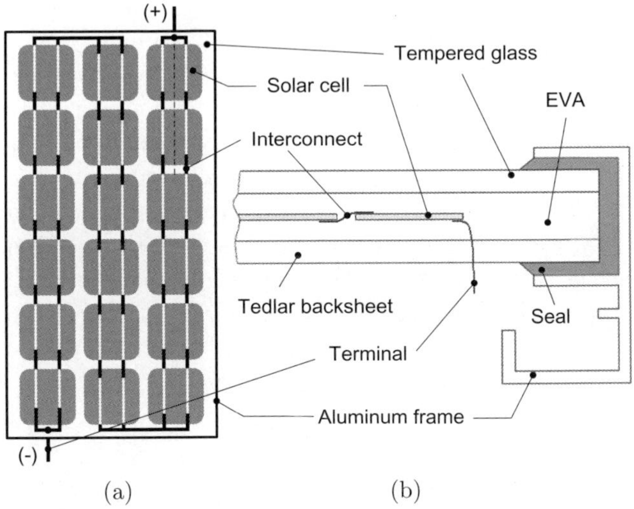

Fig. 9. (a) Schematic representation of the front-view of a solar panel based on silicon solar cells. (b) Cross-section of the panel at approximately the location marked as a dashed line.

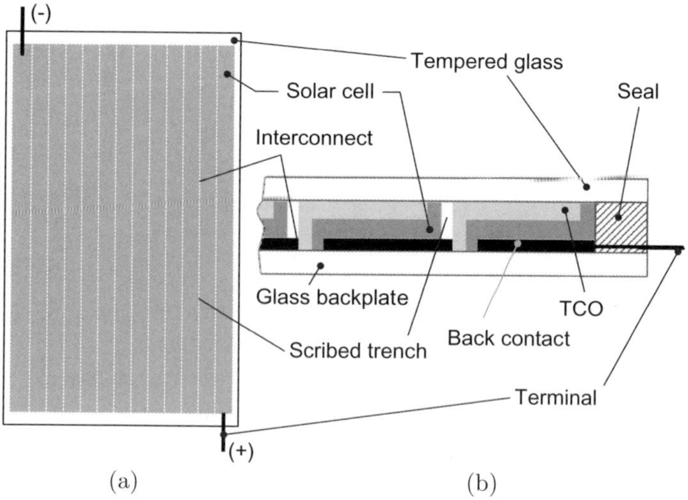

Fig. 10. (a) Schematic representation of the front-view of a solar panel based on thin film technology. (b) Cross-section of the panel at approximately the location marked as a dashed line.

this technology the solar cell material is deposited on the module back-sheet (glass, metal, plastic) or superstrate (glass). Individual cells are manufactured by scribing the solar cell material (either with a laser beam or mechanically) and the cell interconnections are made during the deposition. Typically, each single solar cell is in the form of a long narrow strip and is connected in series with the adjacent strip (i.e. cell). Accordingly, being a series connection of solar cells, the current of the module is determined by the area of each strip and the voltage by the total number of strips (cells). This connection scheme minimizes the inter-cell area giving the module a very uniform appearance, which is even enhanced by the fact that thin film modules are typically frameless — just insulated with edge seal. The aesthetic impact of these two factors has allowed a great penetration of thin film technology in the building integration market.

4.2 *Characteristic Equation and I–V Curve*

4.2.1 *I–V characteristic of a PV module*

Figure 11(a) sketches a generic configuration of a module with an arbitrary number (N_P) of cell strings in parallel, each string consisting of N_S cells in series. Considering that the solar cells in the module are identical and that they operate at the same conditions (irradiance and temperature),

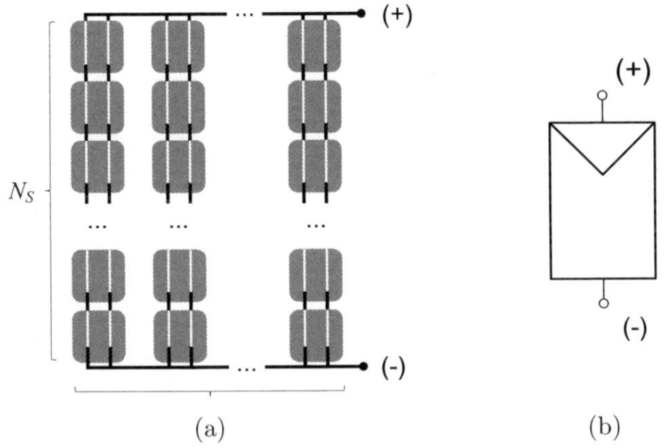

Fig. 11. (a) Generic electrical configuration of a PV module with N_P branches in parallel and N_S cells in series per branch. (b) Electrical symbol for a PV module.

the I–V characteristic of such a connection can be deduced from the I–V characteristic of the individual solar cells presented in Sec. 3:

$$I = N_P I_L - N_P I_0 \left[\exp\left(\frac{V + IR_S N_S}{N_S \cdot nV_t}\right) - 1\right] - \frac{V + IR_S N_S}{N_S R_P}, \tag{15}$$

where I_L is the photogenerated current per each individual cell in Amperes; I_0 is the diode dark current of each cell in Amperes; n is the diode ideality factor of each cell; R_S is the series resistance of each cell in Ohms; R_P is the parallel or shunt resistance of each cell in Ohms; and V_t is the thermal voltage in Volts. Equation (15) can be presented in a more compact solar-cell-like form as:

$$I = I_{Lm} - I_{0m} \left[\exp\left(\frac{V + IR_{Sm}}{n_m V_t}\right) - 1\right] - \frac{V + IR_{Sm}}{R_{Pm}}, \tag{16}$$

where all parameters used are analogous as those of the cell as defined in Table 1.

Accordingly, the equivalent circuit of Fig. 2 can also be used for modules with the parameters calculated as in Table 3.

As can be deduced from Eq. (16), the I–V curve of a solar panel under illumination (Fig. 12) has the same shape as that of the individual cells, except that it is scaled according to the relations expressed in Table 3. If the parallel resistance is high enough to be neglected, the characteristic equation of a solar panel can be also expressed in terms of the parameters

Table 3: Relation between cell and module I–V parameters.

Symbols	Parameters	Eqs.
I_{Lm}	Module short circuit current	$I_{Lm} = N_P.I_L$
I_{0m}	Module equivalent dark current	$I_{0m} = N_P.I_0$
n_m	Module equivalent ideality factor	$n_m = N_S.n$
R_{Sm}	Module equivalent series resistance	$R_{Sm} = N_S.R_S$
R_{pm}	Module equivalent parallel resistance	$R_{Pm} = N_P.R_P$

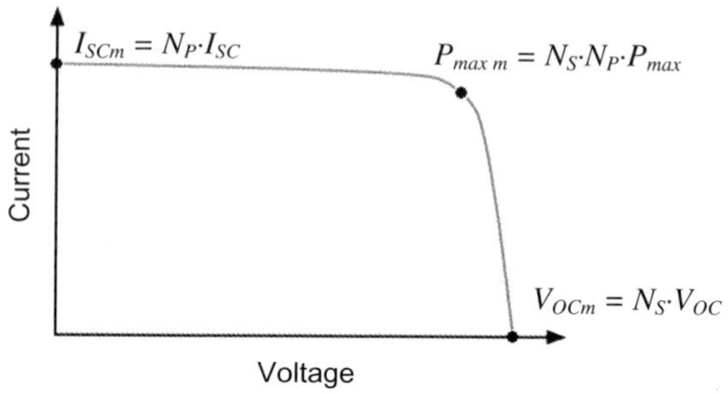

Fig. 12. I–V curve of the PV module of Fig. 10(a).

of the I–V curve as:

$$I = I_{SCm}\left[1 - \exp\left(\frac{V - V_{OCm} + IR_{Sm}}{n_m V_t}\right)\right]. \tag{17}$$

4.3 *Electrical Performance*

4.3.1 *Standard test conditions*

As in solar cells, the electrical performance of PV modules is generally rated under a set of predefined conditions referred to as **Standard Test Conditions** (STC). These conditions are: irradiance of 1.000 W/m^2 impinging perpendicularly on the tested module, a solar spectrum of AM 1.5g and module temperature at 25°C. The maximum electrical power supplied by the module at STC is called the peak power and is given in **Watts peak** (W_p). As STC are very favorable, the peak power of a solar panel may be considered as the maximum electrical power it would deliver in real operation rather than its typical or average output power. STC are very difficult to achieve outdoors in virtually any location so **solar simulators** are used to rate PV modules under these conditions.

4.3.2 *Factors affecting the electrical power of solar panels under real operation*

The operating conditions in the field deviate considerably from STC (negatively), so any solar panel operating outdoors will typically exhibit a lower power output than its peak power at STC. The main factors affecting solar panel power output are:

- *Irradiance*: As shown in Sec. 3.3.3 for solar cells, low irradiance levels degrade the open circuit voltage and make the effects of the shunt resistance more important.
- *Angle of incidence of light*: For solar modules mounted on fixed structures the angle of incidence of the beam radiation (and thus the reflection losses) depends on the position of the sun; in addition, diffuse radiation reaches the solar panels from all angles.
- *Spectrum of light*: It never reproduces exactly the AM1.5 standard since it varies annually with the season, daily with the AM value and locally with altitude, weather and pollution.
- *Module temperature*: As shown in Sec. 3.3.4, an increase in temperature degrades the open circuit voltage.
- *Soiling*: Dust, dirt, debris or other soiling hinder the absorption of light by solar cells causing an effective drop of irradiance.
- *Ageing*: As for any other industrial product, PV modules show signs of wear with time. Typically, manufacturers warrant that the power output will degrade less than 10% during the first 10 years of operation and less that 20% during the first 25 years.

To avoid the deleterious effects that partial shadowing might have on module performance or integrity (i.e. hot spots[24]) sometimes **bypass diodes** are connected in parallel but with opposite polarity to solar cells in a module. The mission of these diodes is to short-circuit shadowed cells. The number of bypass diodes per module varies from one per module up to three or four.

4.3.3 *Modeling the equilibrium cell temperature in a PV module*

The first step of most models describing the performance of photovoltaic modules not operating under STC is modeling the equivalent cell temperature. The most common method used for this purpose is based on an empirical parameter (supplied by the manufacturer) named **nominal operating cell temperature (*NOCT*)**. The *NOCT* is defined as the cell temperature when the ambient temperature is 20°C and the irradiance on the module is 800W/m^2. Typical values of $NOCT$ are around 45°C. Using the $NOCT$, the cell temperature (T_{cell}) for any irradiance and ambient temperature (T_{amb}) can be calculated as:

$$T_{cell} = T_{amb} + \frac{G}{800W/m^2}(NOCT - 20°C). \tag{18}$$

4.3.4 *Electrical power of solar panels at any irradiance and temperature*

Given certain operating conditions (in-plane irradiance and cell temperature), the most precise way to calculate the electrical output power from a PV module is to calculate its I–V curve for those conditions (Secs. 3.3.3, 3.3.4 and 4.2.1), and then determine its maximum power point.[4,9]

Despite being quite accurate, this model is costly in terms of calculations. Therefore, simpler approaches are often used, that implement first order (linear) corrections to the rated module power (P_M) as a function of cell temperature (T_{cell}) and in-plane irradiance (G)[25]:

$$P_M = P_{M,STC} G/G_{STC}[1 + \gamma(T_{cell} - T_{STC})], \tag{19}$$

where $P_{M,STC}$ is the module power rating at STC in W_p; and G_{STC} is the irradiance at STC (=1000 W/m^2); T_{STC} is the temperature at STC (=25°C); γ is a parameter that defines the relative change in module efficiency per °C (γ is typically negative, so power decreases as temperature increases).

4.3.5 *Electrical Energy from a PV module*

Once the instant power from a PV module is known, its electrical energy output ($E_{M,p}$) over a given period of time (p) can be calculated just by integrating:

$$E_{M,p} = \int_p P_M(G, T_{cell})dt. \tag{20}$$

Typically, the continuous evolution of the irradiance and cell temperature over the period p will be not known. On the other hand, a most likely situation is to take a set of discrete values (or averages) of G and T_{cell} at regular intervals (Δt) along period p. Thus, a more convenient formulation of Eq. (20) is:

$$E_{M,p} = \sum_i P_M(< G >_i, < T_{cell} >_i)\Delta t, \tag{21}$$

where $< G >_i$ is the average for the irradiance over interval i in kW/m^2; and $< T_{cell} >_i$ is the mean value for the cell temperature over interval i in °C.

5 PV Arrays and Systems

5.1 *Basic Definitions*

A PV array is the combination of a certain number of PV modules which constitute an electrical power generator providing a desired electrical output.

A **PV system** is a system including a PV array and ancillary elements designed to provide electrical power having the desired performance, safety and reliability.

In PV terminology, all the elements in a PV system other than the PV array are altogether referred to as the **balance of system** or **BOS**. The BOS may include diverse elements such as mounting structures, wiring, fuses, maximum power point trackers, charge controllers, batteries, inverters, electrical safety elements, and meters.

5.2 *Balance of System Components of PV Systems*

5.2.1 *Power conditioning*

Solar panels produce DC electricity while most electrical applications demand AC electricity. The **inverter** is the equipment that transforms the DC electricity from the solar array into AC electricity to power AC-loads or inject to the grid. In addition to DC-to-AC electronics, modern inverters typically integrate circuits intended to force the solar array to operate at optimum conditions by interfacing them with the adequate impedance: these are the so-called **maximum power point trackers** (or **MPP trackers** for short). Other ancillary functions also performed by modern inverters are synchronization with the grid, monitoring, and logging the production of the PV system, and implementing some protections (anti-island operation,...).

5.2.2 *Storage*

In PV systems, as in any other system based on a renewable resource, energy storage is needed to marry production with demand. In stand-alone (i.e. off-grid) systems, the adequate dimensioning of this storage system is indispensable to attain a given quality of service. In grid connected applications, storage is infrequent since the grid provides access to an alternative source of energy when the solar resource is unavailable or insufficient. In PV, the energy reservoir typically consists of a set of **batteries**. When using electrochemical energy storage, special care has to be taken to preserve the life and adequate operation of the batteries. Therefore, dedicated electronics, namely **charge controllers**, are inserted between the solar array, the load and the batteries to avoid overcharging or over-discharging these elements. Many different battery technologies are available today, though the classic lead-acid electrochemical cell remains the dominant choice in the PV market.

5.2.3 *Electric components*

PV installations are typically constituted by many elements and may cover large areas and therefore need many meters of wiring and a large number of connectors. Weatherproof **cables** of the correct cross-section have to be used to ensure safety, reliability and to minimize risks, voltage drops and energy losses. To preserve critical elements (batteries, inverters,...) **fuses** and **circuit breakers** are used. Since electricity generation in a PV array cannot be suppressed as long as there is sunshine, extra **insulation switches** are included in several parts of the system. Finally, in grid-connected systems **meters** are used to record PV electricity production.

5.2.4 *Mounting structures*

PV arrays have to be firmly and securely mounted on rigid structures guaranteeing the maximum exposure to sunlight, as well as ease of installation, access to connectors, ventilation, durability and cleaning. Depending on the location (roof, wall, pole, ground,...) a variety of fixtures exist. For some applications (PV power plants, PV pumping, satellites,...) sometimes solar arrays are mounted on **solar trackers**. These are moving structures that follow the sun during the day, maintaining the angle of incidence between the incoming light and a PV array close to the normal at all times and thus minimizing reflection losses and providing around 30–40% more power than a fixed array.

5.3 *Types of PV Systems*

Depending on their final use, PV systems combine the elements mentioned in the latter section in different topologies. Figures 13–16 show schematic representations of PV systems for grid-connected operation with (Fig. 13) and without storage (Fig. 14), an off-grid system with only DC output (Fig. 15) and an off-grid hybrid system with both DC and AC supply and a backup generator (Fig. 16). For the sake of simplicity, fuses, insulation switches and circuit breakers have not been included in the figures.

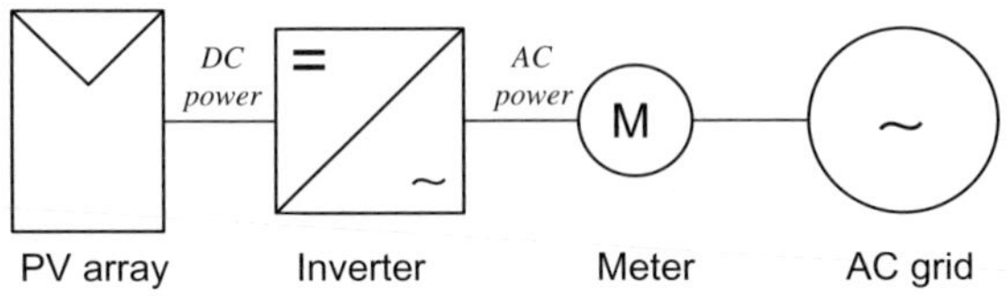

Fig. 13. Grid-connected PV system without storage.

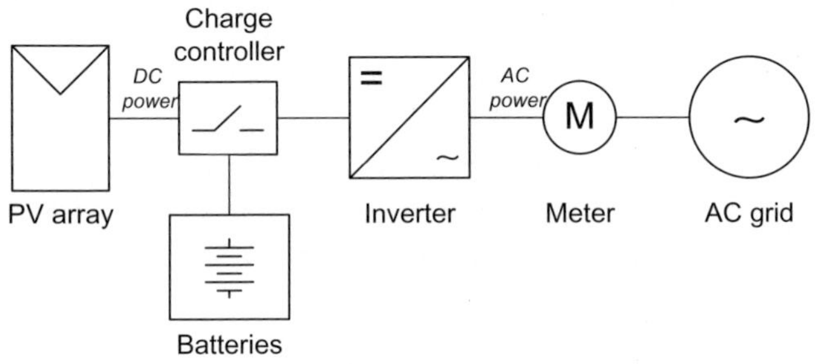

Fig. 14. Grid-connected PV system with storage.

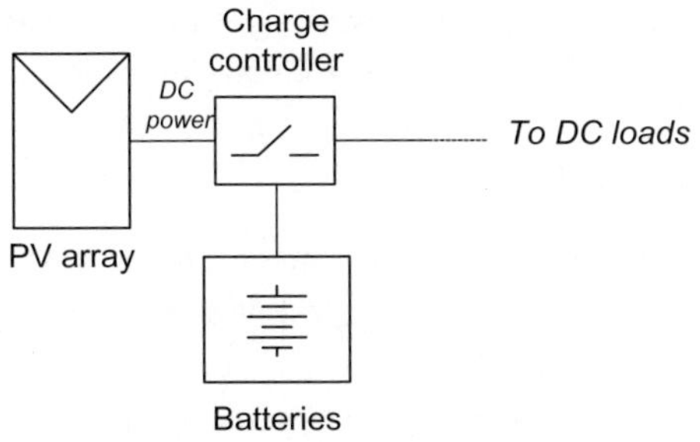

Fig. 15. Off-grid DC PV system.

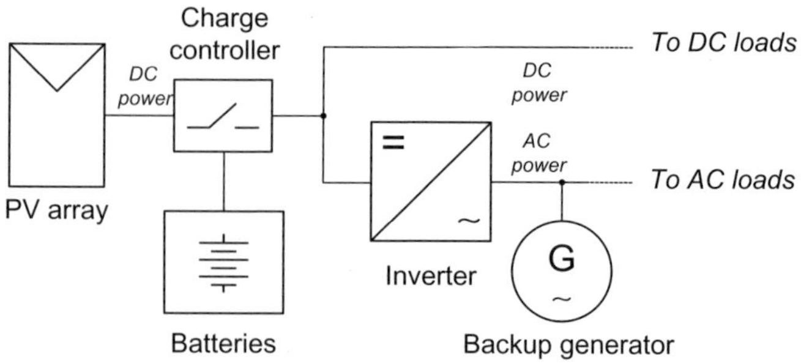

Fig. 16. Off-grid DC/AC hybrid PV system.

5.4 *Designing a PV System*

5.4.1 *Location*

The location, with a certain latitude and climate, determines the available solar resource and thus is a first-order impact factor on the performance of PV systems. The availability of long-term records of meteorological data is a key factor in assessing the suitability of a given location for PV use. PV systems should be located far from obstacles that might produce shading, mostly in the central hours of the day and throughout the year. Snow, dirt or other soiling are location dependent and are also sources of shading. Shading small parts of a PV array may cause severe losses in its energy output. If shading is unavoidable, a good design including adequate module interconnection, protective elements such as bypass diodes and special inverters, may minimize the losses.

In many situations, there is little flexibility in choosing the location of a PV array. The system should be close to the demand in off-grid applications, mounted on the available structures in building integrated PV, or near a high-voltage transmission line for power plants. In all these cases, a careful design is the tool to overcome restrictions associated with the location.

5.4.2 *Orientation and tilt*

The **orientation** of a PV array, also referred to as **azimuth angle** (α), is defined as the angle measured clockwise between true south (or sometimes due north depending on the definition used) and the array plane. The **tilt** of a PV array, also referred to as **elevation angle** (β), is defined as the angle formed by the array plane and the horizontal. Figure 17 illustrates azimuth and elevation angles.

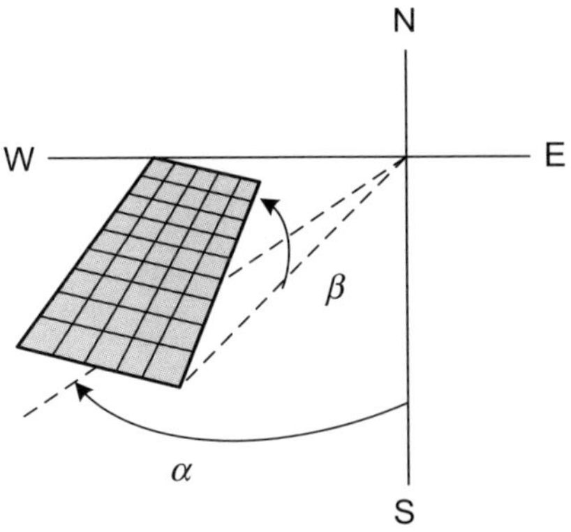

Fig. 17. Orientation or azimuth (α) and tilt or elevation (β) of a PV array.

The optimum orientation of a PV array in the northern hemisphere is true south ($\alpha = 0$), whilst on the southern hemisphere it is due north. For moderate latitudes ($\Phi < 30$), it is generally accepted that the tilt angle that maximizes the annual energy collection is approximately equal to the latitude ($\beta = \Phi$). In higher latitudes, the tilt angles used are approximately equal to the latitude minus 10° ($\beta = \Phi - 10°$).

In some applications, it is important to maximize the energy collection during the worst conditions (i.e. in the worst month in terms of irradiation and not annually). This is the case in off-grid systems, where the PV installation has to meet the user's needs through the year, including the months with less isolation. To maximize the collection in winter the tilt angles used are approximately equal to the latitude plus 10° ($\beta = \Phi + 10°$).

Figure 18 visually illustrates losses resulting from non-optimum orientation and tilt. In this figure, the accessible solar resource has been quantified in a polar chart as a function of azimuth (α) and tilt (β) for a location with latitude $\Phi = 41°$ in the northern hemisphere. It can be seen that for PV arrays with tilt roughly equal to the latitude, even deviations in orientation as high as 30°from the optimum azimuth only cause minor losses (below 10%) in the irradiation collected. Accordingly, installing cumbersome structures to modify the orientation of a preexisting support (roof, façade, wall,...) is not justified in most cases.

5.4.3 *Sizing*

In the most general case, sizing a PV system essentially implies determining the PV array size and configuration (number of modules and

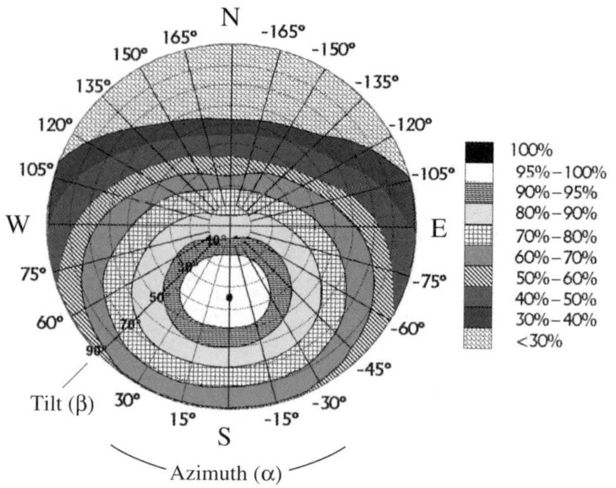

Fig. 18. Polar chart quantifying accessible solar resource as a function of azimuth (α) and tilt (β) for a location with latitude $\Phi = 41°$ in the northern hemisphere.

interconnection), the inverter size (if needed) and the battery bank size (if needed) as a function of location (available resource and statistical variability), load distribution and target **loss-of-load probability** of the system (the latter being the probability that the power demand will exceed the generating capacity of a system during a certain period). The problem of sizing acquires a critical engineering dimension in stand-alone (off-grid) PV systems, since a failure of the system to meet the desired operation may cause an interruption in the electric supply. Accordingly, many sizing methods exist for stand-alone systems.[26–29] Conversely, the design of grid-connected systems is often an exercise more determined by economics (calculating return of investment, as a function of feed-in-tariffs, tax rebates, green certificates and interest rates) than by PV engineering itself.

5.5 *PV System Performance*

5.5.1 *Output power of PV systems*

The maximum output power of an array formed by the connection of an arbitrary number (N_M) of identical PV modules can be calculated as:

$$P_A = N_M \times P_M(G, T_{Cell}) \times \eta_A, \tag{22}$$

where P_M is the actual power supplied by each module as calculated using Eq. (17) and η_A is a correction factor that accounts for mismatch losses between the modules in the array and additional losses in the cables and wiring of the system. The array power is sometimes referred to as the **DC power of the PV system**.

Most PV systems, as shown in Figs. 12–15, will include ancillary elements that may drain some of the energy produced before it is transmitted to the load (i.e. inverters, storage, charge controllers,...). Such losses should be accounted for depending on the configuration of the system including as many efficiencies or correction factors as needed. For instance, in a grid-connected system without storage (Fig. 12) the overall system output power (P_S) would be:

$$P_S = P_A \times \eta_i(P_A), \tag{23}$$

where η_i is the efficiency of the inverter which is typically a function of the input DC power. In systems with inverters, the overall system output power is sometimes referred to as the **AC power of the PV system**.

5.5.2 *Energy rating of PV systems*

Analogously as described for PV modules in Sec. 4.3.5, the electrical energy output of a PV system (E_p) over a given period of time (p) can be calculated as:

$$E_p = \sum_i P_{S,i} \times \Delta t, \tag{24}$$

where $P_{S,i}$ is the average system output power during the ith interval in p.

5.5.3 *Alternative (simpler) energy rating of PV systems*

The use of Eq. (24) needs the determination of the average in-plane irradiance and mean array temperature for each time interval i, as well as many efficiencies and correction factors for other elements in the PV system. These data are not always available or, simply, such a detailed modeling of the system is not needed. For these situations, IEC standard 61724[30] defines a series of parameters for a much simpler determination of the energy output of a PV system. Their definition is as follows:

The **PV system final yield** (Y_f) is defined as the net energy output of the system over a given period of time (E_p) — typically over a year — divided by the nominal power ($P_{A,STC}$) of the PV array:

$$Y_f = \frac{E_p}{P_{A,STC}}. \tag{25}$$

PV system final yield is measured in kWh/kW_p or simply "hours" and thus it is sometimes referred to as the "**equivalent hours**" of the system since it represents the number of hours that the PV array would need to operate at STC to provide the same energy. Typical values for the (yearly) final yield vary from location to location being $Y_f = 700\,h$ for Germany and the Netherlands, $Y_f = 830\,h$ for Switzerland, $Y_f = 1400\,h$ for Spain or California and up to $Y_f = 1600\,h$ for Israel.

The **PV system reference yield** (Y_r) is a magnitude associated with the solar resource available at the system location and is defined as total in-plane irradiation on the system (G_p) over a certain period of time (typically a year) divided by the reference irradiance at STC ($G_{STC} = 1000\,W/m^2$):

$$Y_r = \frac{G_p}{G_{STC}}. \tag{26}$$

PV system reference yield is measured in hours and represents the number of equivalent hours at the reference irradiance. Sometimes the reference yield is also referred to as the **Peak Solar Hours** of the system.

The **performance ratio** (***PR***) of a PV system is defined as the system final yield (Y_f) divided by the system reference yield (Y_r). In other words, the PR is the ratio of the actual and theoretically possible energy outputs of a PV system.

$$PR = \frac{Y_f}{Y_r}. \tag{27}$$

Accordingly, PR accounts for the overall effect of losses in the system compared to the rated (nominal) output power. In addition to the factors affecting the efficiency of PV modules (discussed in Sec. 4.3.2), in a complete PV system the PR may be affected by mismatch between the modules forming the PV array, inverter inefficiency, losses in the wiring or connections and other BOS component failures. Accordingly, the PR is an excellent metric to characterize the performance of a PV system, which, in addition, can be experimentally determined in a quite straight forward way by monitoring the energy output of such given PV system. PR values can be calculated on a daily, weekly, monthly or yearly basis, the latter being the most frequent case. For most systems yearly PR values span a range from 0.6 to 0.8. Because losses in a PV system may change during the year (temperature, soilings, shadows,...), PR values fluctuate in a 10–20% range, being greater in the winter than in the summer.[31]

Combining Eqs. (25)–(27), it is straight forward to obtain a reasonably accurate estimate of the annual energy output of a PV system as a function of its performance ratio, nominal power and irradiation data:

$$E_p = P_{A,STC} \times Y_r \times PR. \tag{28}$$

6 Uses and World Market of PV Solar Energy

6.1 *Overview on the Uses of PV Energy*

Over its first decades of development, PV technology encountered many niche applications where the conventional electric grid did not exist or access to it was uneconomical, impractical or even impossible. Therefore, the initial uses of PV systems were off-grid applications (supplying electrical power to professional equipment, rural electrification or consumer products). In the first decade of the 21st century, this situation has changed and now grid-connected applications (including utility owned or de-centralized rooftop-mounted power plants) account for more than 80% of new systems installed in the field.[33] The driving force for this change has been the implementation of various incentives (in the form of special feed-in-tariffs, tax rebates, green certificates,...) in many countries to encourage the penetration of PV technology into the electric market. Figure 19 summarizes the main application areas and uses of PV systems.

6.2 *World PV Market*

6.2.1 *Size and historic evolution of the world PV market*

Figure 20 shows the recent historic evolution of the global PV installed capacity per world region.[32–34] As of end of 2013, the world cumulated

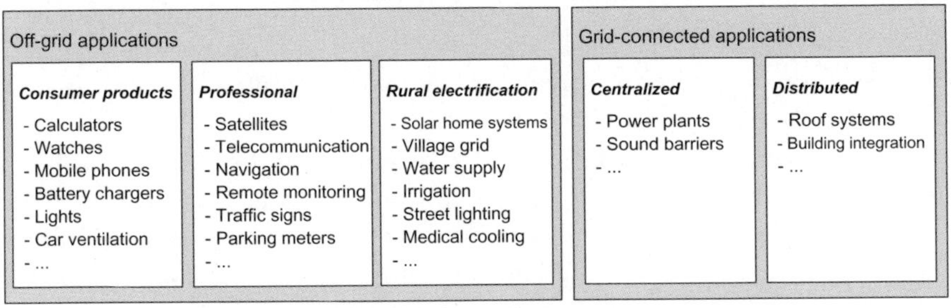

Fig. 19. Application areas and uses of PV systems.

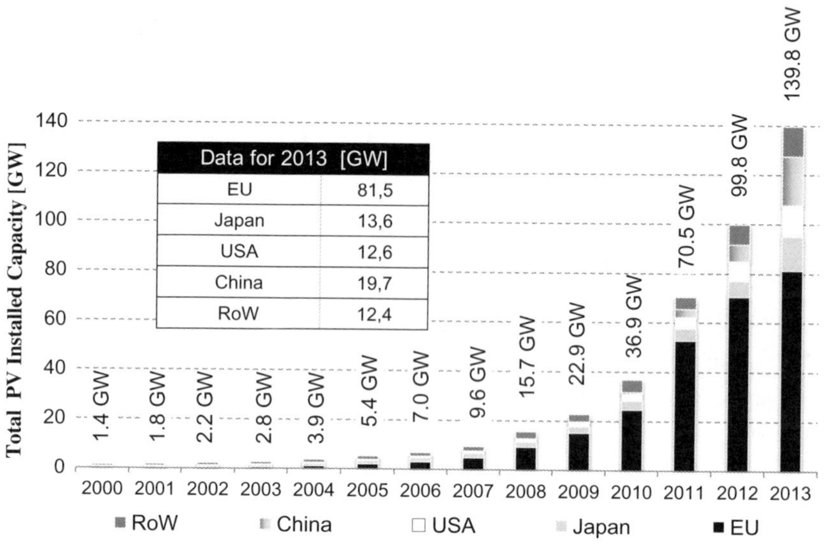

Fig. 20. Evolution of the global PV installed capacity worldwide.

Table 4: Top 5 countries in PV installations by region as of December 2013.

	Country	Installed Capacity [MW]	Installed Capacity [%]
1	Germany	35.765	26
2	China	19.720	14
3	Italy	18.074	13
4	Japan	13.599	10
5	USA	12.079	9
	Total	99.237	71

PV installed capacity almost reached 140 GW, with 60% of such capacity installed in the European Union.

In terms of the distribution of PV installations by country, Table 4 lists the top five countries as of the end of 2013, which constitute more than 70% of the worldwide PV capacity.[33] On the production side, Table 5 summarizes the PV cell production by country, also as of the end of 2013.[33] Tables 4 and 5 show that while market concentration is in Europe, production is in Asia (mostly in China).

Table 5: PV cell production as of December 2013.

	Region	[MW]	[%]
1	China	23.124	58
2	Taiwan	7.176	18
3	Japan	3.189	8
4	Malaysia	2.791	7
5	Germany	1.196	3
	Total	37.476	94

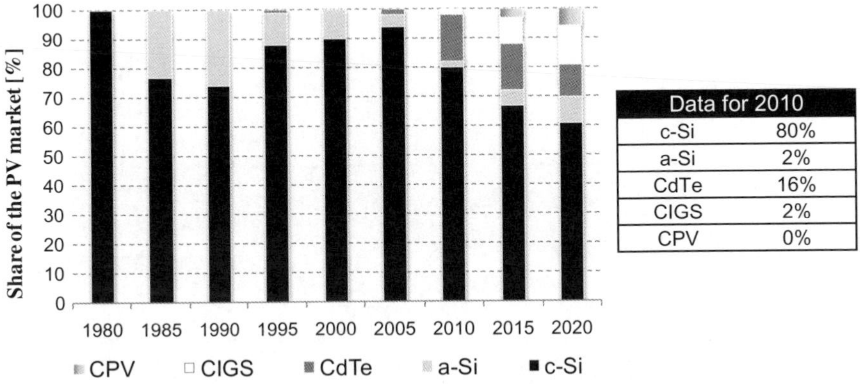

Fig. 21. Historic evolution of the share of solar cell production per technology and expected trends until 2020. c-Si includes mono and multicrystalline silicon; CPV stands for concentrator PV.

6.2.2 *PV cell production by technology*

Figure 21 shows the evolution and prediction of solar cell production by technology from 1980 to 2020.

6.2.3 *Evolution of costs of PV modules and PV electricity*

Over the past decades, the PV industry has achieved remarkable price decreases. The price of crystalline silicon PV modules has decreased by around 20% each time the cumulative installed capacity has doubled (see Fig. 22).[35] This trend seems to be even more marked for CdTe where price reductions close to 30% for each doubling of production have been observed.

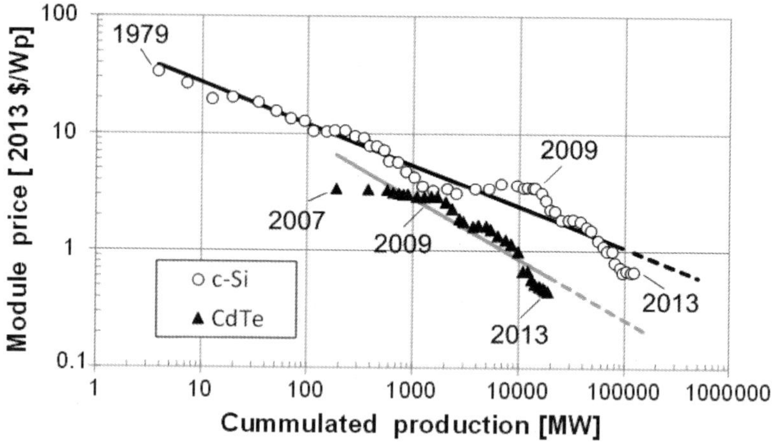

Fig. 22. PV module price experience curve for crystalline silicon and CdTe technology.

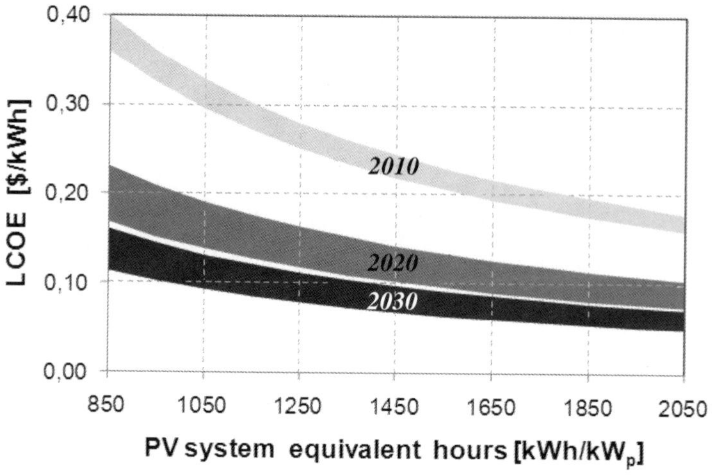

Fig. 23. Evolution of the levelized cost of electricity for PV systems.

Figure 23 shows current and future Levelized Cost of Electricity (LCOE), i.e. electricity generation costs, for large ground-mounted PV systems with PR = 85%. The data is based on the price reductions summarized in Fig. 19 and forecasts that costs of PV per kWh will be below \$0.20 for almost any sunny place by 2020.[34]

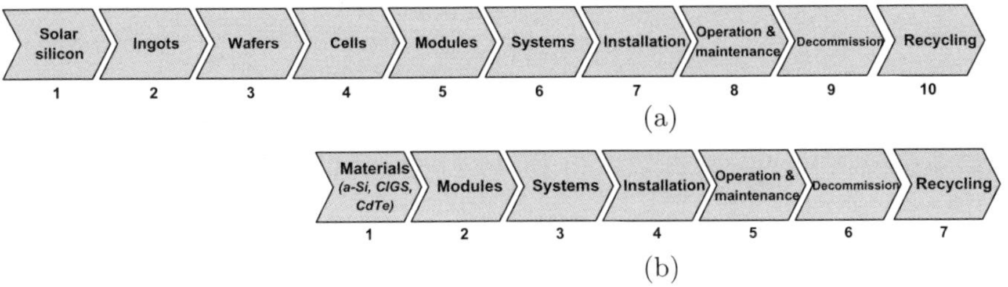

Fig. 24. (a) Value chain for wafer-based PV technology; (b) value chain for thin-film PV technology.

7 Material Usage and Environmental Impact of PV Solar Energy

7.1 *The Value Chain of PV Technology*

Different PV technologies include different processes and services in moving from raw materials to turn-key systems, as shown by Fig. 24. These differences are greatest at the first stages of the manufacturing process. For wafer-based technology (either mono or multi-crystalline silicon) modules are manufactured as assemblies of discrete solar cells, which have been previously fabricated. In thin-film technology, the whole solar cell material in a module is deposited at one time on a substrate and subsequently separated into cells as described in Sec. 4.1.2.

7.2 *Material Usage of PV Technology*

While the solar cell is the essence of a PV system, in terms of weight it accounts for only a small fraction of the total materials required to produce a solar panel. As shown in Table 6, the largest share of the total mass of a PV module is glass followed by the aluminum frame.[36–38]

Some materials used in thin film PV technology are rare (In, Cd, Se, Mo, Te) and this could become an important problem for its large scale deployment.[39] Table 7 summarizes some of the key characteristics, in relation to their scarcity, of these elements.[40]

7.3 *Energy Payback Time of PV Systems*

The **energy payback time (EPBT)** of a PV system is the time needed for the system to produce the energy invested in its manufacturing and installation.[41–44] Accordingly, when calculating the EPBT all energy invested in the value chain (Fig. 24) up to the system level has to be accounted for. A

Table 6: Material inventory for different PV technologies.

Category	Material inventory (kg/m^2 and %)			
	Ribbon-Si	Multi-Si	Mono-Si	CdTe
Cell material	0.9 (*6%*)	1.6 (*10%*)	1.5 (*10%*)	0.065 (*0%*)
Glass	9.1 (*58%*)	9.1 (*55%*)	9.1 (*55%*)	19.2 (*88%*)
EVA	1.0 (*6%*)	1.0 (*6%*)	1.0 (*6%*)	0.6 (*3%*)
Frame	3.0 (*18%*)	3.0 (*18%*)	3.0 (*18%*)	0.0 (*0%*)
Connections and packaging	1.8 (*12%*)	1.8 (*11%*)	1.8 (*11%*)	2.0 (*9%*)
Total	**15.8** (*100%*)	**16.5** (*100%*)	**16.4** (*100%*)	**21.8** (*100%*)

Table 7: Some characteristics of rare metals used in PV.

Material	Scarcity (relative to Si)	2009 spot price ($/Kg)	Max. spot price 2000–2010 ($/Kg)	2010 World primary production (in Tons)	2010 World reserves (in Tons)	Origin
Cadmium	10^{-7}	4	12	22.000	$6,6 \cdot 10^5$	Zn mining
Indium	$5 \cdot 10^{-8}$	350	900	574	?	Zn mining
Molybdenum	10^{-6}	65	70	234.000	$9,8 \cdot 10^6$	Porphyry
Selenium	$5 \cdot 10^{-8}$	75	100	2.260	$8,8 \cdot 10^4$	Cu and Ni mining
Tellurium	$4 \cdot 10^{-9}$	200	220	<200	$2,2 \cdot 10^4$	Cu mining

breakdown of these costs for silicon and thin-film PV technology is included in Table 8.

Obviously, the energy output of a PV system depends on a variety of factors (location, orientation, tilt, design decisions,...) and thus all those issues will yield different energy payback times for the same PV technology. As shown by Fig. 25 EPBT of 0.8–2.1 years are found for South-European or South US locations (irradiation around 1700 $kWh/m^2/yr$).

7.4 *Greenhouse (GHG) Gas Emissions of PV Systems*

During operation PV systems do not produce GHG emissions. It is only at the manufacturing, installation and recycling stages of PV systems when GHG are produced and are mostly (90%) caused by energy use.[45] Figure 26 summarizes the GHG emissions for several PV technologies[42,46]: As shown in Fig. 26, lifecycle GHG emissions are in the 21–54 g/kWh range.

Table 8: Breakdown of the energy cost for manufacturing a multi-crystalline silicon module and an amorphous silicon module using state-of-the-art technology.

	Embedded energy (MJ/m^2)	
Process	Silicon	Thin-film
Cell material	3200	50
Cell/module processing	300	400
Module encapsulation material	200	350
Overhead operations and equipment manufacture	500	400
Total frameless module	**4200**	**1200**
Aluminum module frame	400	400
Total framed module	**4600**	**1600**

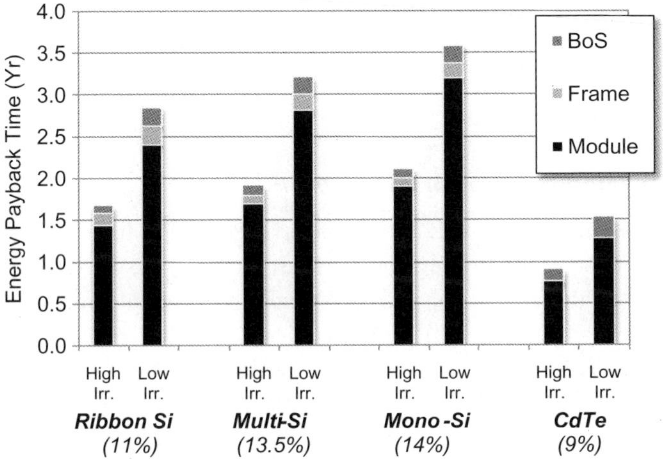

Fig. 25. Energy payback time for different PV technologies installed at two different locations (high irradiation: 1700 $kWh/m^2/year$; low irradiation: 1000 $kWh/m^2/year$).

7.5 *Operational Hazards of PV Systems*

PV systems (as any power station) have the obvious hazards associated with electricity. However, under normal operation, PV systems neither produce any emissions nor is it possible to be exposed to the small amount of toxic substances (some semiconductors or solder pastes) embedded into them.[47] The only possible exposure route is via inhalation if the modules are consumed in the event of a fire,[48,49] which would be itself a much more severe hazard.

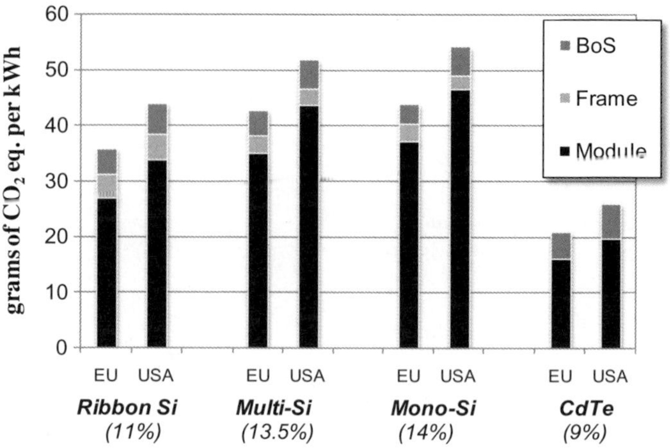

Fig. 26. GHG emissions for several PV technologies for a ground-mounted system, receiving an irradiation of 1700 kWh/m^2/year, with a PR = 0.8, and lifetime of 30 years. Two electrical grids are considered, the European (UCTE) grid mixture and the US grid mixture.

7.6 *PV Module Decommissioning and Recycling*

After their useful life — estimated to be in the 20–30 year range — PV modules will need decommissioning and recycling to both ensure that potentially harmful materials are not released into the environment and to reduce the need for new raw materials. Recent studies show that recycling the useful materials (typically glass and metals) is technically and economically feasible[50,51] and will help to preserve the environmentally friendly nature of PV technology. Accordingly, PV manufacturers are implementing voluntary take-back and recycling programs for end-of-life PV modules.[38,52]

References

1. International Energy Agency, *Key World Energy Statistics 2014* (IEA Press, Paris, 2014).
2. J. Perlin, *From Space to Earth: The Story of Solar Electricity* (AATEC Publications, Ann Arbor, 1999).
3. M. Iqbal, *An Introduction to Solar Radiation* (Academic Press, New York, 1983).
4. E. Lorenzo, Energy collected and delivered by PV modules, in *Handbook of Photovoltaic Science and Engineering*, eds. A. Luque and S. Hegedus (Wiley, Chichester, 2011).

5. J. Page, The role of solar radiation climatology in the design of PV systems, in *Practical Handbook of Photovoltaics*, eds. T. Markvart and L. Castañer (Elsevier, Oxford, 2003).
6. http://re.jrc.ec.europa.eu/pvgis/.
7. http://www.nrel.gov/gis/solar.html.
8. S. J. Fonash, *Solar Cell Device Physics* (Academic Press, Oxford, 2009).
9. M. A. Green, *Solar Cells: Operating Principles, Technology and System Application* (Prentice-Hall, New Jersey, 1982).
10. J. Nelson, *The Physics of Solar Cells* (Imperial College Press, London, 2003).
11. S. M. Sze and K. K. Ng, *Physics of Semiconductor Devices* (Wiley, New York, 2006).
12. ASTM G173-03, Standard tables for reference solar spectral irradiances: Direct normal and hemispherical on 37° tilted surface (2008).
13. R. H. Bube, *Photovoltaic Materials* (Imperial College Press, London, 1998).
14. R. W. Miles, K. M. Hynes and I. Forbe, Photovoltaic solar cells: An overview of state-of-the-art cell development and environmental issues, *Prog. Cryst. Growth Ch.* **51** (2005), pp. 1–42.
15. V. Avrutin, N. Izyumskaya and H. Morkoç, Semiconductor solar cells: Recent progress in terrestrial applications, *Superlattice. Microst.* **49** (2011), pp. 337–364.
16. D. J. Friedman, J. M. Olson and S. Kurtz, High efficiency III–V multijunction solar cells, *Handbook of Photovoltaic Science and Engineering*, eds. A. Luque and S. Hegedus (Wiley, Chichester, 2011).
17. M. K. Nazeeruddin, E. Baranoff and M. Graetzel, Dye-sensitized solar cells: A brief overview, *Solar Energy* **85**(6) (2011), pp. 1172–1178.
18. H. Hoppe and N. S. Sariciftci, Organic solar cells: An overview, *J. Mater. Res.* **19**(7) (2004), pp. 1924–1945.
19. M. Liu, M. B. Johnston and H. J. Snaith, Efficient planar heterojunction perovskite solar cells by vapour deposition, *Nature* **501** (2013), pp. 395–398.
20. M. Grätzel, The light and shade of perovskite solar cells, *Nat. Mater.* **13** (2014), pp. 838–842.
21. M. A. Green, K. Emery, Y. Hishikawa, W. Warta and E. Dunlop, Solar cell efficiency tables (version 44), *Prog. Photovoltaics* **22** (2014), pp. 701–710.
22. NREL Chart on record solar cell efficiencies. Available at: http://www.nrel.gov/ncpv/images/efficiency_chart.jpg. Accessed on 14 August 2015.
23. Solar Generation VI. A report by the European Photovoltaic Industry Association and Greenpeace, Brussels-Amsterdam (2011).
24. W. Herrmann, W. Wiesner and W. Vaaßen, Hot spot investigations on PV modules — new concepts for a test standard and consequences for module design with respect To bypass diodes, *26th IEEE PV Specialist Conference*, Anaheim, pp. 1129–1132 (1997).
25. C. R. Osterwald, Translation of device performance measurements to reference conditions, *Solar Cells* **18** (1986), pp. 269–279.
26. S. Silvestre, Review of system design and sizing tools, *Practical Handbook of Photovoltaics*, eds. T. Markvart and L. Castañer (Elsevier, Oxford, 2003).

27. J. M. Gordon, Optimal sizing of stand-alone photovoltaic solar power systems, *Solar Cells* **20**(4) (1987), pp. 295–313.
28. M. A. Egido and E. Lorenzo, The sizing of stand-alone PV-systems. A review and a proposed new method, *Solar Energy Materials and Solar Cells* **26** (1992), pp. 51–69.
29. T. Markvart, A. Fragaki and J. N. Ross, PV system sizing using observed time series of solar radiation, *Solar Energy* **80**(1) (2006), pp. 46–50.
30. Photovoltaic system performance monitoring — guidelines for measurement, data exchange, and analysis, IEC Standard 61724 (1998).
31. J. A. del Cueto, Comparison of energy production and performance from flat-plate photovoltaic module technologies deployed at fixed tilt, *29th IEEE PV Specialists Conference*, New Orleans (2002).
32. Global Market Outlook for Photovoltaics 2014–2018, European Photovoltaic Industry Association — EPIA, Brussels (2010).
33. Trends 2014 in Photovoltaic Applications: Survey report of selected IEA countries between 1992 and 2013, International Energy Association, Paris (2014).
34. US Solar Market Insight. 2013 Year in Review. Solar Energy Industries Association (SEIA), Washington DC (2014).
35. Photovoltaics Report, Fraunhofer Institute for Solar Energy Systems ISE, Freiburg (Germany). October (2014).
36. Mariska de Wild-Scholten and Erik A. Alsema (2005). Environmental Life Cycle Inventory of Crystalline Silicon Photovoltaic Module Production. MRS Proceedings, 895, 0895-G03-04 doi:10.1557/PROC-0895-G03-04.
37. V. M. Fthenakis and H. C. Kim, Energy use and greenhouse gas emissions in the life cycle of CdTe photovoltaics, Presented at MRS Fall 2005 Meeting, Boston (2005).
38. K. Sander, Study on the development of a take back and recovery system for photovoltaic products, PV Cycle (2007). Available at: http://www.pvcycle.org/fileadmin/pvcycle_docs/documents/publications/Report_PVCycle_Download_En.pdf. Accessed on November 2011.
39. V. M. Fthenakis, W. Wanga and H. C. Kima, Life cycle inventory analysis of the production of metals used in photovoltaics, *Renew. Sus. Ener. Rev.* **13**(3) (2009), pp. 493–517.
40. Mineral Commodity Summaries 2001. United States Geological Survey — USGS, Reston (2011).
41. W. Palz and H. Zibetta, Energy pay-Back time of photovoltaic modules, *Int. J. Solar Energy* **10**(3–4) (1991), pp. 211–216.
42. E. Alsema, Energy payback time and CO_2 emissions of PV systems, in *Practical Handbook of Photovoltaics*, eds. T. Markvart and L. Castañer (Elsevier, Oxford, 2003).
43. V. M. Fthenakis and E. Alsema, Photovoltaics energy payback times, greenhouse gas emissions and external costs: 2004-early 2005 Status, *Prog. Photovolt: Res. Appl.* **14** (2006), pp. 275–280.
44. E. Alsema, M. J. De Wild and V. M. Fthenakis, Environmental impacts of PV electricity generation: A critical comparison of energy supply options, *21st European PV Energy Conference*, Dresden (2006).

45. R. Dones and R. Frischknecht, Life cycle assessment of photovoltaic systems: Results of swiss studies on energy chains, *Prog. Photovoltaics* **6**(2) (1998), pp. 117–125.
46. V. M. Fthenakis, H. C. Kim and E. Alsema, Emissions from photovoltaic life cycles, *Environ. Sci. Technol.* **42** (2008), pp. 2168–2174.
47. V. M. Fthenakis, Overview of potential hazards, in *Practical Handbook of Photovoltaics*, eds. T. Markvart and L. Castañer in (Elsevier, Oxford, 2003).
48. P. D. Moskowitz and V. M. Fthenakis, Toxic materials released from photovoltaic modules during fires: Health risks, *Solar Cells* **29** (1990), pp. 63–71.
49. V. M. Fthenakis, M. Fuhrmann, J. Heiser, A. Lanzirotti, J. Fitts and W. Wang, Emissions and redistribution of elements in CdTe PV modules during fires, *Prog. Photovoltaics* **13** (2005), pp. 713–723.
50. V. M. Fthenakis, End-of-life management and recycling of PV modules, *Energ. Policy* **28** (2000), pp. 1051–1058.
51. J. K. Choi and V. M. Fthenakis, Design and optimization of photovoltaics recycling infrastructure, *Environ. Sci. Technol.* **44**(22) (2010), pp. 8678–8683.
52. http://www.pvcycle.org/.

Chapter 3

Large-Scale Solar Thermal Plants (CSP)

Manfred Becker
Solarberatung,
53797 Lohmar, Germany
manfred.becker@arcor.de

Robert Pitz-Paal
DLR, Institute for Solar Research
Linder Hoehe, 51147 Koeln, Germany
Robert.Pitz-Paal@dlr.de

Wes Stein
Division of Energy Technology, CSIRO
P.O. Box 330, Newcastle, Australia
wes.stein@csiro.au

By concentrating solar radiation a source of heat is provided to generate a hot fluid for a downstream energy conversion process. This can be used to produce electricity, solar fuels or directly use process heat for industry or for chemical applications. The fundamental part of the technology is the solar collector, the most common types being parabolic trough, linear Fresnel, central receiver and paraboloidal dishes. With approximately 4,000 MW of solar thermal power plants are presently operating commercially, and 8,000 MW of plants are presently at various stages of planning and approval (Status, 2014). The majority of existing plants are parabolic troughs using oil to generate steam for steam turbines. Presently, however, there is a strong emerging interest in central receivers (power towers) due to their ability to produce much higher temperatures at higher efficiency power cycles, and ultimately lower cost. The key for concentrating solar power lies in its ability to incorporate thermal storage and thus provide dispatchable solar power, which will also benefit the penetration of other renewables such as photovoltaic and wind. It is likely that the future for concentrating solar power will involve more high efficiency power cycles and high temperature storage for electricity generation, and solar fuels for transport and industry.

1 Introduction

More than 80% of the world's electricity is derived from fuel sources that drive thermal processes (Fig. 1). The single largest source is coal/peat, used for generating steam for steam turbines; nuclear power uses heat from the fission of uranium to generate steam for steam turbines; natural gas is used both for steam generation for steam turbines and as combustion fuel for gas turbines; and oil is used for steam raising and direct combustion in gas turbines and engines.

Concentrating solar power (CSP) provides a solar alternative for generating hot fluids needed to power these cycles. This has the attraction that much of the significant wealth of knowledge, experience and technology in the power industry today is also directly beneficial and transferable to CSP. Indeed, the advances that continue to be made in gas and steam turbo machinery and power technology generally will also be beneficial to CSP.

Solar energy arrives at the surface of the earth as a relatively diffuse source of energy. For example, the solar energy falling on one square meter of land over an average day is similar to the energy contained in 1 kg of coal. Concentrating the solar radiation provides a useful source of heat for use in energy applications based on thermal energy. This is known as *concentrating solar power* (CSP).

This chapter will provide a background to methods of concentrating solar radiation, to basic relationships of concentrating optics and energy collection, to typical applications by thermodynamic cycles for CSP, and

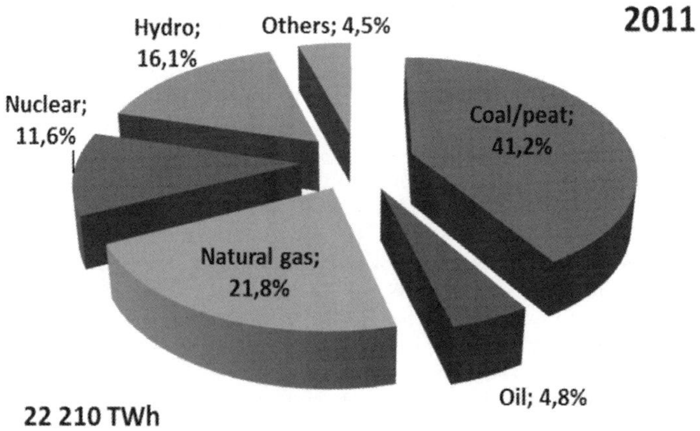

Fig. 1. Fuel use for world electricity supply in 2011.[1]
Note: Other includes geothermal, solar, wind, biofuels and waste, and heat.

giving an overview on the current status as well as future projections for the CSP industry.

2 Solar Radiation and Concentration

The solar irradiance at the outer edge of earth's atmosphere has a value of 1367 W/m^2, known as the solar constant, G. This radiation is scattered and absorbed by atmospheric constituents prior to reaching earth's surface, and arrives as a combination of diffuse and direct radiation. The *direct radiation* is that from the disk of the sun, whereas *diffuse radiation* arrives from all angles. For concentrating radiation from the sun, only the direct radiation portion is useful.

The sun itself is a disk of finite size with a half angle of 0.267° (4.653 mrad). All but the very highest concentrators can also accept *circumsolar radiation*, or light from a narrow area outside of the solar disk. The finite size of the disk means that when concentrating the radiation from the sun, it cannot be assumed that the radiation consists of parallel rays.

Figure 2 shows the principle of concentration.

Assuming that radiant flux is conserved as it passes through the concentrator, then it can be shown[2] that the concentration ratio (CR) is given as:

$$\mathrm{CR} = \mathrm{A/A'} = \sin\theta' / \sin\theta, \qquad \text{(linear focus)} \tag{1}$$

$$\mathrm{CR} = \mathrm{A/A'} = \sin^2\theta' / \sin^2\theta, \qquad \text{(point focus)} \tag{2}$$

assuming the entrance and exit media have a similar refractive index.

With $\theta = 0.267°$ (corresponding to the half-angle subtended by the solar disk), the maximum concentration ratios of point and linear focus

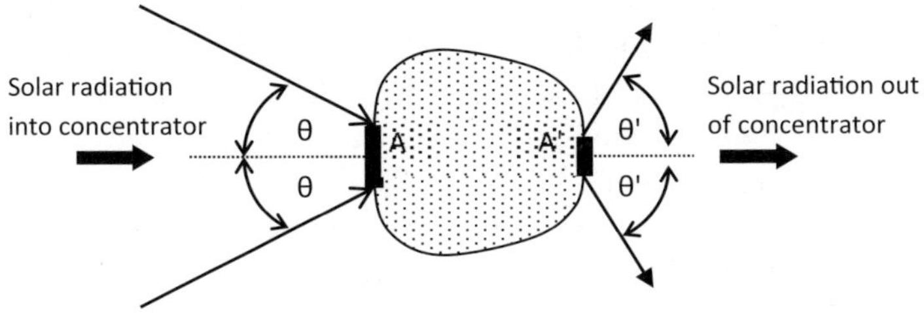

Fig. 2. Geometric quantities of a solar concentrator, where the outlet area A' (with angular spread θ') is less than the inlet area A (which has angular spread θ). An outcome of the second law of thermodynamics is that radiation can only be concentrated by increasing its angular spread.[2]

concentrators can be found by setting θ' to 90°, giving:

— Linear concentrator: $\mathrm{CR}_{\mathrm{max}} = 215$;
— Point focus concentrators: $\mathrm{CR}_{\mathrm{max}} = 46,200$.

In practice, most concentrator concepts use imaging optics to concentrate the solar disk. Imaging optics are limited in their ability to concentrate sunlight by optical aberration effects. Typically one can only concentrate rays of a single incident angle (e.g. parallel to the optical axis). Therefore, rays from the finite sized solar disk, which are not perfectly parallel, cannot be concentrated to the thermodynamic limit. However, other concepts of non-imaging concentrators exist that can overcome such constraints and approach the limit at the expense of a more complex design.

3 Receiving and Absorbing Solar Radiation

At the focus of the concentrator is the receiver located. This is the device which absorbs the incoming concentrated solar radiation, and transferred it as heat into the fluid. Once the radiation has entered the receiver aperture, the aim is to maximize thermal absorption and minimize thermal losses back to the colder environment.

3.1 *Energy Balance*

The performance of a *receiver* is determined by the net *energy balance* of solar radiation entering (gain) against thermal energy leaving (loss). Thermal losses stem primarily from re-radiation and convection (due to wind and natural circulation) and some losses also due to conduction. The re-radiation loss is particularly important in high temperature systems due to the effect of this component increasing in proportion to $\{T^4_{\mathrm{receiver}} - T^4_{\mathrm{surrounds}}\}$. Note that in some receiver designs the radiation entering the receiver will tend to be attenuated if a glass cover is used in order to reduce convective loss.

Figure 3 shows that the receiver's thermal efficiency increases with concentration ratio, but decreases with temperature. The concentration ratio has a clear effect because thermal losses are proportional to the surface area, meaning a system with higher concentration has less area from which losses emanate. Higher temperatures reduce efficiencies because they result in greater re-radiation of energy to the environment.

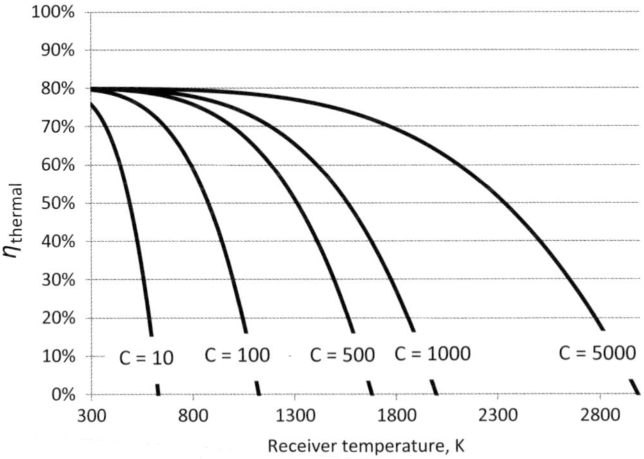

Fig. 3. Receiver thermal efficiency as a function of temperature and concentration ratio (assuming solar irradiance = 900 W/m^2, $T_{\text{ambient}} = 300\,\text{K}$, $F \cdot \alpha = 0.8$, $F \cdot \varepsilon = 0.8$, and zero convective losses.[3]

In high concentration systems it is important to absorb as much energy as possible. A selective surface (Sec. 3.2) can greatly improve the performance of lower concentration systems such as linear troughs but has less relative effect in a point focus system where the concentration ratio is, in practice, 1 to 2 orders of magnitude higher.

3.2 *Selective Surface Theory*

A *selective surface* is a coating which minimizes thermal losses by exploiting the fact that the spectrum of incoming solar radiation is different from that of thermally re-radiated energy.

The performance of a selective surface material depends on its absorptivity and emissivity properties at the operating temperature. A surface can be characterized by Kirchhoff's Law for opaque materials, which states that for a body in thermodynamic equilibrium the absorptivity α is equal to the emissivity ε at a given wavelength and angle of incidence,

$$\alpha(\lambda, \theta) = \varepsilon(\lambda, \theta). \tag{3}$$

In addition, reflectance ρ is related via

$$\alpha(\lambda, \theta) = 1 - \rho(\lambda, \theta). \tag{4}$$

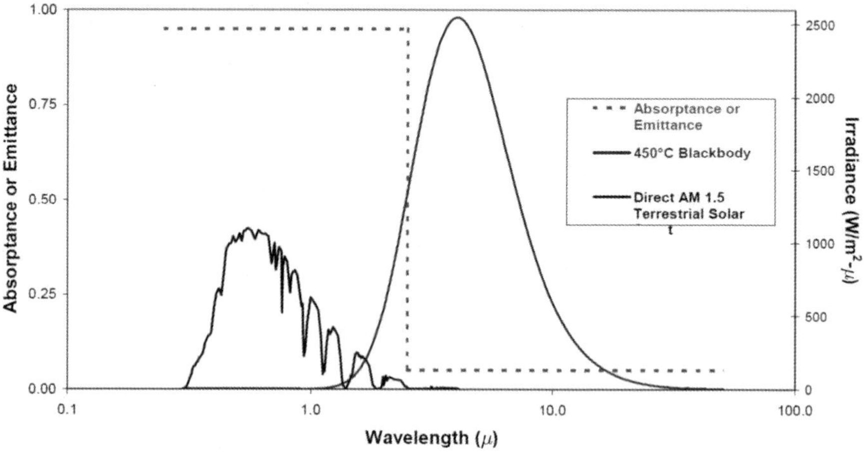

Fig. 4. Spectra for solar irradiance, a blackbody at 450° C and an "ideal" solar selective surface.[4] The solar spectrum can be approximated by application of Planck's spectral distribution corresponding to a blackbody temperature at 5777 K.

The majority of the energy of solar radiation for collector purposes arrives within a wavelength range of approximately 0.3 to 3 μm. The absorbing surface, however, can be treated as a blackbody emitter at its surface temperature which generally results in longer wavelengths for the re-radiated energy. Application of a selective surface treatment to the absorber allows it to have low reflectance and hence high absorptivity (Eq. (4)) at wavelengths less than a critical value — thus allowing maximum absorption of incoming solar radiation — while having high reflectance and hence low absorptivity and emissivity (Eq. (3)) for wavelengths greater than this value, hence minimizing re-radiation. An ideal selective surface would exhibit a vertical crossover as shown in Fig. 4.

This figure shows that at low absorber temperatures there is little overlap between the solar and the blackbody spectra. As the blackbody temperature of the absorbing surface increases, the blackbody curve moves to the left. This increases the degree of overlap with the solar spectrum and the spectrally-weighted ratio of α/ε approaches 1, such that the effectiveness of a selective surface eventually becomes negligible. Selective surfaces therefore offer more significant performance improvement potential for lower-temperature concentrating solar thermal systems than those at higher temperature.

4 Types of Solar Collectors for Power and Fuels

The term *solar collector* refers to a device that collects solar radiation into a fluid in the form of heat. This chapter will generally deal with concentrating solar applications, where the solar collector system is a combination of the solar concentrator (the mirrors) and the receiver (the device that absorbs the solar radiation and converts it to heat). Refractive concentrators such as lenses are not covered — they have application for concentrating PV systems but generally might become too large and unwieldy for larger capacity thermal systems.

This chapter will start with two types of solar technologies without concentration (C = 1), namely the *solar pond* in Sec. 4.1 and the *solar chimney* in Sec. 4.2. These technologies are included here due to the fact that they have been used for solar thermal power generation.

The real concentrating technologies will be outlined in the following. They are:

Line focusing displays the low concentrating technology of a *parabolic trough* in Sec. 4.3. This technology, with concentrations up to C = 100, is one which is simple, economic and has actually been used very successfully. An even simpler concept is that with a linear *Fresnel-type* arrangement of the reflector, discussed in Sec. 4.4.

The *central receiver* or *power tower* technology, described in Sec. 4.5, utilizes medium concentrations between C = 100 to almost C = 1000, resulting in temperatures greater than 1000°C.

The highest *point concentration* is delivered by a *paraboloid*, the three dimensional *dish construction*, shown in Sec. 4.6. It can ideally reach concentrations definitely more than C = 1000 and temperatures higher than 1000°C.

The following Fig. 5 describes the four concepts schematically.

4.1 *Solar Pond*

The most common form of *solar pond* is based on salt water. It is relatively shallow — usually no more than a few meters deep — and has a salt density gradient that increases with depth. At the bottom, the salinity is near-saturation, and it is here in this 1–2 m thick zone that maximum temperatures are reached. The solar radiation penetrates through the upper zones to the bottom, while the salinity gradient prevents natural convection

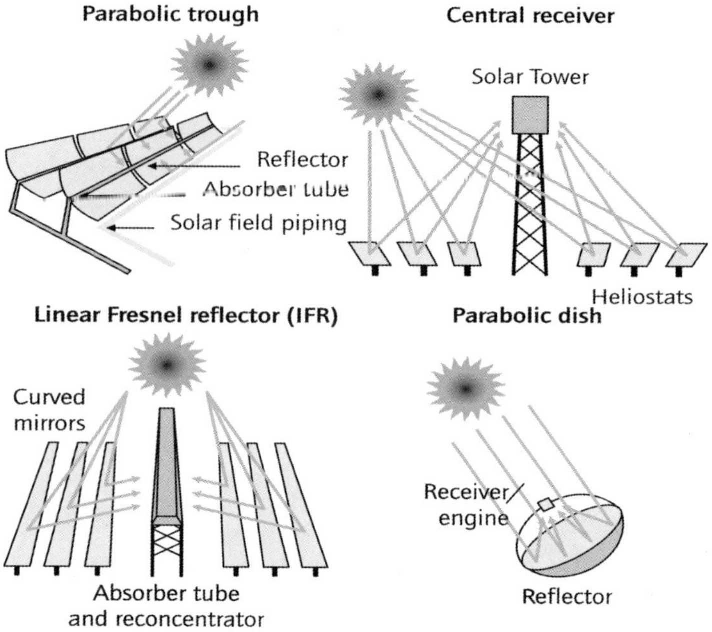

Fig. 5. Concentration Arrangements.[3]
Source: DLR.

upwards. Because the salt tends to naturally diffuse upwards over time, the salt gradient needs regular maintenance or refreshing.

One of the major requirements for a salt pond is in fact fresh or low salinity water for the upper levels due to evaporation. The salt can be common salt (NaCl), able to provide salt water densities of up to 1200 kg/m^3, or other salts such as magnesium chloride which can achieve higher densities (over 1300 kg/m^3) and better performance. The maintenance cost of a salt pond is a critical issue, with salinity gradient, excursion events, clarity, and chemical balance all requiring careful control.

Temperatures above 90°C have been demonstrated in solar ponds, though the potential thermal to electric efficiency is very low — less than 5% in practice. Their application may be most useful for low temperature thermal processes in salinity-affected areas such as desalination or production of some chemicals. An additional attraction of the salt pond is that it provides thermal storage.

A number of small plants have been demonstrated for the production of electricity, including at the Dead Sea, Israel (5 MW); Alice Springs,

Australia (15 kW); El Paso, USA (120 kW), and Pyramid Hill, Australia (60 kW)[5] in the last century. No commercial off-take has happened yet. The main hurdle is the limitation to low temperature heat which results in a low cycle efficiency and the need for a special engine design.

4.2 *Solar Chimney*

The *solar chimney*, also referred to as the *up-draught solar tower*, works on the principle that a large 'greenhouse' solar collector heats air such that its density is less than the surrounding ambient air outside the chimney. The resulting pressure difference causes air to move inside and up the chimney, passing through one or more wind turbines installed at the base of the chimney and generating electricity.

The power generated is a balance between the driving force provided by the pressure drop in the chimney (including frictional effects) and the pressure drop across the turbine. It can be shown that electrical power is optimized when the pressure difference across the turbine is two-thirds of the total available, not considering the transient coupling effects between collector and chimney, allowing maximum power to be estimated for a system with collector efficiency $\eta_{\text{collector}}$, chimney height H_{chimney} and collector area $A_{\text{collector}}$, for given conditions solar irradiance I, ambient temperature T_0 and specific heat capacity of air c_p[6]:

$$P_{\max} = 2/3\, \eta_{\text{collector}}[g/(c_p T_0)] H_{\text{chimney}} \cdot A_{\text{collector}} \cdot I. \tag{5}$$

A typical large-scale solar chimney may have air temperature at the base of the chimney about 18°C above ambient, air velocity through the turbine of approximately 11 m/s, collector thermal efficiency based on global horizontal radiation in the range 40–60% and overall solar to electric efficiency of the order of 1–2% depending on tower height.[6]

An experimental solar chimney (Fig. 6) was constructed and operated successfully from 1982–1988 at Manzanares, Spain.[7] It had a 50 kW rating, 195 m high chimney, 46000 m^2 collector area, 10.2 m chimney internal diameter (at base) and 10 m turbine rotor diameter. The plant was constructed with an intended lifespan of just three years, but ended up operating for eight. The tower collapsed in a storm in 1989 due to a corroded guy wire, though the structure was beyond its design life.

The technology has been developed and advanced largely through the German engineering firm Schlaich, Bergermann and Partner. Commercial projects from 30–200 MW have been proposed for Jordan, India, Australia

Fig. 6. Illustration of a 5 MW solar chimney.
Source: Schlaich, Bergermann and Partner.

and the southwest USA. A 200 MW plant would typically require a 1 km high chimney (with 120 m diameter at the base) and a 7 km diameter collector. With annual solar radiation of 2300 kWh/m^2/yr, such a plant could generate 680GWh/yr of electricity.[7] Very large solar fields with a tower height clearly exceeding 500 m are necessary to allow for a reasonable efficiency and electricity costs. No commercial off-take of this technology has happened yet, mainly due to the perceived risk to the investors who are reluctant to invest several $100 million in a very tall tower project without reasonable scaling steps.

The attraction of the technology is its simplicity, with capital costs able to be minimized in countries with low labor costs. However, operation and maintenance is likely to be an area of concern. If plastic is used as the collector, the lifetime will likely be considerably less than for glass; for either plastic membrane or glass, the cleaning cost could be significant given the essentially flat arrangement.

4.3 *Parabolic Trough*

4.3.1 *Trough technology*

Parabolic trough technology, often referred to simply as a *solar trough*, uses linear parabolic reflectors to concentrate solar radiation to a focal line.

A fluid flows through a receiver tube positioned along the focus and is heated for use in a downstream process.

The concentration ratio of a parabolic trough collector (PTC) is given by the ratio of collector aperture area (planar) and the circumferential area of the absorber tube, with typical values ranging from C = 10 to 100.[8]

Parabolic troughs generally have a single axis of rotation that is aligned in either an east–west or north–south orientation. A north–south axis provides greater annual solar energy collection because the angle of incidence of the sun's rays to the normal of the trough aperture (θ) is less on an annual average basis' leading to $\cos\theta$ being closer to the ideal value of one. Additionally, during the summer months when some three to four times more solar energy falls in than in winter (depending on latitude and seasonal climate), the average daily incidence angle is lowest, meaning collection is greater. An east–west system, by contrast, provides a lower seasonal variation in energy delivery.

For a solar trough at latitude ϕ on day of year n and at time of day t ($0 \leq t < 24$), the angle of incidence between the incoming beam and the normal of the aperture of the reflector, θ, can be found by the following relationships.

For a parabolic trough with east–west horizontal continuous tracking,

$$\cos\theta_{\text{ew.hor}} = (1 - \cos^2\delta \cdot \sin^2\varphi)^{1/2} \tag{6}$$

and for north–south horizontal continuous tracking,

$$\cos\theta_{\text{ns.hor}} = [(\sin\phi \cdot \sin\delta + \cos\phi \cdot \cos\delta \cdot \cos\varphi)^2 + \cos^2\delta \cdot sin^2\varphi]^{1/2}, \tag{7}$$

where the quantities δ and φ are defined by

$$\sin\delta = -\sin(23.45°) \cdot \cos[360°(n + 10)/365.25] \tag{8}$$

and

$$\varphi = 360° \cdot (t - 12)/24. \tag{9}$$

Polar axis continuous tracking

$$\text{Cos } \theta_{\text{ns.polar}} = \cos\delta. \tag{10}$$

For comparison, a dish concentrator with ideal two axis tracking gives

$$\text{Cos } \theta_{\text{two axis}} = 1.$$

In large PTCs for solar power stations, a number of trough modules are connected in series such that the overall length might be up to 150 m with apertures over 6 m wide. In these cases, the collector unit is rotated by hydraulic motors. Due to the large forces exerted, careful position control and high precision is required.

The reflectors of parabolic trough collectors are usually made from rear-silvered low iron glass (92–94% spectral reflectance) which has been thermally sagged to give the correct shape. Alternative reflectors currently use polished aluminum sandwich construction sheets and polymer film reflectors on a substrate. It is noted, however, that rear-silvered glass reflectors have now been in operation in a variety of applications and environments for over 20 years, and have also been subjected to accelerated life tests that indicate negligible drop in reflective performance with time.

The heart of the PTC is the receiver tube. In PTC power stations, these tubes comprise a steel pipe (to contain the fluid pressure) with an annular glass cover. To minimize convection losses, the space between the glass and the steel pipe is evacuated, requiring a metal bellows and special glass-to-metal weld to seal the glass to the metal pipe and allow for differential expansion. The vacuum is maintained through the use of 'getters' — chemical sponges that maintain and indicate vacuum status. The evacuated tube is especially valuable when operation temperatures are above 250°C, where the reduction in thermal losses outweighs the extra cost.

The other mechanism for reducing thermal loss is the selective surface, which has been discussed in Sec. 3.2.

Because the receiver tube rotates with the trough reflector, a suitable means of transporting the fluid to the ground is required. Both flexible hoses and ball joints have been used, with ball joints being used more commonly in modern troughs that use oil as the heat transfer fluid (HTF) due to lower maintenance requirements and reduced pressure drop.

Table 1 provides characteristics of the nine *SEGS* (Solar Energy Generating System) parabolic trough plants built by LUZ that have been in commercial operation for more than 20 years. Figure 7 shows a typical parabolic trough plant layout.

The LS-1, LS-2, and LS-3 collector family was developed by LUZ simultaneously with the implementation of the first large scale solar thermal power plants (SEGS I–SEGS IX, Table 1) in California that have been in commercial operation for more than 20 years. The collector has a modular design based on steel structures with parabolic pre-shaped, silvered glass mirrors and improved efficiency by implementation of evacuated tube receivers. For the first time this series of commercial projects justified the investment into series production facilities for key components like the

Table 1: Characteristics of the SEGS IIX plants.[9]

SEGS plant	1st year of operation	Net output MW_e	Solar field outlet temp. (C°/°F)	Solar field area (m^2)	Solar turbine eff. (%)	Fossil turbine eff. (%)	Annual output (MWh)
I	1985	13.8	307/585	82,960	31.5	—	30,100
II	1986	30	316/601	190,338	29.4	37.3	80,500
III & IV	1987	30	349/660	230,300	30.6	37.4	92,780
V	1988	30	349/660	250,500	30.6	37.4	91,820
VI	1989	30	390/734	188,000	37.5	39.5	90,850
VII	1989	30	390/734	194,280	37.5	39.5	92,646
VIII	1990	80	390/734	464,340	37.6	37.6	252,750
IX	1991	80	390/734	483,960	37.6	37.6	256,125

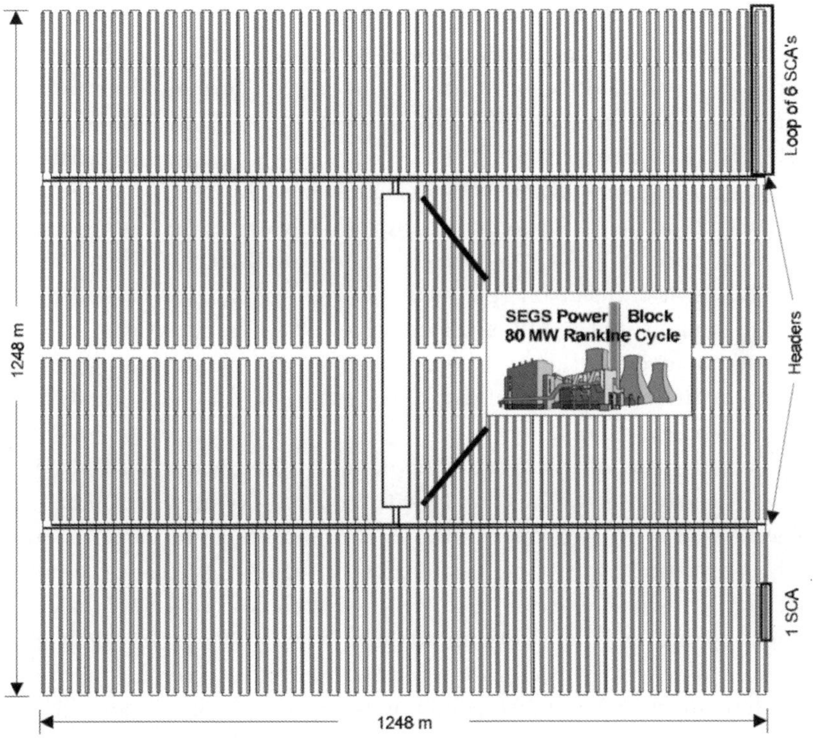

Fig. 7. Typical layout of an 80 MW parabolic trough power plant, no storage, showing a density of mirror aperture to ground covered of approximately 30% and approximately 0.5 MW/ha.[9]

curved mirrors and the evacuated absorber tube, establishing their dimensions as a *de-facto* standard for subsequent developments.

Consequently, the *EuroTrough* designed by a European consortium in the late-1990s was based on the LS-3 concentrator geometry with a focal length of 1.71 m and an aperture width of 5.77 m, but offered advantages in stiffness and costs.[10] Increasing competition in the emerging market started by the attractive feed-in tariff in Spain and led to further developments of individual variants by the different companies of the former consortium for the SKAL-ET collector that was used in the first commercial parabolic trough plant in Spain. Recent developments show a continuing trend towards larger aperture sizes as for the Heliotrough, Senertrough2 and Ultimate Trough (Table 2). These constructions have thick glass/silvered reflector and a steel torque tube or box in common (Table 2).

4.3.2 *Heat transfer field*

The fluid used in the *heat transfer field* of parabolic trough collector power plants being installed today is most commonly a thermal oil, in particular VP-1, an eutectic mixture of 73.5% diphenyl oxide and 26.5% diphenyl. This oil is attractive because it can reach reasonably high temperatures of 395°C before decomposition becomes significant. However it has a relatively high melting point (12°C) requiring auxiliary heating, and at atmospheric pressure it boils at 257°C requiring an inert gas blanket.

Other heat transfer oils are available that extend this operating temperature range slightly, but are expensive. Other heat transfer fluids are under active investigation and demonstration, viz. steam (known as *direct steam generation* — DSG), *molten salt, carbon dioxide* and radiation *on a drizzling particle curtain.*

The attraction of steam based on demineralized water is that it can be used directly in the Rankine cycle with no need for an intermediate heat exchanger. Also, it is cheaper than oil, the environmental hazards are less and higher temperatures are possible. The difficulties are that the pressure associated with useful steam temperatures becomes high, and thus collector field piping and flexible joint fittings increase costs. In addition, the two-phase flow in the evaporation portion of the receiver field can create hydrodynamic flow instabilities and temperature control over long distances is needed under transient solar radiation. Research has been carried out in the USA on DSG since the 1980s. LUZ built an LS4 DSG prototype in Israel to follow their successful LS2 and LS3 PTC's which used oil as the

Table 2: Characteristics of different parabolic trough collectors.[10–14]

	LS-1	LS-2	LS-3	Euro-trough	Helio-trough SKAL-ET	Sener-trough 1	Sener-trough 2	Ultimate Trough
Start of development	1984	1985	1989	1998	2005	2005	2006	2009
Aperture width in m	2550	5000	5.77	5.77	6.78	5.77	6.87	7.51
Length per Module/SCE in m	6.3	8	12	12	19	12.27	13.23	24
SCA length in m	50.2	47.1	99	147.8	191		158.8	242.2
Focal length in m	0.68	1.40	1.71	1.71	1.71	1.71	2	
Torsion force carried by	Torque tube	Torque tube	V-truss Frame-work	Torque box	Torque tube	Torque tube	Torque tube	Torque box

heat transfer medium. A considerable research program was conducted by CIEMAT, DLR, ZSW and Siemens in the 1990s. A European R&D project named DISS (Direct Integrated Solar Steam) test facility was built at the PSA in Spain in 1998 and had been extended and used for considerable testing since then.[15] The current configuration is a single row of length 750 m and 3,822 m^2 aperture area, providing a nominal 1.8 MW_{th} with 1 kg/s of superheated steam at 400°C and 10 MPa pressure. The experimental results concluded that no additional heat transfer enhancement devices are required to keep within allowable temperature gradients. The preferred arrangement to ensure reliable control is recirculation rather than once-through, or two-phase stratification in the evaporator section prevented (or minimized) by water injectors and mass flux control (Fig. 8).

Molten salts already have a long history as heat transfer fluids in solar tower power plants including the French Themis Power Plant (1983),[16] the US Molten Salt Electric Experiment (1983)[17] and in the Solar Two plant (1996).[18] In the last few years, investigations of liquid salts in parabolic trough systems have also emerged. Liquid salts promise to overcome drawbacks of the classic heat transfer medium diphenyl/diphenyl ether. Salts are chosen due to their suitable thermophysical properties; they comprise high boiling/decomposition points, low vapor pressure, large specific heat capacity, large thermal conductivity and high densities at low pressures.[19] A higher process temperature leads to a significant rise in the thermodynamic

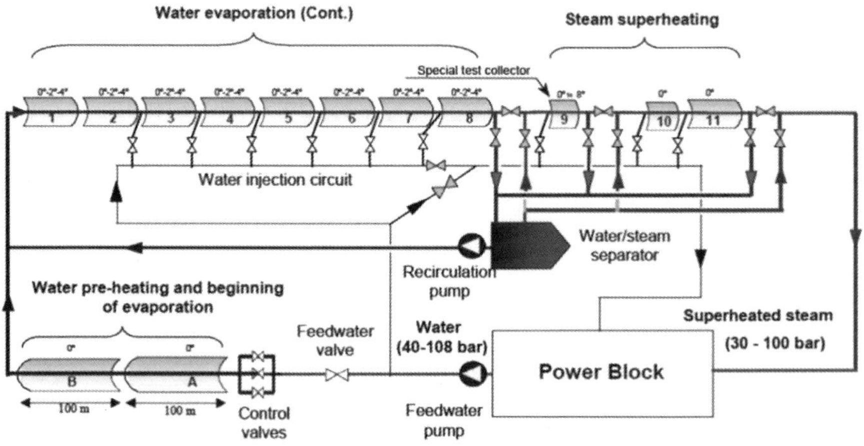

Fig. 8. EU project "direct integrated solar steam (DISS)" on the PSA, Spain. *Source*: DLR.

conversion cycle efficiency. The large capability to store energy in small volumes has two advantages. Solar field parasitics can be significantly lowered due to the very low flow velocity of salt and also the amount of storage mass is drastically lowered. Furthermore, typical salt mixtures are significantly cheaper than synthetic oils. Heat can therefore be absorbed by the salt in the irradiated receiver tubes and directly stored in big flat-bottom storage tanks. The solar field and power block are fully decoupled; such a system is beneficial for satisfying the demands of full dispatchable power plants.

The current main candidates are nitrate salts. Sodium (60 wt.%) and potassium (40 wt.%) nitrate is being used in the 5 MW-Archimede plant.[20] The upper allowed temperature is above 550°C[21]; first crystallization of the non-eutectic melt occurs at 238°C. Adding further substances to the salt allows lower crystallization and freezing temperatures. Very promising future nitrate salts are calcium/sodium/potassium nitrate and lithium/sodium/potassium nitrate; the eutectic mixtures achieve solidification temperatures of 133°C and 120°C, respectively,[19] while still maintaining high maximal temperatures of >480°C and 550°C, respectively.[22] The overall group of currently investigated salt candidates varies widely.[23,24]

Nevertheless, there are also concerns about molten-salt-based parabolic trough plants. Due to the high solidification temperature the process setup needs to be adapted. Thus, the solar plant must be fully equipped with impedance and trace heating systems in order to secure non-freezing of the salt. Salt melts at high temperatures are highly corrosive, but the latest research and process experience with demonstration plants shows that this can be handled with the right choice of stainless steel.[25] Both of these main issues and their handling lead to higher investment costs. In current research projects[22,26,27] it has been demonstrated that the increase in total plant efficiency and lower storage and HTF costs outweigh the higher investment in solar field and piping and therefore lead to a decrease in the electricity costs from solar thermal power plants.

Gaseous heat transfer media like air or CO_2 are also proposed to be used in parabolic trough systems to reach operation temperatures up to 600°C. The Swiss company Airlight has specifically developed a receiver for non-pressurized air with secondary optics for re-concentration and spillage minimization, based on coiled cavity heat exchangers and low thermal inertia radiative shields insulation. The receiver implements a cross-flow design and has inlet and outlet on the same side, which minimizes piping. This solution is considered inexpensive and environmentally friendly. It also makes it possible to use a packed bed storage concept that uses a closed container filled with stone gravel. A 3 MW demonstration plant built close to Agadir (Morocco) is starting up in 2014. Another approach has been proposed by CIEMAT and leads to a test loop that has been erected at the PSA,[28] with the aim of working with CO_2 and other gases at temperatures and pressures up to 525°C and 100 bar, respectively. High pressure leads to low pumping power requirements so that a conventional tube design is feasible. CO_2 has been selected as more advantageous than helium or nitrogen, mainly due to the fact that CO_2 has higher density at the same values of pressure and temperature. Coupling such system directly to closed gas-turbine cycles (Joule–Brayton) has been proposed.

4.4 *Linear Fresnel*

Large reflectors broken down into multiple Fresnel sub-elements can help to improve the manageability of moving components. The concept was first advanced by Baum[29] in 1956 and later important development work in linear Fresnel systems was undertaken in 1962 by the Italian patent No. 6788664 of Francia, described by an Italian group for the history of solar energy by Silvi,[30] and also containing tower concepts. Essentially, the linear focusing Fresnel system is a still simpler variation of the parabolic trough, whilst the central receiver (a quasi point area focus system) could be considered split into many almost planar dishes with various focal lengths.

The linear focusing Fresnel concept was revived for its potential commercial application in 2000.[31] Among the attractions of this technology is that the moving structural components are at ground level with reduced wind loading and cost compared to a trough, and the receiver is an inverted cavity fixed in space. A fixed receiver can be designed without the limitation of flexible joints, making direct steam generation quite feasible, as well as allowing the possibility of very long lengths, thus reducing end losses. Some receiver designs have proposed a simple boiler tube arrangement, and others are using evacuated tubes from trough technology with a secondary reflector.

The disadvantage of the linear Fresnel system is that annual optical efficiency is around 43% (depending on latitude) compared to a parabolic trough field of approximately 57%. The aperture-specific investment cost of a linear Fresnel field in order to break even with a parabolic trough field has been calculated to between 29% and 78% of the parabolic trough costs, depending on optical and thermal performance assumptions, the definition of the aperture, as well as operation and maintenance (O&M) costs.[32]

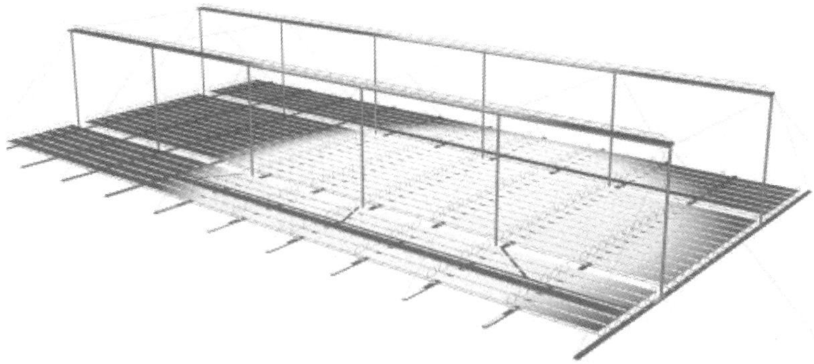

Fig. 9. Illustration of a compact linear Fresnel reflector in which reflectors can target one or another receiver depending on time of day and solar incidence angle.[33]

Over the last 10 years, there has been a rapid rise in linear Fresnel demonstrations and projects (Fig. 9). In Australia, a 1.3 MW demonstration loop was integrated with the Liddell coal-fired power station by Solar Heat and Power (now AREVA) in 2005, and is presently being upgraded to a 9 MW system by Transfield and Novatec Solar.[34] Further, north in Queensland, AREVA are building a 44 MW linear Fresnel as a solar boost to the dry-cooled supercritical Kogan Creek coal-fired power station.[35] More recently, an AU$1.2 billion, 250 MW linear Fresnel and natural gas hybrid project called "Solar Dawn" has been announced for Australia.[36]

Other linear Fresnel projects include the AUSRA's 5 MW Kimberlina demonstration project in California,[37] the Man Ferrostaal 1 MW loop in Spain (2007), and the plants PE1 (1.4 MW, Fig. 10) and PE2 (30 MW) built by Novatec Solar in Murcia, Spain.[38]

4.5 *Central Receiver* (*Power Tower*)

Central receiver systems[39,40] comprise many ground-mounted reflecting panels (*heliostats*) that individually track the sun in two axes and reflect it to a receiver mounted on a tower (Fig. 11).

4.5.1 *Heliostats and field layout*

Heliostats are typically based on rear-silvered glass, though some stretched membrane designs have been tried. Sizes range from 1 m^2 to 150 m^2 each, and prototypes of up to 200 m^2 have also been developed. They are driven around each axis by either a motor and gearbox, linear actuators or a

Fig. 10. Novatecs Solar PE1: A 1.4 MW demonstration plant at Murcia, Spain. PE2 (30 MW) is presently under commissioning on an adjacent site.
Source: Novatec Solar.

hydraulic system. Control is most often accomplished with open loop controllers using well-known algorithms for the solar position. However, closed loop heliostat positioning systems are also used in some designs.

The function of the control and actuation system is primarily to maintain the normal of the reflective surface such that it bisects the angle between the sun, heliostat and receiver, with refined systems enabling target-alignment strategies on the receiver. The most common actuation mechanism in use involves rotation about the azimuth axis with adjustment for the altitude angle (e.g. elevation). However, others such as pitch/roll and target-aligned also offer particular benefits.

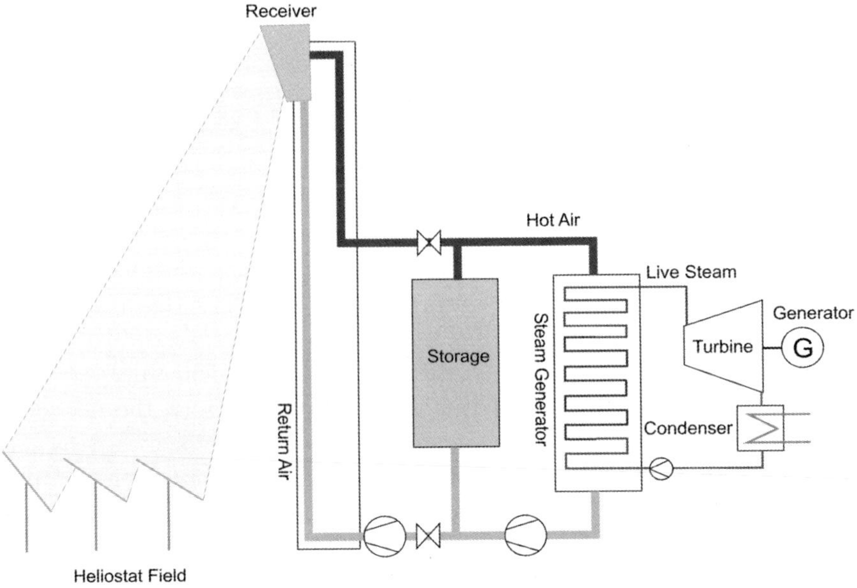

(a)

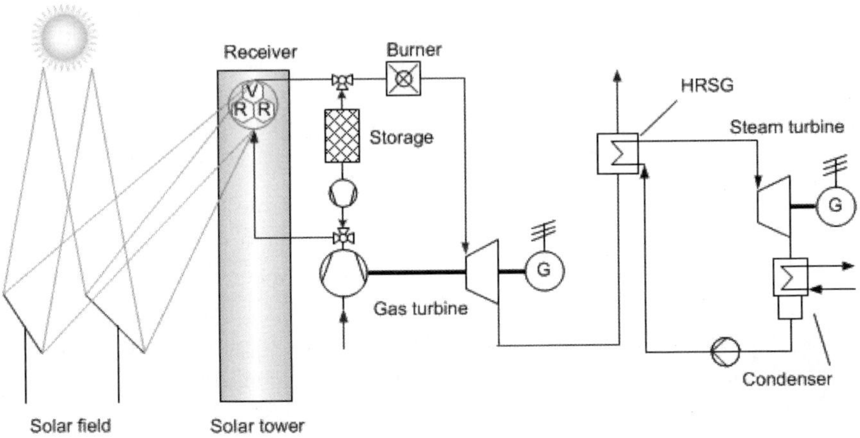

(b)

Fig. 11. Illustration of the central receiver concepts[3] using multiple heliostats to reflect solar radiation to a central fixed receiver. (a) with medium temperature steam cycle; (b) with high temperature gas cycle.

Source: DLR.

4.5.2 *Receivers*

There are generally three types of central receivers in use.

In *external receivers*, the radiation directly heats the tubes that contain the heat transfer fluid with tubes exposed to the atmosphere. They are most often used for medium temperature applications such as steam generation.

Cavity receivers use a tube bank as the absorber surface within a cavity, but the aperture where the concentrated radiation enters is much smaller than the absorbing surface and is useful for minimizing re-radiation and convection losses in higher temperature applications. Note that partial cavities are possible.

Although point focus concentrators are capable of delivering very high concentration ratios to the receiver, the ability to absorb the energy is dependent on the heat transfer characteristics of the interface between surface and fluid. Thus, in many cases, the peak allowable flux on the receiver is heat transfer limited rather than being limited by the solar concentrator. For example, in-tube pressurized air or superheated steam in a tubular receiver might need to be limited to an irradiance of under $200\,kW/m^2$ because of the poor heat transfer properties of a gas. In contrast, the evaporator section of a solar boiler might handle radiation of the order of $500\,kW/m^2$, molten salts up to $1000\ kW/m^2$ (Fig. 12) or more in the case of volumetric receivers (as outlined in the following), and particle receivers even higher.

Volumetric receivers are those where, instead of using tubes to contain the fluid (which may fatigue under a high number of high temperature cycles and which limits the allowable incident flux due to heat transfer limitations), the receiver comprises a porous arrangement of metal honeycomb or ceramic foam material to absorb a very high concentration of flux. This provides a small surface area for high heat transfer to the fluid. Volumetric receivers may be at atmospheric pressure or, if pressurized, require a transparent window (with the absorber in a cavity behind). Both volumetric receiver techniques have been developed by DLR, tested in the solar furnace and operated on the towers of the PSA.[47,48]

Figure 13 describes the potential of the open cycle concept, especially considering the fact that air as a heat transfer medium is always available, free and without any danger. Conduction and convection losses are minimal; exit temperatures range at the very "downstream" end of the material layer at maximum. And even better: the gas exit temperatures nearly equal the material temperatures at the downstream end.

Fig. 12. Solar Two in California — a 10 MW demonstration of a receiver using molten-salt as the heat transfer fluid. Operating from 1996 to 1999, it provided the technology basis for new commercial projects currently operating or under development that promise cost-effective long term thermal storage.

Source: SANDIA.

The details of the DLR Jülich pilot power plant with open cycle volumetric receiver are shown in Fig. 14. The closed cycle solution is given in Fig. 15.

4.6 *Dish Concentrators*

From an optical perspective, the ideal solar concentrator is the *paraboloidal dish.* It tracks the sun continuously around two axes, and because the receiver moves with the dish it is always collinear with the sun resulting in no cosine losses. The dish provides the highest levels of concentration and is useful for applications requiring high solar flux.

As with heliostats, a wide range of solar dish sizes have been developed. Some very small applications (under 1 kW) have been proposed, but generally sizes have ranged from approximately $40\,m^2$ up to the largest ones built of nearly $500\,m^2$ (Fig. 16).

Dish systems have been classified as distributed or centralized. *Distributed dish systems* have generally been based around the Stirling engine, though micro turbines (based on the Brayton cycle) are also under investigation. In each case, electricity is generated directly from the

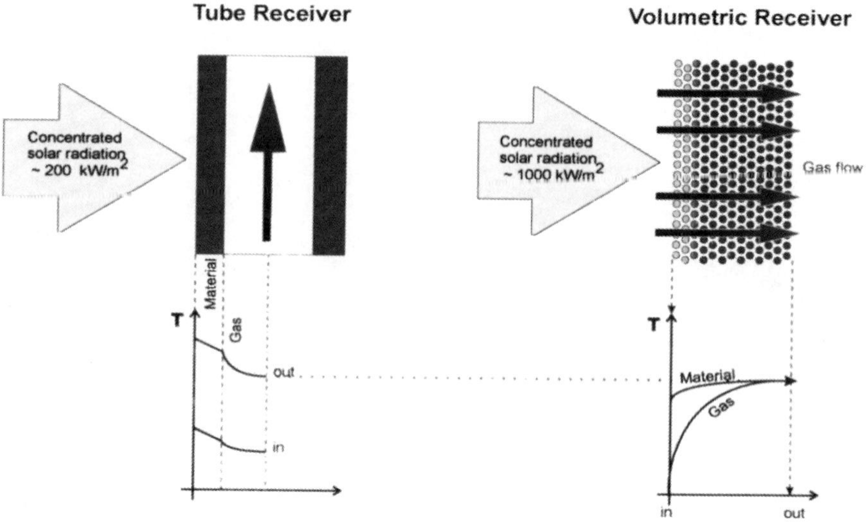

Fig. 13. Principle of heat transfer and temperature distribution, comparing the concepts of "in-tube receiver" with "volumetric receiver".
Source: DLR.

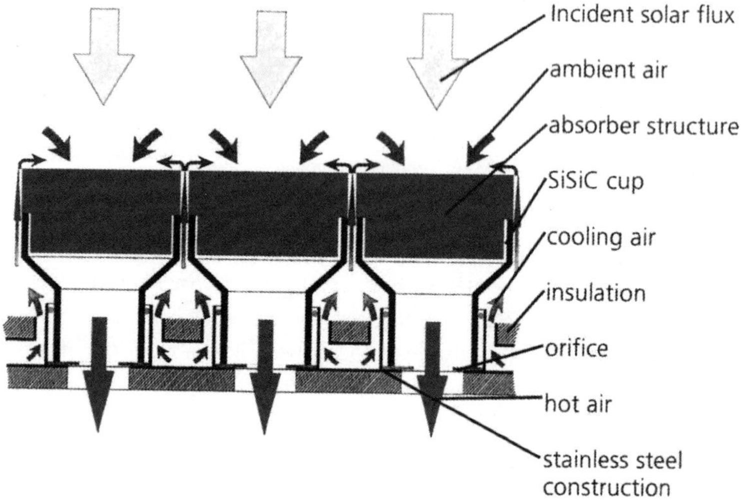

Fig. 14. Open air volumetric receiver at the DLR Juelich Power Tower.[60]
Source: DLR.

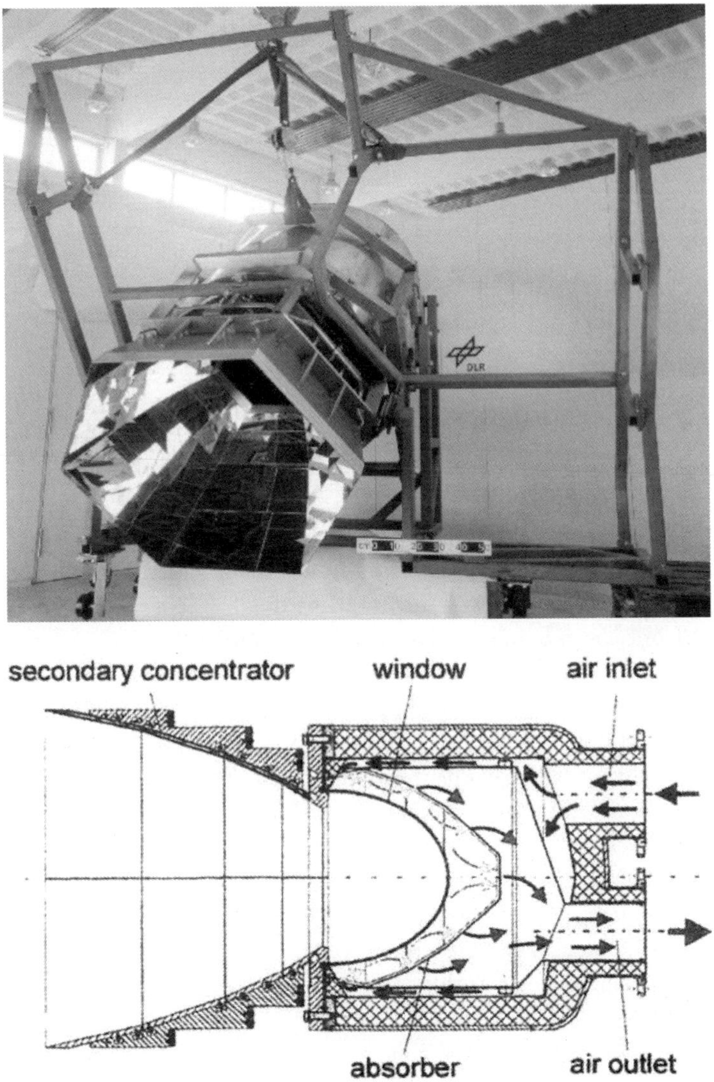

Fig. 15. Closed cycle volumetric receiver arrangement.[48]
Source: DLR.

engine/generator which is mounted at the focus of the dish. This allows relatively simple collection of energy from many point sources.

In *centralized dish systems*, an array of dishes is used to generate a high temperature fluid. This provides high solar to thermal efficiency at the

Fig. 16. Left: One of the largest solar dishes in the world at the Australian National University, Canberra, shown here with a steam receiver. It has a 494 m^2 aperture, 13.4 m focal length, 50.1° rim angle and total mass of 19.1 tons. The reflector uses square panels with 1 mm rear-silvered glass adhered to a pre-formed sandwich substrate. Right: 10 kW dish Stirling units installed at PSA, Spain.

Source: Left: ANU. Right: DLR.

receiver but requires the aggregation of fluid from multiple point sources to the central facility. Each dish also requires flexible joints to allow the fluid to travel from the moving receiver to the ground.

The largest centralized dish demonstration to date was the *Solarplant 1* system built in Southern California by the Lajet consortium in 1983/84. It comprised 700 dishes with a total collecting area of 30590 m^2. The dishes generated steam in their cavity receivers, with 600 of the dishes producing saturated steam and others taking the saturated steam and superheating it to 460°C. The project experienced problems with the reflective film used on the dish facets and long start-up times.[40]

More recently, Wizard Power have proposed a project based on large dish technology generating high temperature steam for a 40 MW steam turbine[41] using dish collector technology developed at the Australian National University.[42]

A variety of dish reflector concepts have been developed. Some early designs used a fiberglass mold to which reflective tiles were glued. An interesting stretched membrane concept was used for a number of the dish Stirling demonstration units. In this design, the dish is shaped into a parabola by plastically yielding the membrane using a combination of uniform and non-uniform loading, with mirrored tiles or reflective film providing the reflection. Membranes for the Schlaich, Bergermann and Partner dish Stirling units in Spain (44.2 m^2, 9 kW) were made from 0.23 mm thick stainless steel sheet. Shape is maintained in practice by drawing a slight vacuum behind the membrane. If a temperature excursion occurred at the receiver, the vacuum could be released quickly to disrupt the flux.

Up to now no real commercial system has been built using a dish Stirling system. On the one hand, this is due to the fact that the effort in operation and maintenance of single units is still significant at the current stage of development. On the other hand, the integration of thermal storage is more complicated than in trough, Fresnel or tower concepts due to the small size of the individual collector-receiver unit. Thus, competition with PV systems is very challenging.

However, in the field of process heat supply, in particular for cooking, a specific dish application has been very successful. W. Scheffler[43] has developed a fixed focus concentrator which not only provides simple and precise automatic tracking but also a fixed focus away from the path of the incident radiation. This design also provides an opportunity to shift the receiver for indoor applications. The versatile reflector rotates along an axis parallel to a polar axis with an angular velocity of one revolution per day from east to west to counterbalance the effect of earth rotation. Therefore, the relative position of the Scheffler reflector with respect to the sun remains stationary and provides a fixed focus on the line of the axis of rotation. The reflector not only provides daily tracking but also a seasonal tracking device to ensure that the focus remains at the same fixed point with changing solar declination. Nevertheless, there is little compromise on the aperture area as compared with the conventional paraboloid concentrator. Scheffler reflectors are used today in different sizes from $2\,m^2$ to $60\,m^2$ capable of delivering temperatures in the range of 300°C. They are used for cooking purposes e.g. by providing steam to large community kitchens and hospitals in India and elsewhere, with several thousand systems in operation today. The use of solar energy for the generation of steam is now an economically attractive possibility since the payback period of such a system lies between 1.5 and 2 years.

5 Thermal Storage

Solar and wind power are generally disadvantaged by the intermittent nature of the source. Not only does this lead to uncertainty for the grid operator, but also invested capital such as the turbine and the balance of plant equipment remains unused for much of the year. For CSP, the solar field requires about one-third to one half of the total capital cost, and of course this cannot be used when the sun is not shining. However, if thermal

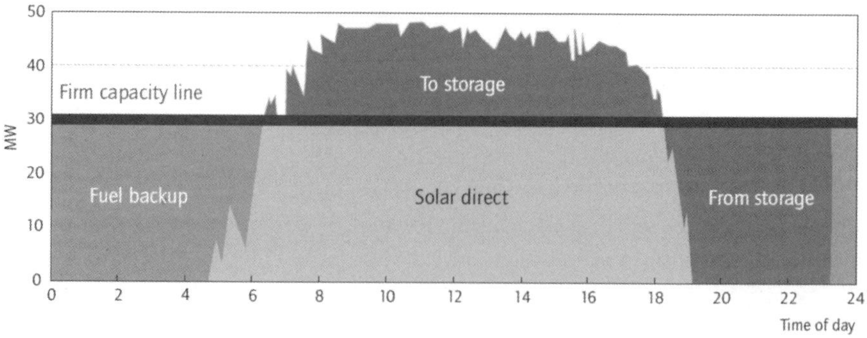

Fig. 17. Schematic of the basic principle of storage in a CSP system. The solar field, storage and turbine may be sized for any desired outcome — from simply charging the storage during solar hours for later turbine operation, through to simultaneously running the turbine and charging the storage during solar hours with the turbine then continuing to operate into the night.

Source: IEA CSP Technology Roadmap, 2011.

storage is employed, the balance of plant equipment, comprising the electricity generating equipment such as the turbine, can be used to continue generating electricity regardless of solar conditions.

The basic concept is that solar energy during the day is used to charge a thermal store. Then, on demand, that stored heat is used to generate steam for a steam turbine or otherwise heat a working fluid or feedstock in a controlled manner for a downstream energy conversion process (Fig. 17).

The US Department of Energy has established a target cost for thermal storage of under \$15/kWh based on round trip efficiency.[44]

5.1 *Two-Tank Molten Salt*

The thermal storage technology of choice in today's commercial CSP systems is the two-tank molten salt system. A number of 50 MW trough projects (Fig. 22), with oil as the heat transfer fluid in the PTC field interfacing with a salt storage loop, are now operating with this approach, though this limits temperatures to that of the heat transfer oil. A new PTC receiver tube designed for molten salt is under early test operation.

A 20 MW tower developed by Gemasolar in Spain (Fig. 24) began operation in 2011 using molten salt as both the receiver heat transfer fluid and storage medium. Based on the pilot scale concept originally developed at Solar Two[45] in California, it uses salt as the HTF in the receiver (565°C) and a two-tank storage system. Designed for 15 hours storage (6500 hours

Table 3: Data for the Andasol parabolic trough power plant near Guadix.[54]

Turbine Capacity (gross)	50 MW_e
Turbine Efficiency (Full Power)	37%
Storage Thermal Rating	1010 MWh, 7.4 hrs
Storage Tank Size, H × D	2 tanks, 14 m × 38.5 m each
Storage Fluid	Solar salt (60% $NaNO_3$, 40% KNO_3)
Storage Heat Exchanger	16,000 m^2
Solar Collector Field Aperture Area	510,000 m^2
Solar Collector Assembly, W × L	5.77 m × 148 m
Solar Collector Row Spacing	16.2 m
Solar Collector Fluid (HTF)	Therminol VP-1

annually), it has demonstrated 24 hour operation on solar energy. Many new CSP projects are incorporating two-tank molten salt storage, and several large projects, both tower and PTC, are under construction.

5.2 *Single Tank Molten Salt*

One way to reduce storage costs may be to use a single tank system with a thermocline, where the hot and cold molten salt is contained in a single vessel, separated vertically by a small high-temperature gradient region. The single tank system has a much lower volume than the total volume of a two-tank system. The two isothermal regions remain vertically stratified through thermal buoyancy forces. A low cost filler material (such as quartzite rock) with high specific heat and low void fraction (<0.25) is used in the tank to further reduce the inventory of the relatively expensive molten salt. Comparative studies have shown that the thermocline approach may offer a 35% cost reduction over the two-tank system.[46] The issues that have been raised are thermal ratcheting (when a filler material is used) and also the risk related to disturbance of the stratified regions when charging or discharging and the consequent loss of temperature difference. This latter effect is managed through well-placed inlet and outlet points, distributor regions and suitable flow rates.

5.3 *Alternative Thermal Storage Options*

Molten salt offers a number of attractions such as relatively low cost, and good specific heat and density, but poses problems both at the low temperature end (due to elevated freezing temperatures) and the upper end (due to instability at temperatures that would be even more attractive for thermal power cycles). Given that point focus solar collectors can reach temperatures in excess of salt upper temperatures, alternative materials are constantly under development to enable thermal storage at temperatures well in excess of the present ≈600°C.

Solids such as graphite, silicon carbide, alumina, high temperature concrete and fluidized sand have been investigated as storage media. Generally, the weight of these materials for any significant storage capacity precludes them being located at the receiver in the case of central receivers or dishes, although they may be useful for overcoming operational transients. Thus, a high temperature heat transfer medium is usually needed to move the heat from the receiver into the solid material. Heat transfer fluids considered above, include supercritical fluids such as steam and CO_2, lower pressure gases such as helium and air, and even solid particles used in a "falling curtain" arrangement in the receiver. Given the solid material will not have freezing point issues and will generally have a much higher upper temperature limit than today's solar salt formulations, the HTF/bulk solid material combination may be a cost effective form of storage particularly for higher temperature CSP applications.

The Jülich Solar Tower project has reported results for temperatures up to 680°C with success, using alumina porcelain as the packed bed material in a regenerative design.[47] High temperature concrete has been investigated and has shown long-term stability for temperatures up to approximately 500°C, suitable for parabolic trough power plants.[49] As an indication of thermal capacity, approximately 50000 m^3 of concrete would be needed for a 50 MW plant with design storage capacity of 1,100 MWh plant similar to that shown in Fig. 22.

The above thermal storage options rely on sensible heat, and thus, as heat is drawn from the storage media, the temperature begins to drop. Phase change materials provide a constant discharge temperature over much of the inventory by operating within the latent heat region. This is particularly valuable for the evaporation section of steam generators. However, this requires the development of phase change materials (PCMs) for the desired operating temperature. Many PCMs have been formulated for low temperatures; however there are few low-cost synthetic options available for CSP at present. The phase change temperature of KNO_3 (333°C) and $NaNO_3$ (308°C) may be suitable for low temperature application, and NaCl (802°C), Na_2CO_3 (851°C) and K_2CO_3 (891°C) for high temperature application but present management difficulties such as corrosion. Liquid metals are a form of PCM though they are relatively expensive. PCMs often exhibit poor thermal conductivity — one option is to encapsulate the

PCM material within a matrix of higher thermal conductivity to improve the surface area available for heat transfer.

5.4 *Thermochemical Storage*

An alternative to sensible heat storage is to store the solar energy through a thermochemical process. A closed loop thermochemical cycle provides storage only, whereas an open loop thermochemical cycle provides a means to store solar energy prior to using it as a fuel. The principle of closed-loop thermochemical storage is that solar energy is used to drive, at constant temperature, a dissociation or reforming reaction which is preferably highly endothermic. The products are then stored, and later recombined in an exothermic reaction as and when the heat is needed. This reverses the endothermic reaction and closes the loop. Detailed information is given in Chapter 9 of the present publication.

5.5 *Cost Reduction of Thermal Storage Through Higher Temperatures*

One method of significantly reducing thermal storage cost is simply utilizing higher temperature storage. Modern parabolic trough plants use oil as the HTF, which limits the maximum salt storage temperature to about 385°C even though conventional nitrate salt formulations are capable of working up to 560°C. In this low temperature configuration for PTCs, the temperature difference is low ($\Delta T \approx 100°C$). This is because cold storage is kept at approximately 290°C in order to keep conditions well away from the salt melting point of 220°C. This low temperature difference results in a much larger storage volume (and hence salt and tank cost) being required for a given storage capacity. A PTC has recently been demonstrated with a receiver tube that uses salt as the HTF, as noted before.

Central receivers can reach the upper operating temperature of today's salt formulations with relative ease, providing a working temperature difference that is two to three times higher than possible with PTC's using oil as the HTF (though the ΔT that is thermodynamically useful will depend on the cycle). Thus, the inventory of salt and the size of tanks required is less, though the temperatures are higher. This is expected to lead to lower costs of thermal storage.

A number of research programs are under way to develop salt formulations that both decrease the melting point (Sandia have demonstrated a salt with a melting temperature below 100°C) and increase the upper operating temperature to make full use of the capabilities of point focus concentrators.

6 CSP Systems

According to the principles of thermodynamics, power cycles convert heat to mechanical energy more efficiently at higher temperatures. However, the collector efficiency drops with higher absorber temperature due to greater heat losses. Consequently, for any given concentration factor there is an optimum operating temperature at which the highest conversion efficiency is achieved. With increasing concentration factors, higher optimum efficiencies are achievable. Figure 18 illustrates this effect assuming an ideal solar concentrator combined with a perfect (Carnot) power cycle. If the spectral absorption characteristics of the absorber are perfectly tailored to maximize absorption in the solar spectrum but avoid thermal radiation losses in the infrared part of the spectrum (selective absorber), additional efficiency gains can be expected, in particular, at lower concentration factors.

In practice, the optimum operation temperatures will be lower than these theoretical graphs, because power cycles with Carnot performance and ideal absorbers do not exist. Furthermore, the impact of frequent operation under part-load conditions throughout the year on the efficiency of the system has to be considered.

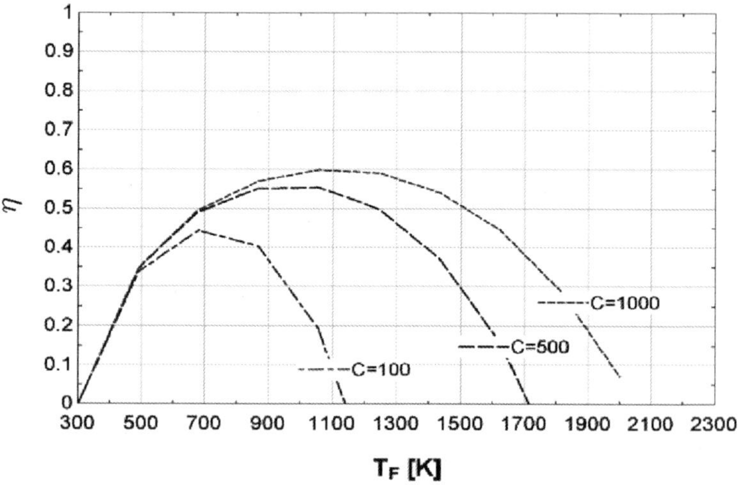

Fig. 18. Efficiencies versus temperature with increasing concentrations. *Source*: DLR.

A CSP power plant is an optimization of the type of collector, its operating parameters, and the thermodynamic cycle employed. In theory, the heat engine is the defining component as concentrating solar collectors can provide heated fluids for most temperatures. For commercial systems, the cycles of choice are Rankine, Brayton, and Stirling cycles. Each presents specific features such as preferred fluids, temperatures, efficiencies and, importantly, part load response curves.

An important point is that each of these cycles is able to operate independently of a solar input which affords the ability for CSP to be hybridized with fossil fuel in such a way as to improve performance and reduce the cost of the solar energy delivered.

6.1 *Rankine Cycle Based*

All of the large-scale commercial CSP projects in the world at present are based on the *Rankine cycle*, the most common form of which is the steam turbine (though it should be noted that a 1 MW parabolic trough power plant using an organic Rankine cycle has been built in Arizona, USA). The majority use the PTC with oil as the HTF. The only real variations are that different developers have applied different trough structures. The majority use receiver tubes available commercially from a small number of suppliers.

The parabolic trough power plants typically operate with steam temperatures of about 380°C at subcritical pressures and steam cycle efficiencies of around 37–38% (gross). Even lower temperature operation has been considered for linear Fresnel systems using saturated steam turbine designs from the nuclear industry. However, at present most interest is in higher temperature turbines (due to higher efficiency) and thus the development of receiver tubes using molten salt or direct steam generation.

Central receivers can easily achieve the temperatures of steam turbines used in many coal-fired power stations today (≈540°C), and even the conditions needed for supercritical steam turbines of around 25 MPa and over 600°C, where cycle efficiencies could increase to greater than 40%.

One way in which CSP may be introduced most readily to the market is through solar steam supplementation of existing or new coal or gas-fired steam plants. This allows the solar steam to benefit from the higher efficiencies of larger steam turbines and utilizes existing power block capital; the downside is that the solar plant is then constrained to the solar irradiance of that site and the solar contribution is generally low. Overall, however such an option provides excellent transitional opportunities for CSP. In addition to integration with a conventional coal or gas-fired Rankine cycle, the solar steam can also be integrated with the steam cycle (bottoming cycle) of a combined cycle plant.

Such plants are called "Integrated Solar Combined Cycle Systems (ISCCS)". They are combined cycle power plants using additional solar heat generated by parabolic troughs in the bottom cycle. Three ISCCS plants are in operation: Kuraymat, Egypt, Hassi R'Mel, Algeria, and Ain Beni Mathar, Morocco. These plants produce saturated steam in the solar steam generator to be fed into the HRSG. The electricity generated from the solar part is in all cases less than 17% at the design point. The annual solar fractions are even lower because the solar field delivers less than the design point heat for a large portion of the year.

Two additional plants viz. Archimede, Italy, and Martin, Next Generation Solar, Florida can also be included in this category.

Table 4 presents for the ISCCS plants an overview of the total design output and of the solar share. Figure 19 gives an outline of the fossil-solar integrated power station at Kuraymat, Egypt.[50]

6.2 *Brayton Cycle*

Central receivers and dishes can be used to provide hot air for *Brayton cycles*, i.e. gas turbines. In the industrial context, gas turbines have hundreds of megawatts capacity and operate with turbine inlet temperatures exceeding 1300°C. Point focus collectors have demonstrated that they can reach these temperatures, though the materials and fluid handling for such temperatures needs to be carefully considered. Transporting such high temperature gas streams to a turbine at ground level incurs either a high temperature loss or expensive insulation; alternatively mounting a large gas

Table 4: ISCCS power plants[50] (Status: June 2014).

ISCCS-Plant	Plant design output (MW)	Design output from solar (MW)	HTF
Kuraymat, Egypt	140	20	Diphenyl-oxide/biphenyl
HassiR'Mel, Algeria	150	25	Diphenyl-oxide/biphenyl
Ain Beni Mathar, Morocco	470	20	Diphenyl-oxide/biphenyl
Archimede, Italy	130	5	Molten salt
FPL Martin Next Generation Solar, Florida	3780	75	Diphenyl-oxide/biphenyl

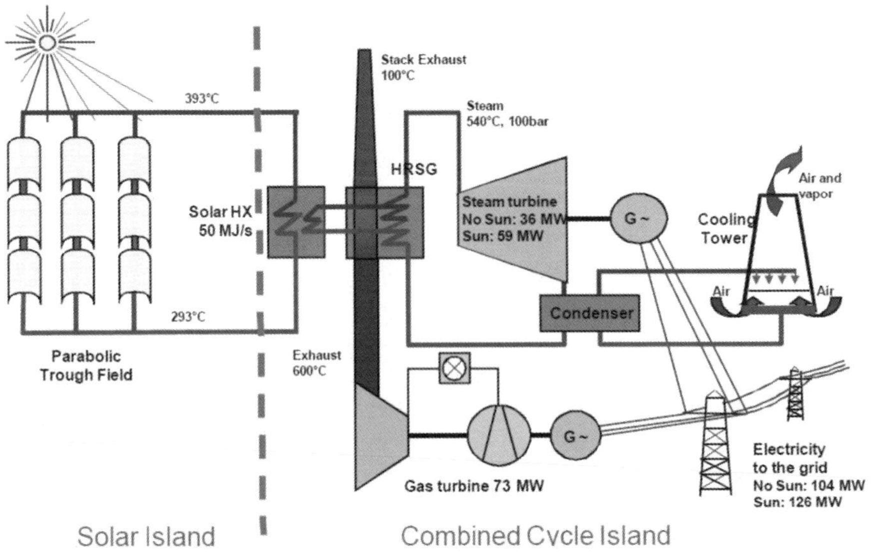

Fig. 19. The cycle arrangement for the integrated solar combined cycle system (ISCCS)[50] at Kuraymat, Egypt.

Source: Fichtner Solar GMBH.

turbine next to the receiver (on the tower top) can be structurally expensive. It is noted that at present, the allowable gas turbine combustor inlet temperature is generally much lower than the solar receiver can generate. If solar/gas hybrid operation is desired, some significant gas turbine redesign is required or the solar air temperature would need to be reduced which would yield a low solar contribution. Though an open cycle gas turbine requires no water for cooling, it is often less efficient than a large steam turbine. Thus the advantage for the Brayton cycle lies in the modularity and lower cost of gas turbines for small capacity systems, or in the very high efficiency (≈60%) of a gas turbine combined cycles at larger capacities (see Fig. 20). A number of research and demonstration programs for the solar tower Brayton cycle are under way or have been previously undertaken.[51,52]

A difficulty with such high temperature cycles is the ability to integrate competitive thermal storage. Whilst electrochemical storage is possible after the generator (as with wind and PV) this prevents the use of the inherent advantages of thermal storage of lower cost and 'spinning reserve'. A cycle under development for CSP is the closed loop Brayton cycle using *supercritical* CO_2 as the working fluid. The particular properties of this

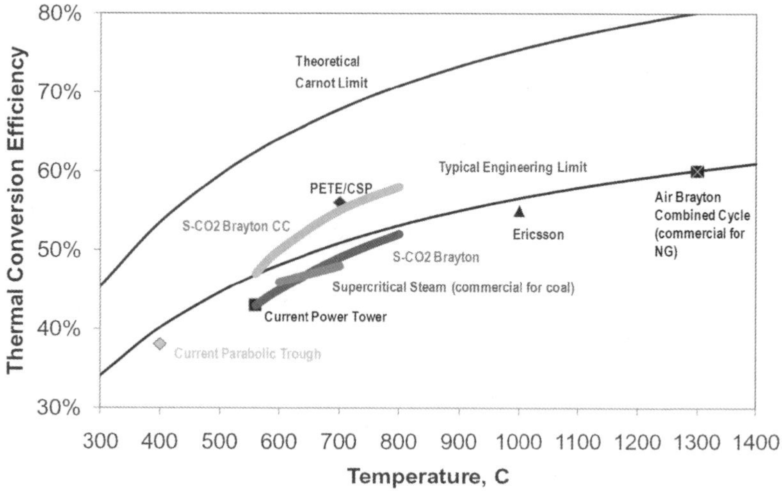

Fig. 20. Plot of comparative Brayton cycle efficiencies versus capacity. *Source*: N. Siegel.

fluid mean that the power used to drive the compressor is significantly less than in the conventional air Brayton cycle. In an optimized configuration, this yields very high cycle efficiencies of over 50% even with dry cooling at high ambient temperature. A bottoming cycle can improve this further. Most importantly, as shown in Fig. 20, this efficiency can be achieved at temperatures of the order of 650–700°C which is a relatively comfortable operating range for central receivers, and within the range of thermal storage developments.

6.3 *Stirling Cycle*

The *Stirling cycle* offers one of the highest efficiencies of the thermal cycles, and has been under development and demonstration for many years. It is a compact device, and a solar dish can provide ideal concentrator characteristics. It is based on a combination of constant temperature and adiabatic processes, with the receiver being externally heated. The working fluid is usually hydrogen or helium. Demonstrated capacities have been in the range of tens of kilowatts, and solar-to-electric efficiencies of approximately 30% have been reliably reproduced — one of the highest demonstrated by solar energy in the field to date.[53]

7 New Commercial Power Stations

The early commercial success of the SEGS plants in the 1990s in the United States was interrupted by a break of around 20 years due to the unfavorable market conditions that ensued. However, after the turn of the century, a new commercial dawn began, starting in Spain and with some delay, also in the US, mainly driven by the concerns about climate change. This was followed by the initiative of several other countries in the sunbelt that have identified CSP as a technology option that fit well with their plans to decrease climate change, provide energy security and create local employment. Table 5 gives an overview on the worldwide installed capacities of parabolic trough power plants in operation, under construction, under development or in an early planning stage.

7.1 *Parabolic Troughs*

Meanwhile after 2007, the development of more efficient 150 m long parabolic trough arrangements, including highly efficient receiver tubes, enabled a successful build up of a large number of units with 50 MW peak performances in Spain. The first 3×50 MW plants were erected as a block with Andasol I, II, and III[54] which can be seen in Fig. 21 and with details in Fig. 22. Technical data is given in Table 3.

Based on this concept, up to now 22 plants have been installed in Spain, each with a normal output of 50 MW capacity and 550000 m^2 of solar field. In addition, 24 systems of the same nominal output without storage are now operating so that presently about 2.4 GW of capacity is available in Spain. However, the economic crisis forced the Spanish Government to stop the incentivizing of solar thermal technology and it has even reduced the revenues for the existing plants so that no further installations are expected in the near future.

In the United States, some trough plants have been put into operation. Among them are the large systems Solana with 280 MW in Arizona and Genesis with 250 MW in California.

In the Middle East North Africa (MENA) region, in South Africa and in China are a few plants under construction.

The first large solar thermal power plant in the Arabian region is the parabolic trough facility Shams I.[55] The construction commenced in 2010 and was set to go online in 2013. The installed capacity amounts, with a field of 768 collectors, to a performance of 100 MW. This plant is located at Madinat Zayet, 120 km away from Abu Dhabi and is shown in Fig. 23. The

Table 5: Capacities of parabolic trough power plants in MW of various countries and in various stages of implementation (Status: June 2014).

	Operational	Under construction	Development	Planned	Total
— Parabolic trough	**3,108**	**1,391**	**980**	**1,065**	**6,543**
Argentina				20	20
Brazi		1		50	51
Chile			360		360
China	0.18	101.5	100	150	35168
Egypt			100		100
India	51	345		10	406
Iran	0.25				0.25
Israel			120	180	300
Italy	0.35	30		230	260.35
Japan					
Kuwait				50	50
Mexico					
Morocco		163	200		363
Oman					
Papua New Guinea					
South Africa		250	100		350
Spain	2,250				2, 250
Thailand	5				5
Tunisia				50	50
UAE	100				100
US	701	500		325	1, 526
— Parabolic trough — Biomass	**23**				**23**
Spain	22.5				22.5
— Parabolic trough — ISCC	**145**	**107**	**60**	**100**	**412**
Algeria	25				25
Canada		1			1
China		92			92
Egypt	20				20
Italy	5				5
Kuwait			60		60
Mexico		14			14
Morocco	20				20
US	75			100	175

Fig. 21. Areal view of the parabolic trough power plant complex of Andasol I, II, and III near Guadix at the feet of the Sierra Nevada, Spain.[54]

goal of this operation is to reduce the emission of CO_2 into the atmosphere by 175,000 tons annually.

7.2 *Central Receivers (Power Towers)*

After the mothballing of the 10 MW tower stations Solar 1 and 2 at Barstow, California, after 1998 no attempt has been made to build new tower power facilities in the USA.

Then, the Spanish company Abengoa built two power tower plants: PS 10 with a rated 11 MW power output which started operations in 2007, and PS 20 with 20 MW output which started in 2009 near Sevilla.[56]

The concentrated solar radiation heats metallic tubes and produces steam at 40 bars and 250°C to drive the turbine. A pressurized water tank is used to provide auxiliary storage for 30 minutes. The 20 MW plant is based on the same technology, but with 1 h of storage available.

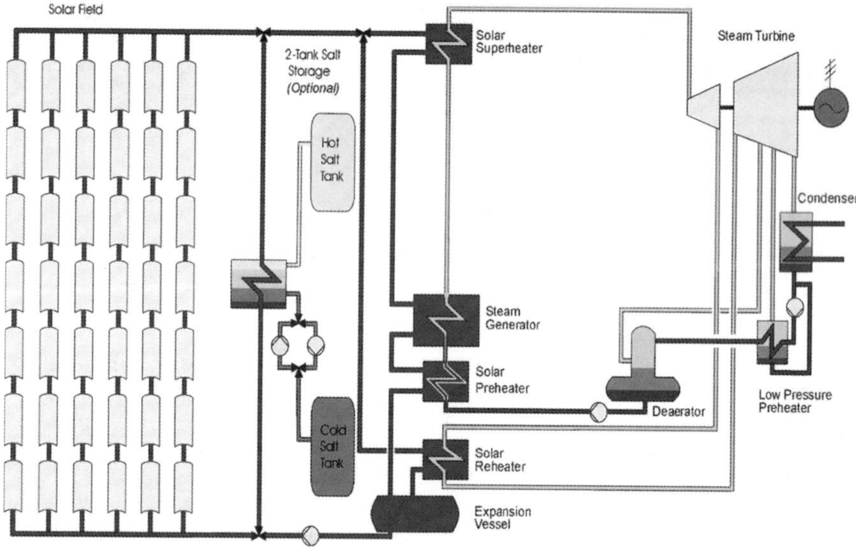

Fig. 22. In the upper part: A 50 MW unit of an Andasol power plant near Guadix, Spain with two storage containers in the middle of the parabolic trough field.

In the lower part: Schematics of the Andasol plants.[54]

Fig. 23. 100 MW SHAMS I — Masdar Clean Energy parabolic trough plant near Abu Dhabi.[55]

More recently, the 20 MW GEMASOLAR tower power plant[57] was erected near Ecija, Spain and went into operation in 2011. The plant uses molten salt as the heat transfer and storage medium. During solar operation, the molten salt is pumped from the cold storage tank (at about 290°C) to the receiver where it is heated to about 565°C. The hot salt is then piped to the hot storage tank. For power generation, the hot salt is pumped to the steam generator where superheated steam at 540°C is generated to drive the turbine and the generator. The cooled-down salt (290°C) is then pumped back to the cold storage tank. This plant concept allows decoupling of solar energy collection and electricity production. The circular heliostat field consists of 2,650 heliostats, each with a reflective area of 115 m^2. The concrete tower has a height of 140 m. With its 15 h storage system the plant can operate in summer for 24 h; the annual capacity factor reaches 75%. GEMASOLAR is the first solar facility which enables non-stop electricity production under summer conditions. Figure 24 gives a panoramic view of the plant with the tower in the middle of the heliostat field and the power block at the outside. A schematic of the plant is shown in Fig. 25.

The Sierra SunTower demonstration system[58] built by eSolar is installed near Lancaster, California. It is a modular system with two identical tower units. In the receivers, superheated steam is generated and piped to a common steam turbine for power production. The steam cycle has a

Fig. 24. GEMASOLAR solar thermal plant, owned by Torresol Energy.[57] © SENER.

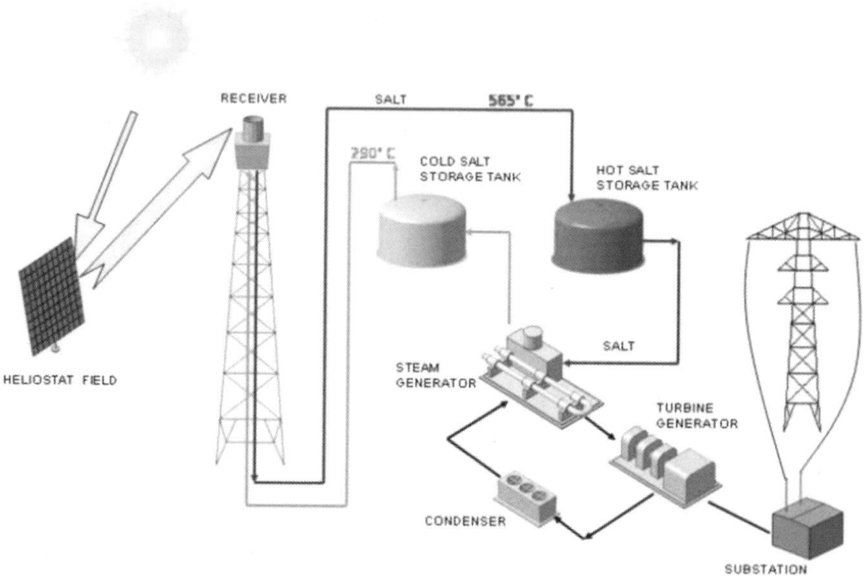

Fig. 25. GEMASOLAR tower system schematics.[57]

power level of 5 MW. Larger power levels can be achieved by combining multiple modular tower units, all with identical layout of heliostat field, tower and receiver.[59] Mainly, the power cycle must be adapted to the changing power level. Since the heliostat sub-fields are relatively small, prefabricated and easily handled, high solar collection efficiencies can be achieved.

The DLR solar tower at Jülich, Germany[60] is a pre-commercial solar tower demonstration plant with a power level of 1.5 MW (see Fig. 14). Air at near-ambient pressure is used as the heat transfer medium. The air is heated with solar energy in an open volumetric receiver to about 700°C. The hot air can be used directly in a heat recovery steam generator (HRSG) to produce power in the steam cycle. Alternatively, in charge mode, hot air is directed to the regenerator type storage system. In discharge mode, the flow direction in the regenerator is reversed to heat the entering air which is then used in the HRSG for power generation. The heliostat field contains over 2,000 heliostats with a total mirror area of about 18,000 m^2.

The above-mentioned solar tower plants are all significantly below economic power levels. But very recently a new large solar tower plant was erected and put into operation in 2014. This is the Ivanpah[61] Bright Source Energy tower power station, representing a total power output of 390 MW, the largest CSP facility worldwide. The plant consists of three independent solar tower units, each with its own power block. The receivers are used to generate superheated steam to drive a steam cycle for electricity power production. In total, 173,500 heliostats are installed, each with a relatively small area of 15 m^2. To cope with the arid area, dry cooling with air-cooled condensers guarantees extremely low water consumption, only 5% in comparison to wet cooling. The plant is shown in Fig. 26.

At the end of 2014, the Crescent Dunes tower power project (SolarReserve) is expected to go online. This solar tower plant is under construction near Tonopah, Nevada (USA) and has an output power level of 110 MW. Molten salt is used as the heat transfer fluid and storage medium. The storage is designed for about 4,500 full load operation hours of the power block annually (capacity factor >50%).

A number of other solar tower plants are in different phases of project development. Abengoa has started construction of a 50 MW solar tower plant in South Africa. SolarReserve is developing two additional projects, one in Spain (50 MW) and one in California (150 MW). Bright Source plans to erect another plant in California with a total power output of 500 MW. Several other tower projects are in preparation.

Fig. 26. Ivanpah[61] solar thermal central receiver power stations of Bright Source Energy, San Bernardino, California. The towers are 140 m high.

8 Economy and Cost Reductions

8.1 *CSP in the Market*

The first commercial CSP plants were those built by LUZ in California from 1985–1991 (see Table 1). Over 2,000,000 m^2 of trough collectors were installed in 9 projects providing 354 MW of power in total, which are still in commercial operation today.

There was little commercial activity until 2006. Since then there has been a strong growth in deployment of commercial CSP plants (see Fig. 27). These plants have been supported financially through various mechanisms, such as up-front grants and loans, through feed-in-tariffs or mandatory targets, reducing over time.

In Spain, CSP industry momentum has been fostered largely through the incentives offered under the Royal Decree. In addition, growth is again underway in the United States, and a range of other countries and regions are now accelerating their own programs — the Middle East and North Africa, South Africa, Australia, Israel, China, India, Brazil, Mexico, and

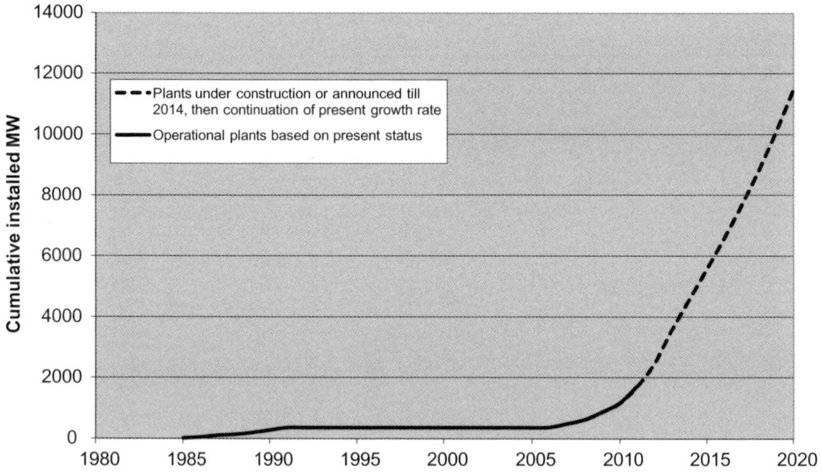

Fig. 27. Installed global CSP capacity as of 2011, plants under construction, and predicted to 2020.

Chile. An exciting concept known as Desertec aims to deploy plants in the sunbelt region of North Africa and export much of the electricity to Europe.

However, since 2010, the deployment of solar thermal electricity from CSP plants has been slower than expected. After the initial deployment of the first commercial plants in California in the 1980s, the renaissance in Spain was limited to 2.3 GW under governmental decisions in the context of the financial and economic crisis. Deployment in the US has been slow due to long lead times and competition from both cheap unconventional natural gas and the rapidly-decreasing cost of PV energy. Deployment in other places was even slower as projects took significant time to be conceived, financed, built and finally connected to the grid.

In Spain, legislation was such that plants needed to be restricted to 50 MW, with the aim of fostering competition and preventing monopolization. This strategy has borne fruit with numerous companies entering the market through both 'bottom up' growth and technology acquisition. Many of these companies now have a global presence. Now, around the rest of the world, plants are typically sized at over 100 MW, with a number between 250 and 300 MW. Thermal storage of several hours is now regularly incorporated to improve grid access and the financial rate of return. In 2014, approximately 4000 MW of CSP plants is in commercial operation,

Table 6: World wide capacity of CSP technology in MW in various stages of construction (Status: June 2014).

	Operational	Under construction	Development	Planned	Total
Algeria	25		7		32
Argentina				20	20
Australia	9.5	44		73	126.5
Brazil		1		50	51
Canada		1			1
Chile		110	370	410	890
China	2.18	307.5	100	150	559.68
Cyprus			50.76	25	75.76
Egypt	20		100	250	370
France	0.25	12	9		21.25
Germany	1.5				1.5
Greece			125		125
India	53.5	445		111	609.5
Iran	0.25				0.25
Israel	6.1		241	181	428.1
Italy	5.35	30	50	230	315.35
Japan					
Jordan				225	225
Kuwait			60	50	110
Mexico		14			14
Morocco	20	163	301		484
Oman					
Papua New Guinea					
South Africa	0.23	300	100	230	630.23
South Korea	0.2				0.2
Spain	2361.6		50		2411.6
Thailand	5				5
Tunisia			2000	55	2055
UAE	100.1				100.1
US	1,192	611.5	1,155	1,120	4078.5
Total	**3,803**	**2,039**	**4,719**	**3,180**	**13,741**

2000 MW under construction and some 7000 MW at various stages of planning and approval as shown in Table 6.

For the CSP industry, there are presently several strategic issues in play, including the consideration of placing power stations close to available grid capacity, the move toward dry cooling (which is well understood by the power industry), and the possibility of large 'solar zones' to centralize infrastructure or, conversely, the development of more modular/compact CSP technologies to expand the range of sites. However, the single most important issue is cost reduction, discussed in Sec. 8.2.

The structure of a commercial CSP project is very similar to other large power plant projects and typically involves several players. An "Engineering, Procurement and Construction (EPC)" contractor and its suppliers provide and warrant the technology to the owner, who finances it through equity investors, banks and eventually public grants. The owner gains revenues from the electricity off-taker (typically the electricity system operator) based on long-term power purchase agreements needed to pay off the debt and operation costs, and to generate a profit. An operation and maintenance company provides services to the owner to operate the plant. This approach results in a complex contractual arrangement to distribute and manage the overall project risk, as the overall project cost of several hundred million Euro typically cannot be backed by a single entity. The perception and distribution of risks, as well as local and regional factors, strongly affect the cost. There is therefore no single figure for the costs of electricity from CSP, nor for similar reasons, for other generating technologies to which it needs to be compared. One approach that is often used to compare costs of electricity generation is to calculate the "levelized electricity cost" (LEC or LCOE for levelized costs of electricity). This also relates average annual capital and operating costs of the plant to the annual electricity production. Recognizing the limitations of this approach, particularly when comparing fossil-fired and renewable technologies where it does not capture differences in value to the customer, it nonetheless gives a useful "first cut" view of comparative costs. For comparisons between fossil-fired plants and CSP plants with storage and/or supplementary firing, the limitations of this approach are less significant as the technologies offer similar services.

Revenues for CSP started with the feed in tariff in Spain of more than 30 €cents/kWh. In 2014, offered revenues for new projects are approaching 15 €cents/kWh. This demonstrates a significant cost reduction over the last seven years.

8.2 *Cost Developments and Cost Reductions*

The US Department of Energy Sunshot initiative (Fig. 28) and various CSP roadmaps have identified the technology key areas that most effectively impact cost, and have established targets that would lead to CSP being a low cost source of dispatchable solar electricity. These generally relate to higher temperature receivers for associated advanced cycles (particularly supercritical cycles), lower cost collectors and higher temperature storage.

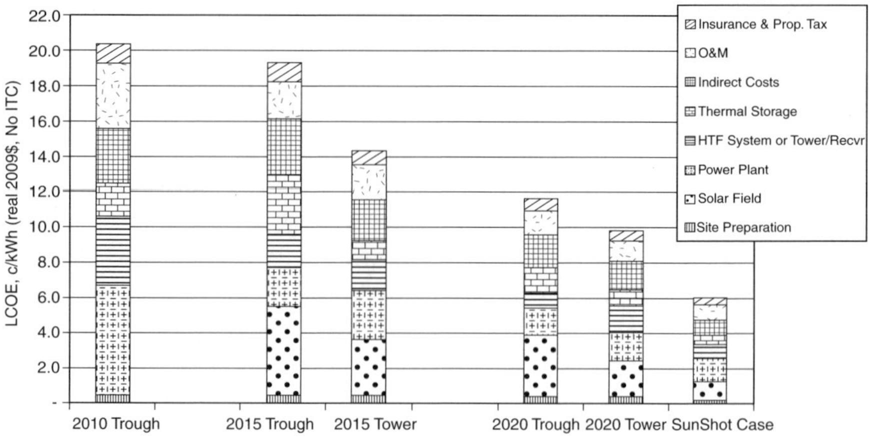

Fig. 28. The targets established by the US DoE Sunshot initiative and corresponding roadmaps.[62]

8.2.1 *Cost reduction*

Three main drivers for cost reduction are scaling up, volume production and technology innovations. As an example, one of the first comprehensive studies of the potential for cost reduction of CSP was undertaken in the framework of the European 'ECOSTAR' project.[63] The study proposed the potential relative reduction of the levelized electricity cost (LEC or LCOE) of trough plants of up to 60%. Half of this potential can be obtained by technical innovations, the other half by scaling-up and mass production effects. Further details of the cost breakdowns and the other costs are given in the listed references.[63–65] If this potential is exploited, the cost for dispatchable solar power from CSP plants will drop significantly below 9 €cents/kWh and can be considered competitive in many commercial markets. The time to achieve this cost reduction is strongly coupled to the deployment rate of the technology. A total installed capacity between 10 and 100 GW is estimated to be sufficient to achieve this target between 2020 and 2030.[66]

8.2.2 *Scaling up*

CSP technology favors big power plant configurations[67] because procurement of large amounts of solar field components can lead to discounts. Engineering, planning and project development costs are essentially independent of the scale of the plant; operation and maintenance costs decrease with plant size, and large power blocks have higher efficiencies than small ones and cost less per kW.

The impact of scaling up on CSP electricity cost is still under discussion. The Kearney report[64] indicates a 24% reduction of capital expenditure for an increase of trough plant size from 50 MW to 500 MW, and Lipman[65] to 250 MW. Finally, the Sargent and Lundy study[68] points to a 14% cost reduction for a 400 MW power block.

8.2.3 *Volume production*

For parabolic trough technology, the Sargent & Lundy study[68] estimates a cost reduction of 17% due to volume production effects when installing 600 MW per year. Cost decreases in the range of 5% to 40%, depending on components, are expected in the Bright Source energy report.[61]

8.2.4 *Technology innovations*

Technological innovations will increase power generation efficiency, mainly through increasing operating temperature. Innovations will reduce solar field costs by minimizing component costs and optimizing optical designs and will, as well, reduce operational consumption of water and parasitic power.

Expected cost reductions and plant efficiency improvements associated with technology innovations are reported by Kearney.[64]

It is not just the cost of CSP electricity generation which determines its economic competitiveness. Rather, the value is represented by three components:

- the value of the kWh's of electrical energy generated by the plant, which will vary over time in a competitive electricity market, reflecting the availability and cost of electricity from other sources;
- the contribution that the CSP plant makes to ensuring that generating capacity is available to meet peak electricity system demand; and
- the 'services' provided by the plant in helping the electricity transmission system operator to balance supply and demand in the short term (typically, on time scales of seconds and minutes).

Potentially, incorporation of thermal storage in a CSP plant can be beneficial to all three components of value. In relation to the first value component, system simulations[69] indicate that as the solar share rises in an energy system there is increasing value in shifting generation to the evenings when the sun is not shining, and hence indicating an incentive to install CSP plants with thermal storage. The availability of such plants in the system means that higher penetrations of solar power can be achieved overall and is an important consideration, beyond just generating costs, in determining the optimum mix of CSP and PV technology.

With regard to the second component of value, the provision of generating capacity to meet peak electricity system demand, CSP with storage can contribute to meeting peak system loads and can provide back-up capacity to cover variable renewable sources. Incorporation of supplementary firing will further increase the capability of the CSP plant to provide capacity at the system peak, though the efficiency of fossil-fuel use for such supplementary firing is likely to be significantly lower than if it is used in a combined cycle power plant. The value of providing capacity to meet the system peak demand will depend on the system, so its quantification needs to be informed by system models.

Turning to the third component, the value of thermal energy storage in enabling the CSP plant to deliver grid services, CSP with storage can provide spinning reserves, being able to ramp up power, if operating at part-load, in less than 30 minutes by drawing on the stored heat (the rate of ramping is limited by the thermal inertia of the equipment). Ramping down is quicker on timescales of around 15 minutes by diverting heat to storage. This is used in Spain to deliver, on demand, 30% power ramps in less than an hour, enabling the plant to be considered dispatchable by the grid operator REE.

9 Conclusions

CSP is in a strong commercial growth phase at present, building on many years of intensive R&D activity. The present trend continues with the theme of larger capacity plants, most of which are based on parabolic trough technology. However, central receivers using molten salt as heat transfer fluid and storage medium are now being increasingly deployed as the technology of choice.

In fact, there is a general move toward higher temperature systems for improved efficiencies, as this leads directly to a reduction in collector area, which is the single most significant capital cost component, as well as lower storage costs.

The industry has developed rapidly since 2006, when it consisted of relatively few companies with in-house technology, to the present, where a diverse range of large multi-national companies has been established, providing strong competition in components, construction and project management, and finance. Risk still exists as the projects are large and strongly dependent on capital that needs to be underwritten by power purchase agreements and, for now, other incentives.

New and advanced technology opportunities are also emerging. These include very high efficiency supercritical CO_2 Brayton cycles, and the opportunity to apply all of the collector and receiver expertise developed for high temperature CSP electricity generation as the basis for developing a new market in commercial solar fuel technology.

The key for the future of CSP lies in its ability to deploy dispatchable solar energy in large quantities, building on much of the knowledge and technology available in the energy industry today.

References

1. International Energy Agency, *Key World Energy Statistics 2011* (IEA Press, 2010).
2. C. J. Winter, R. L. Sizmann and L. Vant-Hull, *Solar Power Plants: Fundamentals, Technology, Systems, Economics* (Springer-Verlag, Berlin, 1991).
3. R. Pitz-Paal, High temperature solar concentrators, in *Solar Energy Conversion and Photoenergy Systems*, eds. J. B. Galvez and S. M. Rodriguez (Eolss Publishers, Oxford, UK, 2007).
4. C. E. Kennedy, Review of mid to high temperature solar selective surfaces, Technical Report NREL/TP-520-31267 (2002).
5. A. Akbarzadeh, J. Andrews and P. Golding, Solar pond technologies: A review and future directions, in *Advances in Solar Energy*, ed. D. Y. Goswami (Earthscan, USA, 2005).
6. D. R. Mills, Solar thermal electricity, in *Solar Energy — The State of the Art — ISES Position Papers* ed. J. M. Gordon (James and James, 2001).
7. http://www.solar-updraft-tower.com.
8. P. DeLaquil, D. Kearney, M. Geyer and R. Diver, Solar thermal electric technology, in *Renewable Energy — Sources for Fuel and Electricity*, eds. T. B. Johansson, H. Kelly, A. K. N. Reddy and R. H. Williams (Earthscan Publications Ltd, and Island Press, USA, 2001).
9. http://www.nrel.gov/csp/troughnet/power_plant_data.html.
10. E. Lüpfert, M. Geyer, W. Schiel, E. Antonio, R. Osuna, E. Zarza and P. Nava, Eurotrough design issues and prototype testing at PSA, in *Proceedings of the ASME Int. Solar energy Conf. — Forum 2001, Solar energy: The Power to Choose*, April 21–25, Washington DC (2001).

11. K. J. Riffelmann, J. Kötter, P. Nava, F. Meuser, G. Weinrebe, W. Schiel, G. Kuhlmann, A. Wohlfahrt, A. Nady and R. Dracker, Helitrough — A new collector generation for parabolic trough power plants, Presented at *15th Int. SolarPACES Symp.*, September 15–18, Berlin, Germany (2009).
12. N. Castañeda, J. Váquez and D. Castañeda, First commercial application of senertrough. High performance at reduced cost, Presented at *17th Int. SolarPACES Symp.*, September 20–23, Berlin, Germany (2011).
13. S. Fernández and A. Acuñas, New optimized solar collector SENERtrough-2, Presented at *18th Int. SolarPACES Symp.*, September 11–14, Marrakech, Morocco (2012).
14. H. Price, E. Lupfert, D. Kearney, E. Zarza, G. Cohen, R. Gee and R. Mahoney, Advances in parabolic trough solar power technology, *J. Sol. Energy Eng.* **124**(2) (2002), pp. 109–125.
15. R. Pitz-Paal, J. Dersch, B. Milow, F. Téllez, A. Ferriere, U. Langnickel, A. Steinfeld, J. Karni, E. Zarza and O. Popel, Development steps for parabolic trough solar power technologies with maximum impact on cost reduction, *J. Sol. Energy Eng.* **129** (2007), p. 371.
16. L. P. Drouot and M. J. Hillairet, The themis program and the 2500-KW themis solar power station at Targasonne, *J. Sol. Energy Eng.* **106**(1) (1984), pp. 83–89.
17. M. Martin, Molten salt electric experiment (MSEE), Sandia National Laboratories, SAND85-8175, Albuquerque, New Mexico (1985).
18. R. W. Bradshaw, D. B. Dawson, W. De La Rosa, R. Gilbert, S. H. Goods, M. J. Hale, P. Jacobs, S. A. Jones, G. J. Kolb, J. E. Pacheco, M. R. Prairie, H. E. Reilly, S. K. Showalter and L. L. Vant-Hull, Final test and evaluation results from the solar two project, OSTI ID: 793226 (2002).
19. N. P. Siegel and R. W. Bradshaw, Thermophysical property measurement of nitrate salt heat transfer fluids, Presented at *ASME 2011 5th International Conf. Energy Sustainability*, Washington DC (2011).
20. M. Falchetta and D. Mazzei, Design of Archimede 5MW molten salt parabolic trough solar plant, Presented at *15th Int. SolarPACES Symp.*, September 15–18, Berlin, Germany (2009).
21. D. A. Nissen and D. E. Meeker, Nitrate/nitrite chemistry in sodium nitrate-potassium nitrate melts, *Inorg. Chem.* **22**(5) (1983), pp. 716–721.
22. R. W. Bradshaw and C. E. Tyner, Chemical engineering, Factors affecting solar central receiver applications of ternary molten salts, Sandia National Laboratories, SAND88-8686, Albuquerque, New Mexico (1988).
23. T. Bauer, N. Pfleger, D. Laing, W.-D. Steinmann, M. Eck and S. Kaesche, High temperature molten salts for solar power application, in *Molten Salts: Fundamentals and Application*, eds. F. Lantelme and H. Groult (Elsevier, Amsterdam, 2012).
24. R. W. Bradshaw, J. G. Cordaro and N. P. Siegel, Molten nitrite salt development for thermal enregy storage in parabolic trough solar power systems, *Proc. ASME Int. Conf. Energy Sustainability*, San Francisco, California, USA (2009).

25. D. Kearney, B. Kelly, U. Herrmann, R. Cable, J. Pacheco, R. Mahoney, H. Price, D. Blake, P. Nava and N. Potrovitza, Engineering aspects of a molten salt heat transfer fluid in a trough solar field, *Energy* **29**(5–6) (2004), pp. 861–870.
26. R. W. Bradshaw and S. H. Goods, Corrosion resistance of nickel-base alloys in molten alkali nitrates, Sandia National Laboratories, SAND2000-8240, Albuquerque, New Mexico, USA (2000).
27. Müller-Elvers and Wittmann, Design and construction of molten salt parabolic trough HPS project in Évora, Portugal, in *Proc. 18th Int. SolarPACES Symp.*, September 11–14, Marrakech, Morrocco (2012).
28. Annual Report 2012, CIEMAT — PSA.
29. V. A. Baum, Prospects for the application of solar energy, and some research in the USSR, *Proc. World Symp. Applied Solar Energy*, Stanford Research Institute, Menlo Park, pp. 289–298 (1956).
30. C. Silvi, The pioneering work on linear fresnel reflector concentrators (LFC) in Italy, in *Proc. SolarPACES Conf.*, Berlin, September (2009).
31. D. R. Mills and G. L. Morrison, Compact linear fresnel reflector solar thermal power plants, *Solar Energy* **68**(3) (2000), pp. 263–283.
32. G. Morin, J. Dersch, W. Platzer, M. Eck and A. Häberle, Comparison of linear fresnel and parabolic trough collector power plants, *Solar Energy* **86**(1) (2012), pp. 1–12.
33. http://www.physics.usyd.edu.au/app/research/solar/clfr.html.
34. http://www.novatecsolar.com/79-1-Australias-Largest-Solar-Thermal-Plant-to-Expand.html.
35. http://www.areva.com/EN/operations-415/concentrated-solar-power-technology.html.
36. http://minister.ret.gov.au/MediaCentre/Speeches/Pages/Round1SolarFlagshipsProjects.aspx.
37. http://www.nrel.gov/csp/solarpaces/project_detail.cfm/projectID=37.
38. http://www.novatecsolar.com/8-1-Projects.html.
39. http://www.volker-quaschning.de/articles/fundamentals2/index_e.php.
40. P. DeLaquil, D. Kearney, M. Geyer and R. Diver, Solar thermal electric technology, in *Renewable Energy — Sources for Fuel and Electricity*, eds. T. B. Johansson, H. Kelly, A. K. N. Reddy and R. H. Williams (Earthscan Publications Ltd, and Island Press, USA, 2001).
41. http://www.wizardpower.com.au/index.php?option=com_content&view=article&id=36&Itemid=35.
42. K. Lovegrove, G. Burgess and J. Pye, A new 500 m^2 paraboloidal dish solar concentrator, *Solar Energy* **85**(4) (2011), pp. 620–626.
43. W. Scheffler, Introduction to the revolutionary design of scheffler reflectors, in *Solar Cookers and Food Processing Conf.*, Granada, July, pp. 12–16 (2006).
44. Funding Opportunity Number DE-FOA-0000471.
45. http://www.nrel.gov/csp/solarpaces/project_detail.cfm/projectID=40.
46. D. A. Brosseau, P. F. Hlava and M. J. Kelly, Testing thermocline filler materials and molten salt heat transfer fluids for thermal energy storage systems used in parabolic trough solar power plants, SAND2004-3207 July (2004).

47. S. Zunft, M. Hänel, M. Krüger, V. Dreißigacker, F. Göhring and W. Wahl, Jülich solar power tower — experimental evaluation of the storage subsystem and performance calculation, *J. Sol. Energy Eng.* **133**(3) (2011), p. 031019.
48. R. Buck, Volumetric receivers for solar-hybrid gas turbine power systems, DLR Forschungsbericht 2001-10, pp. 62–67 (2001).
49. D. Laing, High-temperature solid-media thermal energy storage for solar thermal power plants, *Proc. IEEE* (2011).
50. www.csp-world.com/cspworldmap/kuraymat-iscc.
51. p://ec.europa.eu/research/energy/pdf/solgate_en.pdf.
52. http://sfera.sollab.eu/downloads/4th_presentation.pdf.
53. T. Mancini, P. Heller, B. Butler, *et al.*, Dish-stirling systems: An overview of development and status, *J. Sol. Energy Eng.* **125** (2003), pp. 135–151.
54. F. Dinter and D. M. Gonzales, Operability, reliability and economic benefits with CSP with thermal energy storage: First year of operation of Andasol 3, *Energy Procedia* **49** (2013), pp. 2472–2481.
55. www.shamspower.ae/en/.
56. www.environmental-expert.com/products/abengoa-ps10-andps20-solar-power-plants-334737.
57. J. I. Burgaleta, S. Arias and I. B. Salbidegoitia, Operative advantages of a central tower solar plant with thermal storage system, *15th Int. SolarPACES Symp.*, Paper 11 720, Berlin (2009).
58. P. K. Meduri, C. R. Hannemann and J. E. Pacheco, Performance characterization and operation of eSolar's sierra suntower power tower plant, Presented at *16th Int. SolarPACES Symp.*, September 21–24, Perpignan, France (2010).
59. D. Rogers, M. Slack and B. Cassity, Addressing the challenges associated with esolar's unique approach to central receiver power plants, in *Proc. 18th Int. SolarPACES Symp.*, September 11–14, Marrakech, Morocco (2012).
60. S. Pomp, P. Schwarzbözl, G. Koll, F. Göhring, F. Hartz, M. Schmitz and B. Hoffschmidt, The solar tower Jülich — first operational experiences and test results, in *Proc. 16th Int. SolarPACES Symp.*, September 21–24, Perpignan, France (2010).
61. http://www.brightsourceenergy.com/ivanpah-solar-project (2013).
62. SunShot Vision Study, US Department of Energy, February (2012).
63. R. Pitz-Paal, J. Dersch, B. Milow, F. Tellez, A. Ferriere, U. Langnickel, A. Steinfeld, J. Karni, E. Zarza and O. Popel, Concentrating solar power plants — how to achieve competitiveness, *VGB Power Tech* **8** (2005), pp. 45–46.
64. A. Kearney, *Solar Thermal Electricity 2025* (ESTELA, 2010).
65. E. Lipman, Reducing CSP plant LOEC with precies engineering synergies and R&D, Plenary Lecture presented at *16th SolarPACES Symp.*, September 21–24, Perpignan, France (2010).
66. Easac, Concentrating solar power: Its potential contribution to a sustainable energy future, EASAC policy report 16 (2011).
67. U.S. Energy Information Administration, Updated capital cost estimates for electricity generation plants, US Department of Energy (2010).

68. Sargent & Lundy LLC Consulting Group, Assessment of parabolic trough and power tower solar technology cost and performance forecasts, National Renewable Energy Laboratory (NREL), Golden, Colorado, USA (2003).
69. C. Kost and T. Schlegl, Stromgestehungskosten erneuerbare energien, Renewable Energy Policy Innovation, Fraunhofer ISE (2010).

Chapter 4

Large-Scale Photovoltaic Power Plants

G. Almonacid Puche, P. G. Vidal and E. Muñoz-Cerón
Centro de Estudios Avanzados en Energía y Medio Ambiente
Universidad de Jaén
(Center for Advanced Studies in Energy and Environment, University of Jaén)
Campus de Las Lagunillas, 23071 Jaén (Spain)

This chapter contains an overview of the main aspects that characterize or have a particular impact on large-scale photovoltaic (PV) power plants. The chapter starts with an introduction and an overview of the megawatts installed in PV plants, and then analyses some key issues in the electrical engineering of plants' layout, devices and grid connection. The chapter also goes through important aspects regarding the behavior, commissioning, operation and maintenance of this type of plants. Several socio-economic factors like policies, planning and regulations, financing tools etc. are considered and a forward-looking view is taken on the prospects for this technology.

1 Introduction

Large-scale photovoltaic (PV) power plants (also called utility-scale or megawatt-scale power plants) may be installed on large rooftops or on the ground, and may employ different PV technologies. What distinguishes large-scale solar from distributed generation (DG) is the size of this type of project and the fact that the electricity that they produce is sold to wholesale utility buyers, not to end-use consumers.[1] Large-scale PV plants provide the benefit of fixed-priced electricity during peak demand periods when electricity from fossil fuels is more expensive.

According to the latest data from the PV Power Systems Programme (PVPS) of the International Energy Agency IEA,[2] at least 36.9 GW of PV systems have been installed and connected to the grid in the world during the past year. Likewise, large-scale PV projects have grown rapidly in number and size over the last few years, aided by the fall in the price

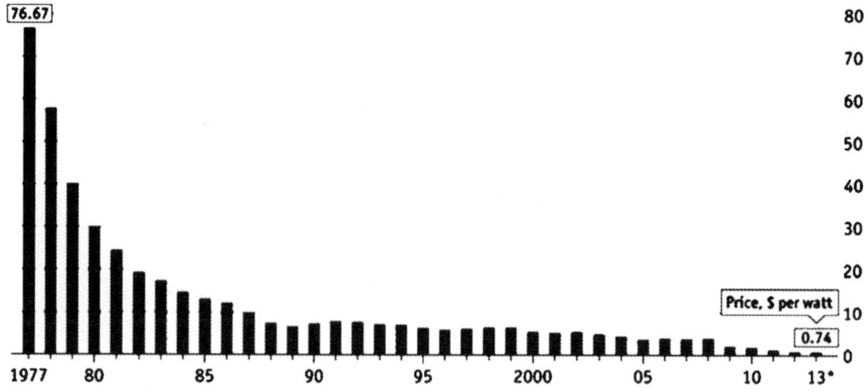

Fig. 1. PV module price (1977–2013).
Source: Bloomberg New Energy Finance.

of the PV components, and, most significantly, by the drop in the price of conventional crystalline silicon PV modules, that has decreased sharply, from 3.5 €/W to 0.5 €/W,[3] as shown in Fig. 1. Due to this incredible drop in price, the installed capacity of solar energy has increased more than tenfold in the last 6 years.

As major markets like the EU, the USA, China, and Japan continue to ramp up installations and new markets like India, South Africa, Chile and others emerge, the promise of large-scale PV electricity production continues to be a bright spot in the PV market.

A large-scale PV project is complex and involves many different parties including developers, landowners, utilities, grid operators, government agencies and financing parties. These projects must pass through a number of different stages grouped as pre-financial and post-financial activities. For instance, Fig. 2 shows an overview of the project time frame for a 20 MW utility-scale PV project.[4]

A successful PV project requires careful consideration of financing options. Financing stands or falls depending on the actual solar power fed annually into the grid. Usually, banks expect a return on project investment of 9% or possibly higher. The strong market for large-scale plants that prevailed at times in Spain (2008), Germany (between 2010 and 2012) and Italy (2011) often created unrealistic expectations of the returns that solar power was able to yield. The Eurozone crisis and the withdrawal of state subsidies are therefore forcing investors to take a more realistic view of PV plants and are enabling them to exploit long-term advantages more

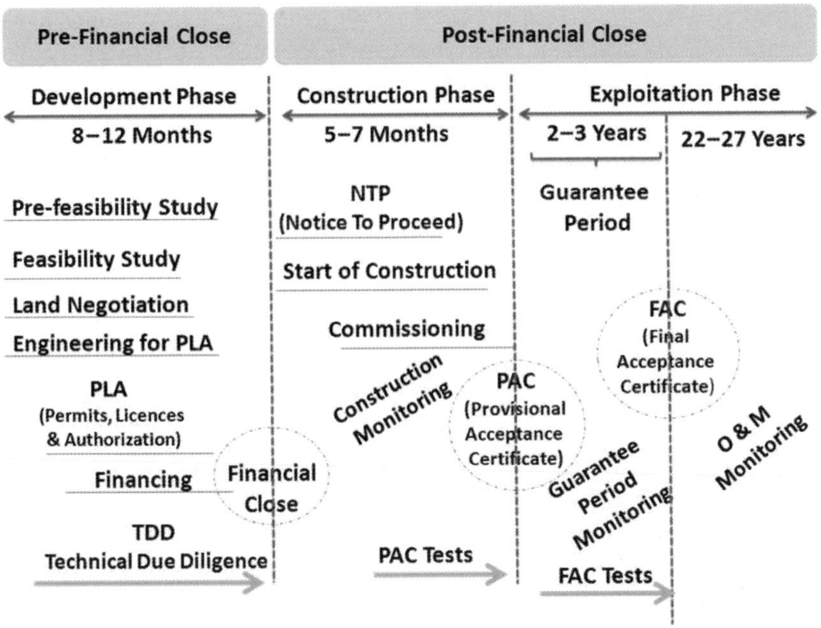

Fig. 2. Project time frame for a typical large-scale PV project.[4]

successfully. To achieve this, large-scale PV plants do not only require an accurate solar energy forecast, but also quality electrical engineering, and careful monitoring, along with other socio-economic considerations. All these are paramount to generate a successful project. A growing number of investors are recognizing the fact that the technology has matured and the risk of default is low. The investment, and consequently the price achievable on the power market, is not linked to the uncertainties of the oil market. The sun supplies its energy for free, which gives PV a fundamental advantage.

2 Electrical Engineering: Key Points in Large PV Generators

2.1 *Layout of a Large-Scale PV Plant*

The choice of the architecture employed in a PV generator is key when designing large-scale PV plants. Such architecture is determined by the configuration of the modules connected to a given inverter, by the number of branches, by the number of modules per branch and by the number of inverters required to obtain the total power generated by the plant, as well as by the number of transformers that will be necessary to connect the plant to the medium voltage grid. In general, there is not one particular indicator to know *a priori* which is the best module configuration or the best inverter topology since the best possible solution depends on a variety of factors linked to the power of the plant, the characteristics of the geographical area of the installation, the characteristics of the connection grid and on applicable regulations.

Any design process oriented to obtain the best possible solution must aim at minimizing the initial cost of the installation and should consider important factors like the value of the energy produced, the efficiency and reliability (performance ratio (PR)) along with the cost of the electricity produced.

The design of large-scale PV plants has traditionally been based on centralized installations with powerful centralized inverters. In such designs, a large number of PV modules is connected in series and/or parallel to one inverter. Then, one or two of these inverters are connected to the medium voltage grid through a transformer (see Fig. 3). These centralized inverter-transformer blocks may surpass 1 MW and have been used in PV plants beyond 250 MW. However, this configuration of centralized blocks is only effective when the structure of the plant is homogeneous, that is to say, when it uses the same modules, with no partial shading and with the same slope and orientation.

In the last few years, large-scale PV plants have also been built through configurations of small inverters of reduced power connected in parallel in a low voltage grid. Then, this block is connected to the medium voltage grid through a transformer that creates a generation block yielding an amount of power similar to that of the previous configuration.

These distributed inverters can be arranged in two different ways: in string inverters, the first type, each string of the generator is wired to one string inverter; in multistring inverters two or three strings are wired in parallel to one inverter. Each string of the first type or group of strings of the second has an independent (maximum power point) (MPP) tracker. As a consequence, higher efficiency is obtained in non-homogeneous installations or installations with shadowed lines at dawn or dusk.

Given the impact that the cost of inverters has on the initial investment, it might be logical to conclude that any configuration based on a centralized generator is the best possible solution. However, these results may vary if

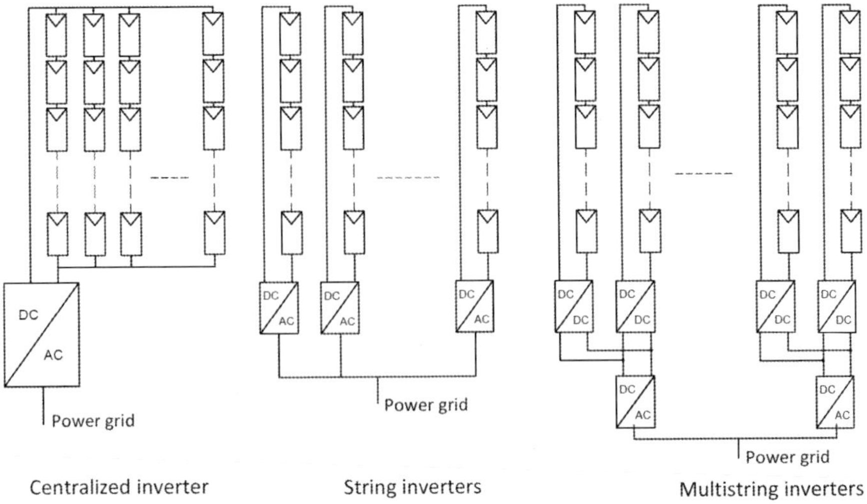

Fig. 3. Different configurations of PV inverters.

all costs and energy efficiency of the system throughout its operating life are also analyzed.

Both installation and operation costs should be included in a detailed analysis. The specific cost for the investor (the cost per kW) diminishes as the power increases. While plants with high-power centralized blocks favor lower cost as well, small-decentralized inverters may be more suitable if factors like wiring, junction boxes, building costs, transportation or installation are taken into consideration. Maintenance is easier and repair costs more economical in plants with standard small-decentralized inverters, which can be easily found on the market and whose replacement parts are easy to find as well. This availability facilitates their substitution by local installers in the event of a malfunction. As a consequence, the usual maintenance contracts would not be completely necessary.

PR are also better in installations with decentralized architectures. When large-scale plants are divided into smaller subgenerators with an MPP tracker, inverters can be adapted to the subgenerators more easily and shading losses can be reduced by using different modules with high fabrication tolerances and groups of modules with different orientation or slope. In the event of a malfunction of the inverter, production losses are also reduced since only a small part of the installation is affected, which can be rapidly substituted if replacement inverters are in stock.

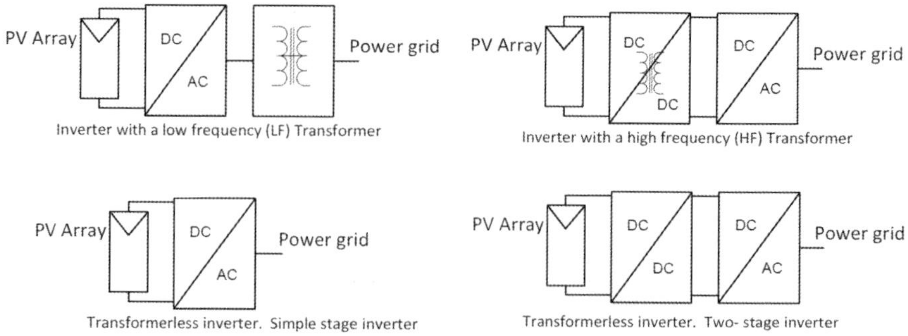

Fig. 4. Solar inverter topologies.

2.2 *Inverter Topology*

Until a short time ago, all inverters used in PV systems included a low-frequency transformer which was situated between the power stage and the grid (see Fig. 4). This transformer, the specifications of which are determined by national regulations in some countries, provides the PV system with galvanic isolation from the grid at the same time that it serves to protect operators, to prevent capacitive leakage, to avoid the flow of direct current to the grid and, in some cases, to increase the output voltage of the inverter. On the downside, these transformers have disadvantages like cost increase, weight, size and, most importantly, efficiency losses (an average of 2%) due to transformer losses.

More recently, high-frequency transformers have substituted for low-frequency ones at the DC side to provide galvanic isolation. This is obtained through a remarkable reduction in inverter size, weight and cost. However, this option has not gained market penetration since it does not improve efficiency and makes the power stage more complex.

Using transformerless PV inverters combined with floating DC systems (in which the positive and negative terminals are isolated from the ground) reduces costs and improves efficiency and, consequently, is becoming a popular alternative. When compared to the previous configurations, the combination of transformerless PV inverters and floating DC systems boosts efficiency by 2% and reduces costs by 25%. The main problem with this configuration is to attain galvanic isolation from the grid. Without galvanic isolation, DC currents can flow to the grid, saturating the magnetic nucleus of the distribution transformer, which may cause overheating and a malfunction. Further shortcomings include capacitive current leakage, safety issues for operators, eventual tripping of the devices that protect the inverter or the AC power line, and issues in the quality of power and system losses.

In order to prevent a malfunction of the inverter, galvanic isolation through a transformer is necessary in those PV systems with ground connection in the DC input (positive or negative terminals connected to ground) and in the AC output (neutral of the grid connected to ground).

Galvanic isolation in these systems greatly depends on national regulations. It is mandatory in countries like the USA, the UK and Italy, and can be built in the two configurations described above (low frequency and high frequency). In countries like Spain or Germany, galvanic isolation can be omitted as long as the inverters offer an alternative solution to separate the PV generator from the grid, which also guarantees the safety of operators and the quality of the power supplied to the grid (see Fig. 4).

2.3 *Capacitive Leakage Current*

Capacitive leakage currents are a key factor to consider when planning PV systems without transformers. These currents are produced by the existing capacity between the PV array and the ground and the common mode voltage fluctuations across this capacitance. Such capacitive currents, depending on the country, are limited for grid-safety reasons and to protect the operators who work at the systems.

All PV generators have some "parasitic" capacitance formed between the surface of the PV panel and the grounded frame. The value of this capacity depends on the size of the PV generator, on the materials used and on the type of module used. Under bad weather conditions like those produced by rain, frost or high humidity, the value of such capacity increases, ranging in crystalline silicon modules between 60nF and 120nF per kWp.[a]

As mentioned above, inverters without transformers lack galvanic isolation between the PV matrix and the grid, which favors the circulation of the leakage currents originating in the fluctuations of the potential of such parasitic capacity. These currents are closed by connecting the neutral of the electric grid to ground and may reach values of hundreds of mA. Large-scale PV plants with distributed inverters usually have a great length of low voltage grid. In such cases there is a very high parasitic capacitance in

[a]kWp stands for kilowatt peak, defined in "Fundamentals of Photovoltaic Cells and Systems", Sec. 3.3.

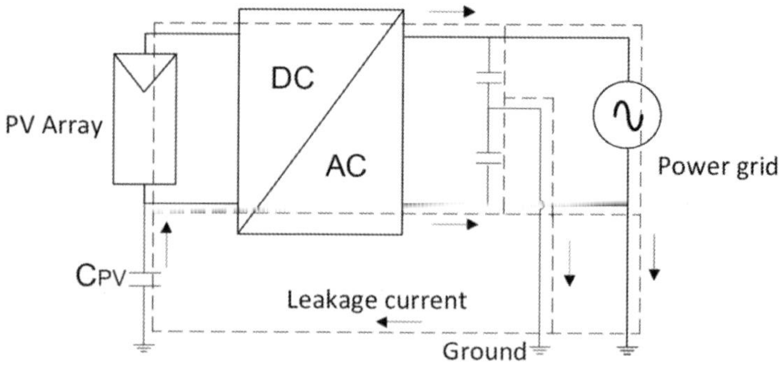

Fig. 5. Leakage current in a PV transformerless inverter system.

the aforementioned low voltage grid. This parasitic capacitance offers an alternative path to leakage currents (see Fig. 5) This current can cause grid current distortion, losses in the system and safety problems.

The new topologies used in transformerless systems aim at minimizing voltage fluctuations in the PV matrix to reduce leakage currents. Three-phase inverters can reduce such voltage fluctuations in large-scale PV plants built with distributed transformerless inverter systems.

In the case of PV plants with big centralized inverters, with compact inverter-transformer solutions for the connection to the medium voltage grid, the low voltage grid is part of the compact block can have any voltage value and its neutral is not connected to ground (if the low voltage winding of the transformer is Y-connected). In this situation, the transformer that is connected to the medium voltage grid performs the role of galvanic isolator from the medium voltage distribution grid (see Fig. 6).

2.4 *Grounding*

There are two possible solutions for the grounding of PV generators. In grounded arrays, for example, one of the conductors of the generator (either positive or negative) is grounded. In ungrounded arrays, on the other hand, none of them is grounded. In both cases, it is necessary to ground all non-current-carrying metal parts. For safety reasons, in grounded arrays, only one grounding electrode can be used both for the grounding of the metal frame and for the grounding of the generator.

As previously mentioned, one of the shortcomings of grounded arrays is the fact that these require galvanic isolation and, as a consequence, inverters with transformers, which reduces efficiency and increases costs.

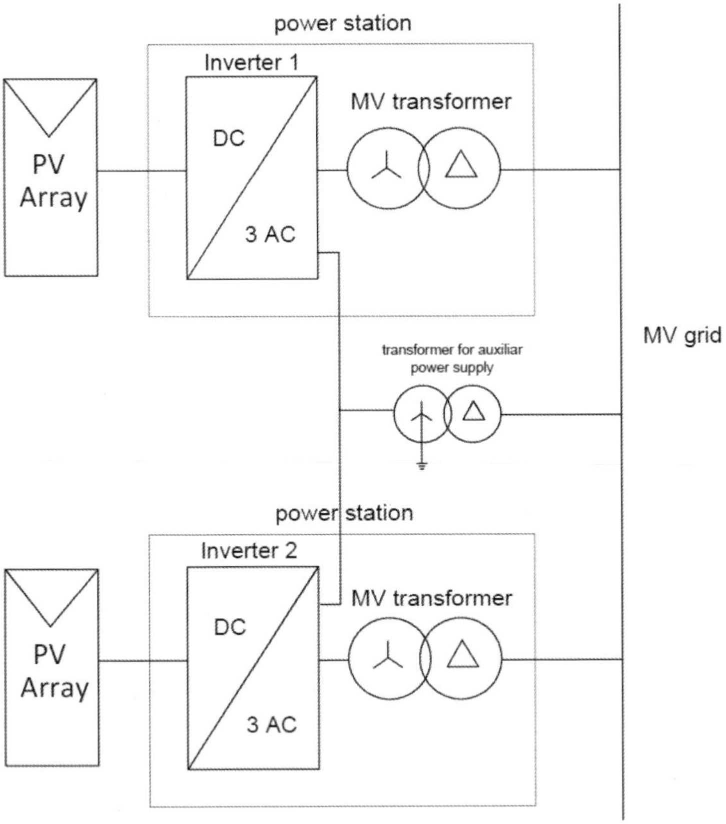

Fig. 6. PV plants with big centralized inverters.

Ungrounded arrays allow, with the right topology, the use of transformerless inverters. Besides the be positive aspects of eliminating the transformer, ungrounded arrays offer safety improvements, equipment-protection enhancements and minimize the risk of fire thanks to the detection of grounding failures in the DC components. For all these reasons, ungrounded arrays are the most common choice in Europe. In the last few years, after the changes introduced in the National Electrical Code (NEC) regulations, these arrays are being used in large-scale PV plants in the USA as well.

Some ungrounded arrays may not be compatible with certain types of modules. In the case of thin-film modules, for example, there is the risk of oxidation and damage in Transparent Conducting Oxide (TCO) layers if the negative conductor is not grounded. In modules with contacts in the back, a polarization effect can occur, which causes a progressive diminution of module efficiency. This effect can be avoided by grounding the positive pole.

2.5 *Insulation Resistance*

A PV generator has non-infinite insulation resistance (R_{iso}) distributed along the installation that creates resistive leakage currents. The value of R_{iso} is determined by the parallel sum of the insulation resistances of the modules, the CC wiring and the inverter, and it diminishes as the size of the installation increases. The R_{iso} of a given installation is also influenced by weather and has its minimal values under wet weather conditions. A drastic drop of this resistance may originate in insulation issues, which should be watched over to prevent short-circuits and fire. The safety of operators is thus linked to high R_{iso} values.

Nowadays, most inverters measure the value of R_{iso}. It is permanently measured in ungrounded arrays and in inverters with transformers. In the case of ungrounded arrays and in inverters with transformers, insulation resistance is measured before the inverter is connected to the power grid. Measuring R_{iso} values in grounded arrays is more complex since the grounded conductor should be insulated first.

The value of the R_{iso} is specified by different regulations, for example:

In the case of modules, international standard IEC 61215 establishes a minimum R_{iso} which is calculated through the formula $R_{\text{iso}} \times A \geq 40\,\text{M}\Omega$, where A represents the area of the module.

For grid-connected PV systems, standard IEC 62446 establishes a minimum value of $1\,\text{M}\Omega$ for R_{iso} in the insulated PV generator.

For inverters without galvanic isolation, the norm DIN VDE 0126-1-1 requires the following R_{iso} values at the point where the inverter is connected to the power grid: $R_{\text{iso}} > 1$ kΩ/V, with a minimum of 500 kΩ. It does not establish any requirement for inverters with a transformer.

The last two regulations do not explain the reasons why such limits are set, limits which are not compatible with large-scale PV plants. In the case of a 100 kWp plant that mounts modules with efficiencies of 16% that abide by IEC 61215 standards, for example, the modules alone yield R_{iso} values of 62.5 kΩ, far beyond the indicated values.

2.6 *Electrical Hazards*

Electrical hazards may be caused by three different types of situations: direct contact with one active conductor; indirect contact with voltages on masses due to insulation issues; and electric shocks originating in capacitive currents if an operator touches the surface of the panel. In large-scale PV plants, the risk is even higher due to the existing high voltage (which in some plants has already surpassed 1000 V) and due to the big area of modules and the length of their wiring (which originates in larger capacitance loads at lower R_{iso} values). The difficulty of avoiding voltage in PV plants makes it compulsory to design facilities in which the risks are reduced to the minimum and calls for preventive measures if such limits are surpassed.

Ungrounded arrays with inverters that use galvanic isolation provide top levels of safety. Ideally, these should also have displays that constantly monitor the R_{iso} values of the PV generator, all the masses of which should be connected to the same ground electrode.

With such a configuration, direct contacts are free from risk under single fault conditions since the voltage on the masses is far below the safety voltage. If the fault is not corrected and a second fault condition occurs, since there is a common ground, there is no risk either since both issues result in a short-circuit.

In the case of direct contact with one active conductor, the current that flows through the person during the electric shock can be limited by controlling the R_{iso}. If the limit of current that a human body can stand is set at 100 mA, without considering the resistance of one particular person, R_{iso} must be maintained at a value 10 times over the maximum generation voltage.[5] In the event of a transitory shock from the parasitic load through direct contact, voltage values in the range of 1500 V have also proved to cause minimal risk for operators.[6] Finally, for these systems, capacitive leakage currents (see Sec. 2.3) are minimal due to the fact that galvanic isolation reduces voltage fluctuations.

In ungrounded arrays without galvanic isolation and with grounded low-voltage AC neutral there is a direct path, closed by the grounding, for the flow of capacitive and resistive leakage currents. As a consequence, safety levels are lower in the event of a direct or indirect contact. In such cases, the differential current device installed in the AC low voltage grid would measure the sum of both currents and insulate the inverter of the AC grid, leading for the system to a situation similar to that described in the previous case.

In grounded arrays with galvanic isolation, the use of one single electrode for the grounding of the masses and of the generator may cause short-circuits originating in insulation issues that prevent the return of currents to the ground. In such situations, the masses do not have potential energy and therefore there are no risks derived from indirect contacts. If direct contacts occur, the worker suffers an electric shock the value of which will depend on the voltage of the generator and on the impedance of the body of the person. Since there are not any differential devices in the grid which are able to detect this situation and to eliminate the voltage, it poses a real electrical hazard.

3 Grid Connection

In recent years, the number of grid-connected renewable energy systems has increased enormously up to the point where, in some countries like Spain, these sources are now the main source of energy generation. This type of generation, geographically dispersed and generally connected to the network, is commonly known as DG.

Grid-connected PV systems are one of these types of energy generation. Their development in recent years has been based on two types of installation:

- Grid-connected low-voltage small installations devoted to residential self-consumption.
- Large PV installations normally connected to medium voltage networks.

The power flow in these systems works in both directions: The grid absorbs the excess of energy produced by the PV installation when the sun is shining and provides the required energy in the hours without radiation.

The growing number of large-scale grid-connected plants with high installed capacities involves important technical challenges and demands new interconnection requirements to maintain the safety, reliability and power quality level in the connected area. For this reason, some countries like Spain[7] and Germany[8,9] (leaders in production, installation and integration of PV technology) have developed connection regulations which require these generators to support the operation and stability of the network. This has raised the possibility of integrating a great amount of PV power into the distribution network.

The international standard IEEE 1547[3], and their respective interpretive standard IEEE 1547.2[4], together establish criteria and requirements for interconnection of DG with Electric Power Systems (EPS). This document provides requirements relevant to the performance, operation, testing, safety and maintenance of the interconnection. The criteria and requirements are applicable to all DR technologies, with a total capacity of 10

MVA or less at the point of common coupling (PCC) (i.e. the grid connection point of a generation plant), interconnected to an EPS at typical primary and/or secondary distribution voltages.

3.1 *Grid-Connection Requirements for Large-Scale PV Plants*

3.1.1 *Voltage control under normal operating conditions — Static voltage support*

According to the Standard IEEE 1547, the DG shall not actively regulate the voltage at the PCC in order not to interfere with the local EPS service regulations, which will also avoid islanding[b] in the generator. This does not mean that the generator cannot regulate the voltage flow by injecting or absorbing reactive power in order to meet the demands of the area EPS operator.

In accordance with the previous recommendations, the Spanish[7] connection regulations establish the power factor limit to be between 0.98 capacitive and 0.98 inductive. Likewise, plants with an installed capacity equal to or greater than 5 MW (0.5 MW in the case of electrical systems for non-mainland territories) must follow the instructions issued by the network operator to modify the mentioned limits depending on the needs of the system. Following the voltage set points of the network operator, the facilities will be allowed to participate voluntarily in the adjustment service of voltage control.

On the other hand, German[9] connection rules require that PV plants connected to medium voltage networks be able to participate in voltage controls when the voltage falls within acceptable limits. For this, the plant reactive power may be adjusted to any power factor limit within 0.95 capacitive and 0.95 inductive on demand from the network operator.

3.1.2 *Dynamic grid support in fault operation*

This requirement makes it compulsory for PV plants to provide dynamic voltage control, that is to say, support under certain fault conditions in the grid. The aim of this requirement is to avoid collapse in the system produced by deep voltage dips due to the sudden disconnection of large PV

[b]Islanding can occur when a part of the grid is electrically isolated from the power system but the part with islanding is energized by distributed generators.

plants. The generation plant must remain connected and provide support during the period that the network fault lasts.

Spain has implemented an operational procedure, as part of connection norms, which states the appropriate response in the event of voltage dips in PV plants. All PV plants with power above 2 MW are to comply with this procedure. For this reason, the PV plant and all its components must be able to remain connected in the event of voltage dips at the connection point produced by three-phases, two-phases or one-phase short-circuits, with the profiles and magnitude indicated in Fig. 7. This means that the installation will not be disconnected due to the voltage dips that fall within the shadowed area of the figure. The functioning conditions during the time that the fault lasts are also specified as regards the consumption-generation of active and reactive power.

In Germany, connection norms make it compulsory for all PV plants connected to medium voltage networks to participate in the dynamic support of the grid. In technical terms, this means that the PV plant must be able to:

- Remain connected in the grid during the presence of a fault.
- Support the grid under fault conditions by injecting reactive current in the grid.

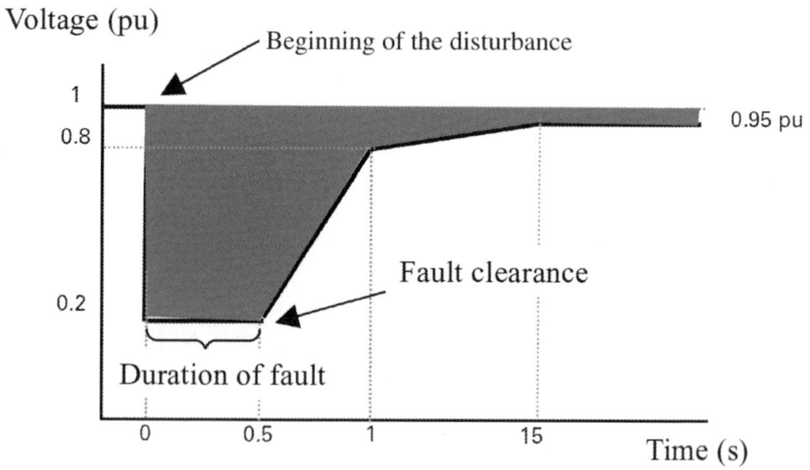

Fig. 7. Voltage–time curve which defines the area of the voltage dip at the connection point which the PV plant must withstand. Ground-phase voltage corresponds to the phases affected by the fault.

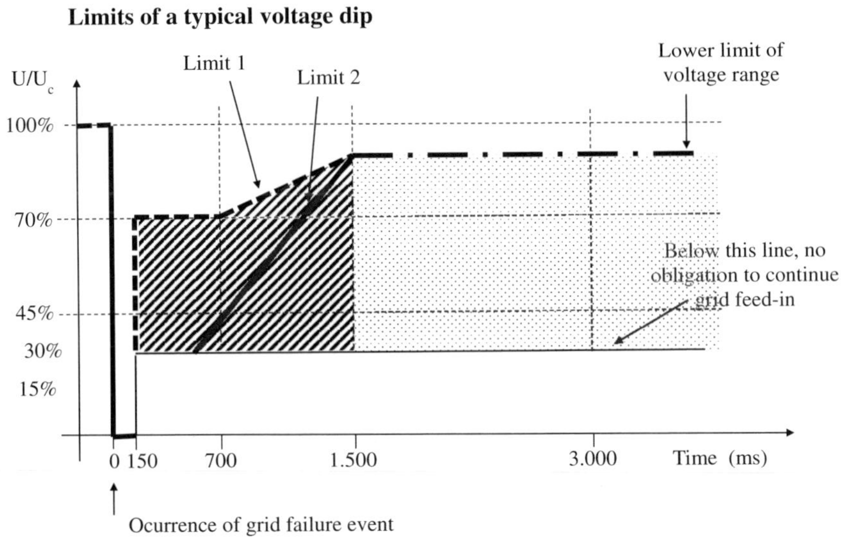

Fig. 8. Borderlines of the voltage profile of a type-2 generating plant at the network connection point.

- Not consume more reactive power after the fault than before the fault occurred.

These requirements are applicable to all types of short-circuits. Figure 8 shows the requirements of the first point. According to this figure, the required behavior for the PV plant is: the plant must remain connected to the network and inject reactive power during the first 150 ms of the fault; over limit point 1 the plant is under continuous and stable operation conditions; between limit points 1 and 2, according to the grid operator, the plant can be disconnected; below limit point 2 and under 30% of nominal voltage, the plant can be disconnected.

3.1.3 *Active power output*

Grid-connected large-scale PV plants are likely to pose significant stability risks in the system if the network operator is not able to limit temporally its output power or able to disconnect the installation.

To avoid this, the German connection regulations make it compulsory for PV plants to be able to limit their output power either by controlling the signals provided by the network operator or automatically in the event of over frequency. In automatic responses, when the grid frequency is over

50.2 Hz, generation units must reduce their active output power with a slope of 40%/Hz with respect to the instant value of active power provided with frequency under 50.2 Hz. The power cannot increase again unless the frequency is less than 50.05 Hz.

In Spain, all installations or groupings of PV installations with power above 5 MW must be linked to a generation control center that will interact with the system by sending real-time information about the installation, guaranteeing its reliability in accordance with the instructions given.

3.1.4 *Unintentional islanding*

Unintentional islanding takes place when the PV generator energizes a part of the grid, which is separated from the EPS, through the PCC. In such cases the generator shall detect the island and disconnect from the grid as soon as possible. According to Standard IEEE 1547,[10] the interconnection system shall cease to energize the EPS within 2 s of the formation of an island, but this time can be reduced in accordance with the local and NECs.

Unintentional islanding is not a desirable operating condition for PV generators. If a zone of the electric system remains energized by the PV system after being islanded from the rest of the power electric system, the safety of the workers operating on the grid is potentially compromised. In addition to this, the voltage and frequency of the islanded zone may reach inadmissible values. This could affect the quality of the power delivered to customers in the area. Other factors to take into account under islanding conditions are the risk that grid equipment may be damaged or the possibility of interfering in a quick restoration of the service through automatic reclosing.

Therefore, protection against islanding is an important requirement for the connection of PV generators to the network. Different detection methods have been developed in recent years. These fall into three categories, namely passive methods, active methods and methods based on the use of communication between the grid and the inverter.

The most common one is the passive method, which monitors grid voltage and grid frequency. If these values exceed the programmed range of values, the inverter is disconnected from the network. All inverters require over/under-voltage and over/under-frequency protection methods. The range of values and time responses that trigger them may vary depending on the country.

Active methods for detecting the island introduce deliberate changes or disturbances to the connected circuit and then monitor the response to determine if the supply network, with its stable frequency, voltage and impedance, is still connected.

Some countries, like Spain, require over/under-voltage and over/under-frequency passive protection measures programmed with the indicated adjustments. Other countries, like Germany, also demand a specific method based on sudden impedance shifts.

3.2 *Power Quality*

One important requirement for grid-connection is the quality of the power delivered by the PV generator. The following factors must be considered when evaluating power quality:

3.2.1 *Voltage fluctuations at the PPC*

These fluctuations are produced during the connection, disconnection, and load variation of the PV generator. The magnitude of such voltage variations depends on the relationship between the size of the generator and the grid impedance at the PPC. Short-circuit power at this point is a reference value for the impedance. Voltage drop variations in this impedance during the connections and disconnections of the PV generator cause fluctuations in the voltage delivered to customers.

The value of the delivered voltage is an important reference when it comes to the quality of the service delivered to customers. This value must be maintained within the allowed limits. These limits may vary from country to country, 5% being a typical value for the maximum deviation allowed from the nominal voltage. Different regulations, such as ANSI C84.1[11] in the USA or EN 5160[12] in Europe, define the main characteristics of the voltage delivered by a distribution network and the limits within which it must be maintained.

German connection rules[9,13] determine that, under normal network operating conditions, the magnitude of the voltage changes caused by all generating plants with a point of connection to a medium voltage network must, at no junction point within this network, exceed a value of 2% as compared to the voltage without generating plants. In the event of disconnection of one generating plant or of several plants simultaneously at one network connection point, the voltage change at every point in the network is limited to 5%.

Spanish connection norms included in Royal Decree 413/2014[7] do not establish a maximum percentage of variation but do limit the connection capacity of non-manageable DG, as is the case for PV generation, in 5% of the short-circuit power at the PPC. This requirement limits indirectly the maximum voltage deviation in medium-voltage networks to values not higher than 2.5% of the nominal voltage at this point of the network.

3.2.2 *Limitation of DC injection*

Injecting DC current into the grid may entail risks for the equipment and the electric system at the same time that it may saturate the distribution transformers turning them into a source of harmonics. Standard IEEE 1547[10] establishes that the DG system and its interconnection system shall not inject DC current greater than 0.5% of the full rated output current at the point of DG connection.

3.2.3 *Harmonics*

Injecting harmonic currents may produce negative effects in the EPS service such as equipment overheating, unnecessary triggering of protection systems, and resonances and may also contribute to the distortion of the voltage delivered at the connection point. To prevent these effects, interconnection regulations limit the harmonics that PV generators can inject into the network.

Standard IEEE 1457[10] defines, for each harmonic order, the maximum current distortion that the generator can inject at the PPC and limits the Total Demand Distortion to 5%. Such limits are applicable to connections to distribution systems with voltages lower than or equal to 69 kV. These emission limits match the most restrictive harmonic current injection limits of the regulations established by IEEE 519-1992 "Recommended Practices and Requirements for Harmonic Control in Electrical Power Systems".

The European standard EN 61727 "Photovoltaic (PV) systems — Characteristics of utility interface" establishes a limit of 5% for total current harmonic distortion.

The German regulations included in the guideline BDEW[9] for the connection of generators to the medium voltage grid set the admissible limits for harmonic and inter-harmonic currents related to the network short-circuit power which may be fed into the medium voltage grid. If several generators are connected to one point, the guideline also establishes the procedure for the distribution of the admissible harmonic currents in each one of the generators.

3.2.4 *Limitation of flicker induced by the PV generator*

Flicker is a power-quality issue predominately associated with noticeable changes in light caused by changes in voltage levels. The severity of this phenomenon is assessed through two indexes, short-term severity (Pst) and long-term severity (Plt). In general, PV generation does not produce annoying flicker fluctuations in the network. Changes in PV generation produced by clouds are not enough to produce flicker.

4 Energy Yields and Efficiencies

In this section, we will deal with factors that characterize the performance of a PV plant, by means of which its energy production can be estimated. Such factors include efficiency, efficacy or the different losses that may reduce the productivity of a given ideal system. In addition, we will outline, through different examples, the software used for the design, simulation and estimation of figures of merit in PV plants.

4.1 *Efficiency and Productivity*

Efficiency and efficacy are two figures of merit in various types of systems (not only in PV plants) which can help to describe the characteristics and quality of a system. Both efficiency and efficacy provide information on the quality of the system, being different and complementary factors. **Efficacy**, the ratio between the theoretical and the actual output of the system, provides information on the performance of the system, at least comparatively. **Efficiency**, the ratio between the output and the input of the system, provides information on the system from a technological rather than comparative perspective, and at the same time provides a quantitative estimate of the production of the system.

In the PV sector, and more particularly in large-scale PV plants, two figures of merit related to the aforementioned concepts are generally used.

On the one hand, the PR is a parameter that measures effectively the real behavior of the system (produced energy) when compared with the hypothetical production of an ideal plant, with the same characteristics as the PV plant being studied but with its cells working at a temperature of 25°C (like at STC[c] [Standard Test Condition]) and with no losses at all.

[c]Irradiance: 1000 W/m^2; cell temperature: 25°C; solar spectrum AM1.5; STC have been defined in chapter: "Fundamentals of Photovoltaic Cells and Systems", Sec. 4.3.1.

The PR describes the relationship between the actual and the theoretical energy outputs of the PV plant. It thus shows the proportion of the energy that is actually available for export to the grid after subtracting energy losses (e.g. due to thermal losses, collection losses, conduction losses, etc.) and after subtracting energy consumption for operation.

The PR indicates the overall effect of losses on the rated output due to array temperature, incomplete utilization of the irradiation and system component inefficiencies or failures.[14]

It is defined analytically by the following formula:

$$PR = \frac{E_{AC}}{E_{th}} = \frac{E_{AC}}{\frac{P_{STC} \int G_i(t)dt}{G_{STD}}} = \frac{E_{AC}}{\frac{P_{STC}}{G_{STC}} H_i}, \tag{1}$$

where E_{AC} is the actual energy produced by the PV system in a given period of time (typically one year),

P_{STC} is the nominal power of the plant under STC (also called peak power),

$G_i(t)$ is the irradiance on the PV generator surface,

G_{STD} is the irradiance at the STC (1000 W/m^2),

H_i is the irradiation on the PV generator along the time considered for E_{AC} and

E_{th} is the energy produced by the same PV system working under STC cell temperature and without losses.

It is clear that this parameter provides information on the behavior of the system. The closer this parameter approaches one, the better the plant will operate. This is a generalized indicator, virtually universal, since it provides useful and comprehensive information on the behavior of the system and is both easy to obtain and to understand: simply by knowing the energy exported to the grid, the PV generator irradiation during the same period and the nominal power of the plant. It also provides a measure of the global losses of the system (L_G) which prevent the maximum theoretical value of the energy output from being reached. So,

$$L_G = 1 - PR. \tag{2}$$

However, this parameter has shortcomings that need to be taken into account. It is important not to focus only on its value but also to seek additional information from other quality parameters. As regards the losses of the PV system, for example, it does not differentiate between unavoidable losses and the losses that a well-engineered plant can prevent.

In this respect, we can distinguish other factors by including only some of the losses in the denominator of Eq. (1). If we only consider unavoidable losses, we would obtain the Performance Index (PI). This parameter does not provide precise information on the engineering of the plant nor on the compliance with the specifications of the equipment. For this reason, its use is less common and the information that it yields may be confusing due to the fact that it is difficult to determine the standard for what can be considered as "avoidable" or "unavoidable" losses.

Another characteristic of PR to be taken into account is the fact that it tends to "favor" low-productivity systems. This may sound paradoxical as regards the profitability of the system. In general, highly productive systems work under high radiation and temperature, which increases the generation losses. In other words, under Nordic weather (low radiation, low temperatures) PR values are higher than under high isolation (and consequently higher temperature) despite the fact that the former produces less energy per unit of nominal power.

This last aspect brings into play a third consideration about PR. For a hypothetical investor, this parameter does not provide information on the profitability of the system. That is to say, its value does not (directly) indicate anything about the production of the system. To obtain such values we must examine efficiency-based indicators. Perhaps, the most widely used in large-scale PV plants is Final Yield Factor (FY), which defines the energy produced during a given period of time (which usually stretches over a year) normalized to the nominal power.

$$FY = \frac{E_{AC}}{P_{STC}}. \tag{3}$$

4.2 *Losses in Large-Scale PV Plants*

The actual energy production of a PV power plant, E_{AC}, is obtained from the expression:

$$E_{AC} = P_{STC}\frac{H_i}{G_{STC}}PR = E_{th} \cdot PR. \tag{4}$$

In Eq. (4), PR reflects the effects of losses in the system which reduce the theoretical energy production. Some of these losses are linked to the transformation of the solar input and the final amount of energy supplied to the network. A brief overview and evaluation of the main factors which account for the total losses in a PV power plant are included in the following paragraph.

Losses can be classified into four main groups, three of which are related to the functioning of the PV generator, the so-called capture losses: collection losses, losses due to functioning under conditions other than STC and losses derived from dispersion and module specifications. The fourth group is associated with the PV Balance of System (BoS).

In the first group, collection losses (L_{coll}), we include several factors that reduce the effective light that will be converted into electricity by the solar cells. These factors include shading due to obstacles in the path of sunlight impinging on the solar cells, reflection due to an off-normal axis incidence angle, module soiling losses and spectral losses due to the actual solar spectrum being other than the 1.5 Air Mass spectrum (AM1.5).[d] Most of the time, cells are not working under STC (which are very unlikely outdoors) and, as a consequence, the efficiency of the module is usually lower than the nominal one under STC. The main factor that reduces efficiency is cell temperature, which is usually higher than 25°C. Taking into account that cell efficiency has a temperature coefficient of about $-0.5\%/\mathrm{K}$ (Si-x) and that the temperature of the cell is usually 20 or more degrees over STC, the temperature loss can easily be above 10%. Another effect included in these losses is the low irradiance effect that reduces the open circuit voltage of the cell. These are the second group we called deviation from STC losses, L_{STC}.

Finally, as regards capture losses, we must take into account the current mismatch of the module in one string that limits the overall current in each string of the array. A string is a set of connected modules, all of which have the same current. Not all the modules can produce the same current under the same operating conditions (dispersion). As a result, the string current is limited by the module with the lowest value. Another factor to take into account in this block is the deviation from the manufacture's specifications (tolerance parameters). The most important is the module power. The actual power of any one module, due to the tolerance, can be different (lower) than the nominal one. These are all called specification losses (L_{Spec}).

[d]AM1.5 refers to the irradiance traversing an atmosphere length 1.5 times its vertical length. The solar spectrum AM1.5 is the spectrum considered in STC. See chapter: "Fundamentals of Photovoltaic Cells and Systems", Sec. 2.3.

In PV power plants there are devices and components other than PV modules. These are known collectively as the BoS equipment, whose non-ideal operation causes additional losses. These include wiring, DC/AC inverter and LV/HV transformer losses. In DC and AC wiring there are ohmic losses due to the resistance of the cables. The DC/AC inverters usually include a Maximum Power Point Tracker (MPPT). This device produces some losses while searching for the Maximum Power Point (MPP) of the PV array that changes along with the environmental conditions. There are also additional losses due to the non-ideal efficiency of the inverter (DC/AC conversion process). Taken together these are all known as the BoS losses, L_{BoS}.

There are various methods and algorithms to calculate some of these loss factors. However, some of these factors can only be estimated through empirical procedures based on experience. This falls beyond the scope of the present chapter and we recommend the simulation tools described below to determine, calculate or estimate these loss factors.

Every device reduces the efficiency of energy production by a factor of $(1 - L_i)$, where the subscript "i" represents the different devices. So, the overall losses of a PV power plant (see Fig. 9) can be calculated as follows:

$$L_G = 1 - (1 - L_{coll})(1 - L_{STC})(1 - L_{Spec})(1 - L_{BoS}). \tag{5}$$

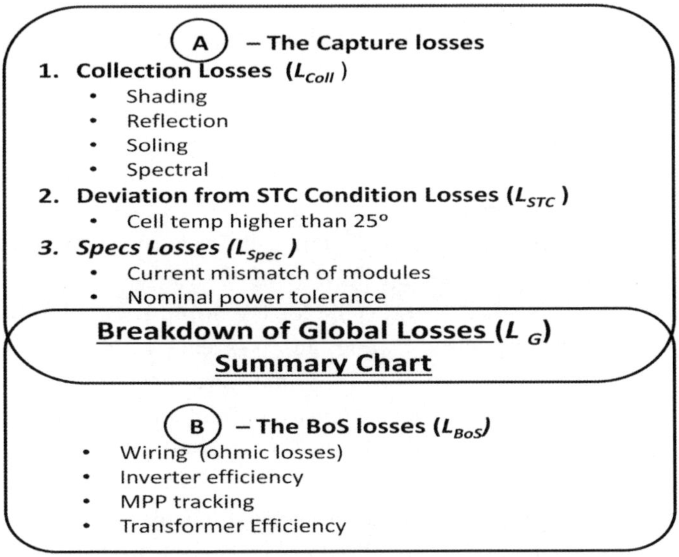

Fig. 9. Overview of the overall losses in a large PV power plant.

From Eq. (2), we can consider PR as a product of the different losses, i.e.

$$PR = (1 - L_{coll})(1 - L_{STC})(1 - L_{Spec})(1 - L_{BoS}). \tag{6}$$

4.3 *Software Simulation Tools*

The huge investment needed to set up a PV plant demand an accurate prediction of PV plant behavior as well as the amount of energy that it can produce to guarantee profitability and minimize financial risks. The analysis of the losses and parameters included in the previous sections leads us to estimate energy production and to optimize the behavior of the PV plant through a detailed study of the elements of the PV plant along with the factors that produce losses.

There are powerful software tools available to perform sound simulations of the behavior of a PV plant and accurate predictions of its energy yield, its productivity, profitability and reliability. Such tools rely on comprehensive and detailed databases of the different devices (modules, inverters, etc.) and of the amount of radiation as well as other environmental variables that impinge on the calculation.

By this means, an initial estimation of the behavior and productivity of the PV plant can be carried out using a few, not very complex initial data like location, orientation, the type of PV technology, the type of inverter and the local azimuth. Nevertheless, it goes without saying that properly designing and engineering a large-scale PV plant (which entails optimal behavior and productivity analysis, a precise scouting of its siting and an appropriate choice of modules, orientation, layout and other BoS devices) is not an easy task and demands powerful simulation tools to predict the best possible energetic, economic and environmental solutions.

Sizing tools and calculation programs are the most straightforward ones. These allow dimensioning PV plants taking into consideration energetic, siting and cost requirements.

Besides the tools which are specific for PV applications, PV plants can also be modeled and simulated through more extended software like Matlab or Pspice. Such tools are the ones that many engineers are familiar with and allow an advanced mathematical treatment aimed at simulating specific devices and systems. Such tools are useful for research, development and education purposes.

However this is not the case for large-scale PV plants, where energetic needs, cost limitations and detailed studies of devices are less relevant than accurate and reliable estimation of behavior, productivity and energy yield.

As a consequence, specific software simulation tools are required in the case of large PV installations.

The interested reader can find a comprehensive review of PV system design, sizing and simulation tools in the existing literature.[15,16] Here, we will outline briefly some of the results of two of these programs, with different characteristics but which have been used extensively.

Among the handy tools that allow an initial estimation of a given investment project, PV Geographic Information System (PVGIS) software can be easily accessed.[17] This easy-to-implement, open access and free piece of software has been developed by the Joint Research Centre (JRC) of the European Commission. The application provides radiation data from different sources, chiefly from the European Solar Radiation Atlas as well as from satellites. Likewise, it offers the possibility of estimating the solar electricity production of a PV system under different variables, since it allows the selection of different parameters for module technology, mounting options, etc. by means of which users can obtain an initial estimate of energy yield data, which can be exported as text or graphs.

On the downside, this tool does not allow the possibility of simulating large PV power plants and, consequently, it does not provide good estimates for large-scale plants. Overall, the software is easy to use, free to access and a great starting point for early productivity estimations for PV plants set in Europe, Asia and Africa.

More powerful tools will be required for a detailed estimation of the behavior and results for large-scale PV plants. Such tools must rely on sounder theoretical grounds and provide a wider range of uses. The PVSyst program is regarded as one of the most powerful and comprehensive tools of its kind. It uses large databases of radiation, modules and devices that are likely to be mounted in a large-scale PV plant. Currently it displays two user levels. The first level assists in the pre-sizing of the PV system and requires limited input data. The second level involves the choice of a wide variety of options that provide a complete calculation of all the mechanisms and processes involved in the performance and results for a large-scale PV plant.

5 Commissioning. Operation & Maintenance

Most of the projects regarding PV systems are turn-key plants in which the owner of the installation asks the installing company to design and to implement the plant. In order to accept the plant, validate its appropriate installation and transfer the responsibility of the system to the owner, all the components must be properly installed and operating as expected, and the system must comply with applicable standards and norms. Beyond this point, the PV plant is commissioned and the starting point for the operation of the installation can be scheduled.

Although the utilities which are receiving the electricity from the PV plants normally have their own set of procedures for the compliance with inspecting, testing and commissioning requirements (in order to guarantee the safety performance of the plant and also the absence of any sort of hazards) there are codes and regulations stipulated by local, regional and national governments.[18]

Among all the codes and regulations, the International Electrotechnical Commission standard (IEC 62446) is remarkable, as it defines the commissioning tests, inspection criteria and documentation expected to verify the safe installation and correct operation of the system. This document can also be followed during the periodic retesting which is necessary for maintenance and operating protocols.[19] Normally, national governments have adopted this standard as a general rule, with some specific modifications adjusted to the singularities of the region.

The list below includes some basic aspects to consider during the commissioning of a plant. These aspects may be complemented and/or modified to abide by local standards and regulations, utility constrains and installer-owner agreements.

5.1 *List of Procedures for the Commission of a Large PV Plant*

Most of the procedures gathered either in the IEC standard or in the protocols defined by the utilities and national or local regulators can be divided into the following:

- Visual inspection,
- Testing and initial startup,
- System documentation.

Although there may exist some differences in the number and ways of implementation of the procedures applied for the commissioning of a plant, the following steps can be used as a general reference.

5.1.1 *Visual inspection*

The visual inspection, also known as the final check-out prior to the initial startup, is the first verification procedure. During the visual inspection, the

Table 1: Visual inspection checklist.

— Verification of the supply and installation of the entire equipment and devices
— Verification of the quality of materials (elements, structures and civil works materials, etc.)
— Proper labeling and identification of PV modules
— Front and rear inspection of PV modules (surface damage, delamination, fastening, grounding connection, electrical frame insulation and encapsulant)
— Verification of the peak power installed (laboratory calibration versus flash test)
— Shading of modules (self-shading or from the surrounding elements)
— Optimal tilt and orientation of PV modules
— Verification of the supporting structure for PV modules
— DC Electrical connection (proper environment isolation of the boxes and electrical connection in accordance with wiring diagrams)
— Verification of the buried cable installation
— Verification of the proper installation of the earthing system and overvoltage protection
— AC Electrical connection (inverter and electricity meter)
— Proper installation and connection of the monitoring and control systems
— Proper installation and connection of the weather station (irradiance and temperature sensors mainly)
— Cleanliness of the installation

correspondence between the planned design and the executed (civil and mechanical) work is checked along with aspects related to electrical and instrument issues.

This visual inspection also comprises the verification of administrative constraints and the compliance with safety standards depending on the system documentation, which should be handed to the owner.

A brief checklist of the most important visual inspection items is collected in Table 1.

5.1.2 *Testing and initial startup*

During the commissioning and prior to the startup of any PV system, several measurements have to be carried out in order to prevent damage to the equipment and to protect the users or operators from safety hazards.[19–21]

Once the previous steps have been accomplished, the PV system can be switched on and its electrical characteristics measured under real operational conditions. This is done to ensure that the system is operating as expected and also to detect any possible sources of future malfunctions in the system.

The procedures and measures that should be applied before and after the startup are collected in Tables 2 and 3.

Table 2: Pre-startup testing procedures and measurements.

— Verification of the polarity of the PV strings
— Short-circuit current measurement of each PV string
— Open-circuit for each string and the entire PV generator
— Insulation resistance of the DC main cable
— Insulation resistance of the PV generator
Insulation resistance of the AC section (wiring and transformer)
— Ground resistance measurement of the PV field (continuity of protective earth)
— Ground resistance measurement of the inverters and transformer centers

Table 3: Post-startup testing measurements.

— Operating current for each string
— Peak power output measurement
— Voltage drop above each safety device
— Voltage drop above DC main cable
— Voltage drop above each safety device
— Verification of the tripping of the protection devices
— Operational state of the inverter (alarms, monitoring parameters)
— Operational state of the electricity meter
— Operational state of monitoring and control systems
— Operational state of the weather station
— Thermography of the PV modules (hot-spot detection)
— Thermographic study of the electrical connection boxes (hot-spot areas detection)

5.1.3 *System documentation*

Once the installation is finished and the initial tests reinforce the proper and safe operational state of the PV plant as contracted and designed, the installer should provide the owner of the plant with all the documentation related to its system. A collection of the essential documents needed for the acceptance and commissioning of the plant is included in Table 4.[18,19,21]

5.2 *Operation and Maintenance of a PV Plant*

As a general rule, grid-connected PV systems require low maintenance, especially if the modules are installed on fixed structures. Nevertheless, it is mandatory to undertake some basic maintenance tasks in order to assure the operational behavior, the optimal energy production and the durability of the plant.

Normally, the installer of the plant should provide the owner with a document with the Operation and Maintenance (O&M) procedures adapted to the particularities of the plant as it was described in the previous section.

Table 4: System documentation recommendations.

— Part list, datasheets and equipment's manufacturer manuals
— Wiring diagrams
— Mechanical design information (also drawings)
— System performance estimate
— Operation and maintenance information (procedures for troubleshooting)
— Safety requirements
— Test Results and commissioning data
— Permits
— Certificates of inspection
— Utility interconnection agreements
— System warranty info

Depending on the severity of the maintenance tasks, two different types of maintenance protocols can be defined:

- Preventive Maintenance
- Corrective Maintenance

Preventive maintenance involves a set of actions whose aim is to maintain the plant under an optimum state of conservation and operation. Their application frequency varies from daily actions to others more extended in time. The objective is to prevent (or to reduce to a minimum) any malfunction in the plant over time.

On the other hand, corrective maintenance involves the substitution and reparation actions that will allow the plant to operate correctly throughout its operating life.

Table 5 shows a list of the main actions corresponding preventive maintenance, gathered in any standard O&M procedure, and recommendations regarding their inspection frequency.

It is easily observable that the O&M procedures defined in the preventive maintenance list include many of the items from the testing and inspection checklist of the commissioning protocol.[21] In this case, with the exception of the triennial activities, these tasks can be carried out by a non-technical operator.

It is important to mention that maintenance requirements could be reduced if the proper monitoring system is installed. This monitoring system has proved to be very useful in daily checks and chiefly during monthly inspections of energy production.

Table 5: Preventive maintenance protocol actions.

Activity	Frequency
— Inspection of the operational state of the inverter and electricity meter	Daily
— Inspection of the operational state of the monitoring and control system	Daily
— Measurement of energy generation (yield)	Monthly
— Inspection of the operational state of the weather station	Monthly
— Module Inspection (surface damage, delamination, burned connections, fastening, corroded grounding connection, electrical frame insulation)	Biannually
— Inspection of the mechanical security of the structure	Annually
— Revision of the connections, wiring and circuit boxes (mechanical damage, loose connections)	Annually
— Verification of the tripping of the protection devices and main switch	Annually
— Inspection of shading on the PV generator's surface	Annually
— Repetition of insulation and ground measurements (same as commissioning)	Triennial
— Repetition of voltage drop measurements (same as commissioning)	Triennial
— Repetition of operating current measurements in each string (same as commissioning)	Triennial
— Repetition of peak power output measurement (same as commissioning)	Triennial
— Repetition of thermographic studies of the modules and the electrical connection boxes (same as commissioning)	Triennial
— Cleaning of the modules (also include pruning to prevent possible fire hazards)	(as required)

Table 6: Most common Failures in PV plants.

— Defects in the connection and cabling
— Faults on modules (delamination, soldering joints, moisture ingress, disconnected modules, reverse polarity, hot spots)
— Isolation and grounding faults
— Defects in protection devices
— Stops and faults of the inverter (hardware defects, decreased efficiency, high level of harmonics)

Corrective maintenance deals with the typical faults and breakdowns of any PV system, and qualified technical personnel should always conduct them.

Although a troubleshooting guide is beyond the scope of the present manual, Table 6 gives a brief summary of the most common sources of faults in large-scale PV power plants that should be covered by corrective maintenance.[20,22,23]

Corrective maintenance should be included and specified in detail in the warranty documentation received when commissioning the plant.

6 Promotion Policies — Planning and Regulation

Nowadays, the different sources of energy in the electricity generation mix is having an expanding impact on the development of the balance of energy sources and on policies within countries. In this scenario, renewables are responsible for a major change in the conception of electrical generation. PV technology stands out in the Renewable Energy (RE) supply, as it represents a realistic alternative in the energy market and is particularly appropriate in countries which are changing the technologies of their electricity generation mix.

Despite the increased deployment of installed PV technology in recent years, its development still remains subject to a policy-driven market and greatly depends on electricity savings and on the potential sales in the wholesale electricity market.[24] In this sense, public policies can not only support but also accelerate PV energy cost reduction. Such is the main objective of government promotion policies, namely, the encouragement of certain technologies in their early-stages so that these can achieve large-scale deployment.[25]

Nevertheless, the definition of several rules or directives is not enough for the effective promotion of RE. A supporting economic mechanism must be designed to minimize both the cost to consumers and generation costs.

The distinguishing approach in the promotion policies proposed by some authors comes from the coupling of energy and economic analyses with the optimization of the place for the installation of a certain RE source. It is intended to strengthen the philosophy that the installation of PV energy should not be made at any price in a massive intervention, regardless the suitability of the location. The principles of equi-marginality and lowering the cost of governmental support should be combined to adapt the PV plant to the places with best renewable energy resources, thus minimizing both the cost to consumers and generation costs.[26]

The regulation tools that governments have been using since the deployment of PV technology in recent years can be grouped according to their energy or finance policy focus. The most representative ones among the large number of supporting mechanisms are listed in the following paragraphs.[27–29]

Feed-in Tariff (FiT): Price-driven regulation which consists in offering PV electricity producers a guaranteed constant price for the energy supplied to the grid during a determined period of time. This unitary price is independent of market electricity prices.

Premium Feed-in Tariff (PFiT): Variation of the previous regulation in which the electricity supplied to the grid will be dependent on price variations of the electricity market. It also includes a floor and a cap unitary price, in order to guarantee a balance between risk and profits.

Net Metering (also Net billing): Access-type regulation consisting in the compensation for the electricity consumed by one user in his residence with the electricity generated by a PV system installed in his domain whose energy is injected to the local distribution grid. The total electricity metered is the balance of the PV electricity delivered to the grid minus the electricity consumed from it. Therefore, there is compensation between the consumption and the production profiles and it is normally applicable during a specific billing period.

Self-Consumption: Access type regulation consisting in connecting a PV installation to the consumer's own grid, allowing him to consume the energy self-produced with the PV system installed in his domain and also receiving value for the non-consumed electricity (excess) that is fed into the general grid.

VAT reduction: Fiscal incentive that reduces the taxes for the production of renewable electricity and that also includes any economic activity related to this technology. It may also consider tax exemption.

Tender and auctions: Quantity-driven mechanism consisting in the remuneration, above the market electricity price, of the electricity supplied by a RE source, whose right to feed into the grid has been obtained through a public tender of a certain quota of PV supply.

Investment Grants: Fiscal subsidy which does not have to be returned with the aim of reducing investment cost when installing any PV system.

Fiscal credit: Fiscal incentive based on the annual tax credit provision to the investor of a PV system depending on the amount invested and the renewable electricity produced. It allows tax deductions.

Beneficial Credit Terms: Public finance mechanism by means of which the investor of a RE plant is financed through lower interest rates and less barriers and requirements.

Although there are more supporting mechanisms like storage incentives, indirect sale of the electricity produced, energy production payment by the government, green certificates, etc. the tools previously defined are the most common. It is important to point out that some regulations do not use one single supporting mechanism, but a combination of them.

Among the regulation tools identified, the FiT has been the one most commonly used, causing several countries to reach PV grid parity, as the price of producing a unit of electricity through PV can be compared to their local retail electricity prices in a competitive way.

Moreover, these FiT policies have boosted the industry to a level where, in certain cases, this parity has been reached without any subsidy or FiT incentives. A favorable regulatory framework has also been necessary[30] as well as the application of reduction mechanisms, where FiT have been progressively reduced based on the connected power in previous periods.[25]

Despite the aforementioned promotion of PV energy and as the share of PV increases, grid-market integration is facing new challenges.[24] In this sense, Europe, as the main PV installer up to date, is facing an enormous competitiveness challenge due to the fact that the priority imposed on renewable energies in the short-term has increased energy generation costs,[31] particularly when there is some skepticism concerning the feasibility of running Europe just on the basis of RE.[32] This technical and economic disbelief is also translatable to the rest of the world.[33]

Beyond the extra cost charges, a critical issue that should be faced is the security of supply, its sustainability and the economic competitiveness, as they seem to be unbalanced in some energy policy targets.[34] In that sense, DG should be favored by policies that increase its market penetration[35] in order to avoid the boom and subsequent bust that the market of PV technology has recently experienced in several countries, especially within the European Union. The lessons learned should be used as an example for the proposal of future policy recommendations for solar PV deployment.[36,37]

Either way, any RE policy should consider its impact and effectiveness[38] and the means through which it can be adapted to dynamic market conditions.[39] At this point, the trending concepts of self-consumption and net-metering are becoming a promising alternative.[40,41] According to some reports, around 4% of the world PV market has been installed according to self-consumption mechanisms.[42]

7 Economic Analysis: Financing and Due Diligence

Large-scale PV projects have experienced tremendous growth in the world in the last few years, above all in the US, China, South Africa, Chile and India. The large budget for such projects requires special development and financing methods. The key consideration during project development is the balance of expenditure and risk. Different financing mechanisms can be implemented in such installations, but the most commonly used tool is project finance, where the aim is to financing the PV plant without risk to the stakeholder. Such projects must pass through a number of different stages, categorized as pre-financial and post-financial activities. Financial institutions require independent advisors to carry out the prior and post-financial closing. In the pre-financial close stage, the Full Due Diligence process is important, consisting mainly of Legal & Tax Due Diligence, Technical Due Diligence (TDD) and Insurance Due Diligence (IDD). It provides support to lenders and investors in ensuring that projects are viable and contractually sound, focusing on identifying and controlling technical risks throughout the project life-cycle.

From a financial point of view, investing in large-scale PV projects is attractive in many countries around the globe, using conventional PV (Si-x) as well as other new PV technologies (e.g. concentrating PV, thin-film technology, etc.).[43,44] In addition, large-scale, ground-mounted PV power plants are largely benign with regard to wildlife and other environmental indicators compared with fossil-fuel based power generation.[45] More investors are recognizing that PV technology has matured and the risk of default is low. The investment, and thus the prices achievable on the power market, is not linked to uncertainties in the oil market. And the sun's energy is free, which gives PV projects a fundamental advantage.

7.1 *Economic Parameters: Grid Parity, Life-Cycle Cost (LCC) and Levelized Cost of Electricity (LCOE)*

Regarding the economic analysis of large-scale PV power plants there are several concepts that need to be defined: Grid Parity, Life-Cycle Cost (LCC) and Levelized Cost of Electricity (LCOE). They play the role of competitiveness and cost referees.

Grid Parity occurs when the cost of producing a unit of electricity by PVs is the same as the average end-price for electricity to consumers. Grid parity should not be confused with generation-parity, as the latter concept includes the transport and distribution of the electricity which was generated off the grid.

The **LCC** of the PV system, includes the present worth of all the expenses incurred throughout the life of the plant (N years) and will depend on the initial investment (PV_{IN}, €) and the annual operation and maintenance ($O\&M$) cost (PVA_{OM}, €) associated with the selected PV technology.[46,47]

Unless the project is not financed, the initial investment will be affected by the annual interest rate for borrowing money during a period of time, and the O&M will extend throughout the lifetime of the plant, so the sum of these annual costs has to be added to the present worth of the initial investment in the PV system ($PW[PV_{IN}]$, €) and to the present worth of the operation and maintenance cost over N years ($PW[PV_{OM}(N)]$, €) associated with the PV technology, being the present worth the current worth of the future investment.

Here, like in any other investment, it is necessary to establish the present worth of the capital invested. Therefore we have taken into account the nominal discount rate, r (%).[48]

$$LCC(€) = PW[PV_{IN}] + PW[PW_{OM}(N)]. \tag{7}$$

The investment can be financed either with company capital (own capital, OC) ($PW[PV_{OC}]$, €) or external capital ($PW[PV_{EC}]$, €). The assumption has been made that there is an annual retribution, for own capital in the form of a dividend (d_i) and the investment will be amortized at the end of the life-cycle of the system (N years). As a result, to calculate the present worth of the owned capital, the following equation is used, considering the factor $q = 1/(1+r)$:

$$PW[PV_{OC}] = PV_{OC}\left[d_i \cdot \frac{q\cdot(1-q^N)}{(1-q)} + q^N\right]. \tag{8}$$

The rest of the investment, $PV_{EC} = PV_{IN} - PV_{OC}$, may be financed with an annual loan interest i_l (%) and loan term N_l (years), so its present worth is,

$$PW[PV_{EC}] = \left[(PV_{IN} - PV_{OC})\cdot i_l \frac{(1+i_l)^{N_l}}{(1+i_l)^{N_l}-1} \cdot \frac{q\cdot(1-q^{N_l})}{(1-q)}\right]. \tag{9}$$

Finally, the present worth of investment cost can be expressed as:

$$PW[PV_{IN}] = PW[PV_{EC}] + PW[PV_{OC}]. \tag{10}$$

The annual operation and maintenance cost is assumed to be proportional to the initial investment. Additionally, an annual escalation rate (ε_{PVAOM}) of the operation and maintenance cost of the system has been defined, so $PW[PV_{OM}(N)]$ may be re-written as:

$$PW[PV_{OM}(N)] = PV_{AOM} \cdot \frac{K_{PV} \cdot (1 - K_{PV}^{N})}{1 - K_{PV}}, \tag{11}$$

where

$$K_{PV} = \left[\frac{(1 + \varepsilon_{PVAOM})}{(1 + r)}\right]. \tag{12}$$

The LCOE [also called LEC], is defined as the cost of a unit of electricity (€·kWh^{-1}) produced by a given system over a specified number of years, normally throughout its whole operational lifetime. This cost is expressed in current monetary units and it is levelised for all the years that the system is intended to be generating electricity.[49] In the LCOE analysis, the cost of transport and maintenance of the network is not considered, so that it is possible to use this number to obtain the value for grid parity. This parameter calculates the price of the electricity that the system generates by dividing the project's total cost into the energy produced during its complete operating life. The LCOE can be defined by the following expression[50,51]:

$$LCOE(€ \cdot kWh^{-1}) = \frac{LCC}{\sum_{i=1}^{N} \frac{E_{PV} \cdot (1 - \varepsilon_{pl})^{i}}{(1+r)^{i}}}. \tag{13}$$

In the previous equation, E_{PV} is the annual PV electricity yield (kWh·year^{-1}), where an annual decrease of the power generated of ε_{pl} (%) has been assumed.

The parameter *LCOE* is a widely used term in the PV industry and it is often used as a marketing tool in discussions regarding large-scale projects. The economic feasibility of PV projects is increasingly being evaluated using that parameter in order compare it to investment in other electricity generation technologies. *LCOE* can also be used for economic comparison between different PV energy system configurations.[52] In addition, LCOE is the widely accepted criterion to fairly compare the cost of energy generated by different power plants.

Table 7 shows the values of the parameters involved in the LCOE for a possible actual scenario in 2014, in wide areas (horizontal irradiation over 1600 kWh/m^2), with Si-x technology and standard economy.

Table 7: Possible scenario in 2014 of the parameter values involved in calculation of LCOE.

		LCC								Present worth	Life
Energy		Financing			Own capital		Q&M				
Final yield	De-rating	Initial invest-ment	Extemal captital	Interest loan	Loan years	Own capital	Dividends	Q&M Cost $\%(PV_{in})$	Growth rate of Q&M	Discount rate	Years of useful life
E_{PV} (kWh/ kWp/ year)	ε_{pl} (annual %)	PV_{in} (€)	x% of PV_{in} (%)	i_l (annual %)	N_l (Year)	$(1-x)$ % of PV_{in} (%)	d_i (annual %)	PV_{old} (%)	ε_{PVOM} (annual %)	r (annual %)	N (year)
1350	0.5	1900	80	4.13	20	20	3	1.5	2.8	3.9	20

With these values we have a present worth (i.e. the current worth of future investments) of the PV system of 2395 €/kWp and a LCOE of 13.5 c€/kWh.

7.2 *Financing*

A primary objective in financing a power plant is to ensure that the plant operates continuously and reliably, thereby generating the maximum economic and energy performance returns. It is therefore important to analyze the risk and uncertainties involved in the project before assessing how much collateral is required or whether project financing is possible. This requires a thorough study of the individual factors and associated risks and by means of mitigation of any identified risks to quantify and to reduce, if possible, the project uncertainties.

A number of financing mechanisms are available for infrastructure projects. They include the following: cash purchase, government funding, corporate or on-balance sheet finance and project finance.

Project finance is the typical tool for financing very big projects in different sectors such as energy and water. In project finance, the viability of projects is solely based on their own capacity to produce future cash flow, which makes it feasible to develop projects with high levels of leverage due to project developers' shortage of funds.[53–55]

Project finance is the usual financing tool for large-scale PV plants. Financial institutions require independent advisors to use this tool. The working tool to carry out the recommendations of the independent advisors is known as due diligence.

7.3 *Technical Due Diligence*

The bankability of a PV plant is not only addressed through the modelling of its energy yield followed by an on-site measurement campaign,[56,57] it is necessary to analyze more aspects. To achieve this, the most frequent resource is the aforementioned Technical Due Diligence (TDD). It comprises all the technical aspects of a project, from the production estimation to the contractual agreements and costs. The objective of the TDD is to understand and mitigate a variety of technical, legal and socio-environmental risks before valuable time and resources are allocated to the project.

Although bankability assessments vary from bank to bank and are not based upon a single criterion or universal formula, the factors by which it is most frequently determined fall into four main categories: company, product, processes and service.

TDD entails a multi-disciplinary analysis in which risk assessments are made for all individual aspects involved in the PV plant. Typical areas assessed in the TDD are[4,58]:

- Solar resource assessment,
- Losses and yield estimation,
- Major component technology review,
- Equipment supplier evaluations: references, previous experience and delivery periods,
- Evaluation of the base case: price, expenses, revenue, etc.,
- Engineering review: technological evaluation of equipment and plant design,
- Contract evaluation (EPC, O&M, PPA, etc.),
- Evaluation of authorizations, licenses, permits, etc. and
- Environmental evaluation.

8 The Future of Large-Scale PV Power Plants

The future of large-scale PV power plants is location dependent and is also influenced by the promotion policies applied. In recent years, the trajectory of the PV industry has been ambiguous. Whereas there has been a pullback because of the recession in Europe, Asian regions have driven up demand. Therefore, in order to objectively assess the possible future development and installation of new large-scale PV power plants, it is necessary to divide the analysis according to the following classification:

8.1 *European Stand-by Future in Large-Scale PV Power Plants*

The EU, which has been the major driving force behind the installation of large PV power plants, has moved from an extraordinary boom, overdeveloped according to some authors, mainly caused by a clear interest and policy support from several national governments, to a stand-by or even recessionary state in certain regions where there has been a reduction in the development of new projects and even the revoking of some economic support mechanisms.[59] The uncontrolled expansion of these systems has led to several measures at a national level to restrain their development. A moratorium on new plants, extra fees and also the retroactive application of some of these measures has not only meant a deceleration in the installing of new PV power but has also led to instability and a lack of viability of existing PV installations.[60]

Beyond those policy measures, the current trend in European PV support schemes is to move to net-metering and self-consumption scenarios,[27] where rooftop small PV systems play a major role, thus promoting local distribution grids.[61] In 2013, around 10% of the installed PV capacity in Europe was designed for self-consumption, and mid-term trends go in this direction as well. Nevertheless, the segmentation in the European PV market is quite varied depending on the country analyzed. The share of ground-mounted systems is around 34% while the commercial and industrial rooftop segments represent 44%[24] altogether.

In this scenario, the future of large-scale PV plants in the European market is uncertain in the short and mid-term, but is stable long term. Some analysts assess that the EU will play only a minor role in the global PV market capacity. The forecast in accumulating PV power is intended to shift from around 80 GW in 2013[42] to around 229 GW by 2050.[62]

8.2 *The Asia Pacific and Middle East and North Africa (MENA) Countries Boosting Future in Large PV Plants*

In 2013, Asia surpassed Europe in the annual increase in PV capacity, with around 56% of the PV power installed worldwide. China and Japan were the leaders, where 11.8 GW and 6.9 GW were installed respectively.[24]

In contrast, in the other regions of the world, centralized grid-connected systems are by far the most common PV systems installed[42] and short-term future projects are expected to reinforce this large-scale grid-connected tendency. The number of new projects forecast is large, on top of those already being installed in China, India, Japan, Kazakhstan, Korea, and UAE among others.[63]

Therefore, the forecast in all reports suggests an optimistic trend and growing opportunity for the future installation of large PV plants, in which China will retain the leadership gained in 2013. China is expected to increase its global capacity from 13% in 2013[24] to around 40% by 2050, producing around 21% of its electricity generation through PV sources.[62]

Despite the great position of China regarding the installation of large-scale PV systems, India and the MENA countries are expected to play a major role in the mid-term, as they represent an untapped potential.[24] In this sense, India, which now accounts for around 2% of the global capacity, will experience, in a favorable scenario, an exponential growth up to 575 GW by 2050, reaching the third highest position in installed PV capacity, just after the USA, and having similar figures to those of China in the percentage contribution to its national electricity generation.[62]

In view of the above facts and trends, the most promising future for large-scale PV systems in the short and mid-term seems to be located in these regions.

8.3 *North and South American Deployment of PV Plants*

America has a great potential in large-scale PV systems, where Latin American countries are expected to lead the deployment of this technology.[24]

In 2013, the USA represented around 9% of cumulative PV systems, but a slight slowdown has been detected in its growth, possibly caused by the solar trade dispute between the USA and China.[63]

Nevertheless, the deployment of these systems is a reality and analysts forecast a cumulative PV capacity or around 811 GW by 2050, positioning the continent as the second largest PV power region.[62] This may represent around 14–15% of PV capacity worldwide.

8.4 *Africa's Increasing Development of Large PV Plant Projects*

Africa has great potential in the installation of large-scale PV power plants, not only because of its favorable radiation conditions and the availability of land area, but also because some countries are experiencing a rapid economic growth. Energy consumption has risen around 45% in some African countries and they are unlocking their huge renewable energy resources. Therefore, the expectations are that half of the electricity generated by 2040 will be produced through renewable sources.[64]

The share of PV produced electricity in Africa is still low due to the lack of access to an electric grid. In fact, most of the systems already installed correspond to decentralized grid-connected ones.[42] Some activity is taking place in those regions through off-grid and mini-grid PV systems,[62] where a noticeable trend is the combination of diesel generators and solar PV microgrid systems.[64]

Nevertheless, in order to absorb the rapid energy growth mentioned, some countries like Nigeria, Uganda or Namibia, among others, are showing great interest in large-scale plants and are aiming to produce several MWs using large PV systems.[63]

In an optimistic scenario, the PV capacity estimated for Africa in 2030 will be around 85 GW, with 165 GW of accumulated PV energy planned for 2050, which makes around 11% of share in the generation mix.[62]

Acknowledgments

We would like to thank our colleague Joaquín Cruz Trapero, from the University of Jaén, for his invaluable assistance in the translation and review of the present chapter. We are also grateful to Pablo Valera, who has been willing to provide us with data derived from his PhD thesis work.

References

1. A. Goodrich, T. James and M. Woodhouse, Residential, commercial, utility-scale photovoltaic (PV) system prices in the united states: Current drivers and cost-reduction opportunities, Tech. Rep. DE-AC36-08GO28308, *Nat. Renew. Energy Lab.* (NREL), Golden, CO, USA (2012).
2. PVPS Report Snapshot of Global PV 1992–2013, IEA PVPS 2013, IEA–EU.
3. C. Winneker, *EPIA Global market outlook for Photovoltaics 2013–2017. European PV Insdustry Assosiation* (EPIA, Brussels, 2013).
4. P. Valera, O. Pereles, A. Trabant, A. Payán, A. Pastor, L. Almonacid, V. Salas and G. Almonacid, State of the technique of the methods implemented in independent technical advisory services, including the technical due diligences (TDD), related with the project finance in photovoltaic energy projects, in *EUPVSEC*, Amsterdam (2014).
5. P. Vidal, G. Almonacid, P. Pérez and J. Aguilera, Measures used to protect people exposed to a PV generator: 'Univer Project', *Prog. Photovolt.* **9** (2001), pp. 57–67.
6. J. Hernandez and P. Vidal, Guidelines for protection against electric shock in PV generators, *IEEE Trans. Energ. Conv.* **24**(1) (2009), pp. 274–283.
7. Ministerio de Industria, *Real Decreto 413/2014, de 6 de junio, por el que se regula la actividad de producción de energía eléctrica a partir de fuentes de energía renovables, cogeneración y residuos* (BOE, Madrid, 2014).
8. BDEW, Bundesverband der Energie-und Wasserwirtschaft e. V., Grid code — extra high voltage, April (2009).
9. BDEW, Bundesverband der Energie- und Wasserwirtschaft e.V., Guideline for generating plants' connection to and parallel operation with the medium-voltage network, June (2008).
10. Institute of Electrical and Electronics Engineers — IEEE, IEEE Standard 1547-2003 — IEEE Standard for Interconnecting Distributed Resources with Electric Power Systems (2003).
11. ANSI C84.1-2006., American National Standard for Electric Power Systems and Equipment — Voltage Ratings (2006).

12. EN 5160, Voltage characteristics of electricity supplied by public distribution systems.
13. A. Notholt, Germany's new code for generation plants connected to medium-voltage networks and its repercussion on inverter control, Valencia (Spain) (2009).
14. IEC 61724, Phtovoltaic system performance monitoring. Guidelines for measuring, data exchange and analysis (1998).
15. T. Marwart and L. Castañer, *Practical Handbook of Photovoltaics* (Elsevier, Amsterdam, 2003), pp. 543–565.
16. German Solar Energy Society, *Planning and Installing Photovoltaic Systems* (Routledge, 2013), pp. 299–335.
17. European Commision — PVGIS, Joint Research Centre, [Online]. Available at: http://re.jrc.ec.europa.eu/pvgis/. [Accessed 16 July 2014].
18. G. Stapleton and S. Neill, *Grid-Connected Solar Electric Systems. The Earthscan Expert Handbook for Planning, Design and Installation* (Earthscan, UK, 2012).
19. IEC 62446, *Grid Connected Photovoltaic Systems — Minimum Requirements for System Documentation, Commissioning Tests and Inspection* (International Electrotechnical Commission, Switzerland, 2009).
20. Deutsche Gesellschaft für Sonnenenergie (DGS), *Planning and Installing Photovoltaic Systems. A Guide for Installers, Architects and Engineers* (Routledge, UK, 2013).
21. J. P. Dunlop, *Photovoltaic Systems* (American Technical Publishers, USA, 2010).
22. IEA — International Energy Agency, *Review of Failures of Photovoltaic Modules* (IEA, Photovoltaic Power System Programme, París, 2014).
23. J. Muñoz Díez, *Contribución a la evaluación de la calidad de sistemas fotovoltaicos conectados a la red de medio y gran tamaño* (Universidad de Jaén, Jaén, 2012).
24. EPIA — European Photovoltaic Industry Association, *Global Market Outlook for Photovoltaics* 2014–2018 (Brussels, 2014).
25. IRENA — International Renewable Energy Agency, Rethinking Energy: Towards a new power system (2014).
26. P. del Río and E. Cerdá, The policy implications of the different interpretations of the cost-effectiveness of renewable electricity support, *Energy Policy* **64** (2014), pp. 364–372.
27. EPIA — European Photovoltaic Industry Association, European PV support schemes overview, Brussels, Q4 (2014).
28. T. D. Couture, K. Cory, C. Kreycik and E. Williams, *A Policymaker's Guide to Feed-in Tariff Policy Design* (NREL, USA, 2010).
29. IRENA — International Renewable Energy Agency, *Evaluating Policies in Support of the Deployment of Renewable Power* (Abu Dhabi, United Arab Emirates, 2012).
30. D. Pérez, V. Cervantes, V. Báez and J. González-Puelles, PV grid parity monitor. Residential sector, *Ecclareon* 2 (2013).

31. D. Helm, The European Framework for energy and climate policies, *Energy Policy* (2014), pp. 29–35.
32. T. Trainer, Can Europe run on renewable energy? A negative case, *Energy Policy* **63** (2013), pp. 845–850.
33. T. Trainer, Can the world run on renewable energy? A revised negative case, *Humanomics* **29**(2) (2013), pp. 88–104.
34. S. Spiecker and C. Weber, The future of the European electricity system and the impact of fluctuating renewable energy — a scenario analysis, *Energy Policy* **65** (2014), pp. 185–197.
35. F. P. Vahl, R. Rüther and N. Casarotto, The influence of distributed generation penetration levels on energy markets, *Energy Policy* **62** (2013), pp. 226–235.
36. P. del Río and P. Mir-Artigues, Support for solar PV deployment in Spain: Some policy lessons, *Renew. Sust. Energ. Rev.* **16** (2012), pp. 5557–5566.
37. A. Ciarreta, C. Gutiérrez-Hita and S. Nasirov, Renewable energy sources in the Spanish electricity market: Instruments and effects, *Renew. Sust. Energ. Rev.* **15** (2011), pp. 2510–2519.
38. IRENA — International Renewable Energy Agency, Evaluating renewable energy policy: A review of criteria and indicators for assesment, UK (2014).
39. IRENA — International Renewable Energy Agency, Adapting renewable energy policies to dynamic market conditions, Abu Dhabi (2013).
40. D. Talavera, J. de la Casa, E. Muñoz-Cerón and G. Almonacid, Grid parity and self-consumption with photovoltaic systems under the present regulatory framework in Spain: The case of the University of Jaén Campus, *Renew. Sust. Energ. Rev.* **33** (2014), pp. 752–771.
41. E. Muñoz-Cerón, *New Trends in Photovoltaic Economy and Development of a Different Technological Spectral Approach in Concentrator Photovoltaics* (University of Jaen, Spain, 2014).
42. IEA — International Energy Agency, *Trends 2014 in Photovoltaic Applications. Survey Report of Selected IEA Countries Between 1992 and 2013* (IEA. Photovoltaic Power Systems Programme, 2014).
43. D. Feldman, G. Barbose, R. Margolis and N. Darghouth, *Photovoltaic System Pricing Trends: Historical, Recent and Near Term Projection* (US Department of Energy, 2013).
44. M. Mendelsohn, T. Lowder and B. Canavan, *Utility-Scale Concentrating Solar Power and Photovoltaics Projects: A Technology and Market Overview* (NREL, 2012).
45. D. Turney and V. Fthenakis, Environmental impacts from the installation and operation of large-scale solar power plants, *Renew. Sust. Energ. Rev.* **15** (2011), pp. 3261–3270.
46. G. Nofuentes *et al.*, Tools for the profitability analysis of grid-connected photovoltaics, *Prog. Photovolt* (2002), pp. 555–570.
47. D. Talavera *et al.*, Sensitivity analysis on some profitability indices for photovoltaic grid–connected systems on buildings: The case of two top photovoltaic European areas, *J. Sol. Energy Eng.* **135**(1) (2012), p. 011003.

48. Centre for Social Impact Bonds (2013). [Online]. Available at: http://data.gov.uk/sib_knowledge_box/discount-rates-and-net-present-value. Accessed 30 August 2014.
49. W. Short *et al.*, A manual for the economic evaluation of energy efficiency and renewable energy technologies (NREL, Golden, Co, USA, 1995).
50. E. Muñoz, Trends in the policy and economy of the PV technology, in PhD Thesis, Universidad de Jaén (2014), pp. 11–25.
51. K. Branker, M. Pathak and J. Pearce, A review of solar photovoltaic levelized cost of electricity, *Renew. Sust. Energ. Rev.* **15**(9) (2011), pp. 4470–4482.
52. M. Campbell, J. Blunden, E. Smeloff and P. Aschenbrenner, Minimizing utility-scale PV power plant LCOE through the use of high capacity factor configurations, in *Proc. 34th IEEE PVSC* (2009).
53. C. Groobey, J. Pierce, M. Faber and G. Broome, *Project Finance Primer for Renewable Energy and Clean Tech Projects* (Wilson Sonsini Goodrid & Rosati, 2010).
54. M. Dailami and D. Leipziger, Infrastructure project finance and capital flows: A new perspective, *World Develop.* **26**(7) (1998), pp. 1283–1298.
55. G. Pollio., Project finance and international energy development, *Energy Policy* **26**(9) (1998), pp. 687–697.
56. W. Lange, Bankability: The ticket for successful solar energy project business, *Global Photovoltaic Business & Resources* (2011).
57. J. Morgenson, Defining bankability for utility-scale PV, *Power Engineering* (2011).
58. P. Valera, J. Tejada, A. Payan, O. Pereles, A. Trabant, L. Almonacid, V. Salas and G. Almonacid., Technical Aspects of Project Finance in Solar Energy Projects — Methodology to Improve the Technical due Diligences, in *EUPVSEC*, Paris (2013).
59. Foresight group, The bigger picture — What is the future of large-scale solar in Europe? *Green Power Conf.*, 11 September (2014). [Online]. Available at: http://www2.greenpowerconferences.co.uk/EF/?sSubSystem=Prospectus&sSessionID=utp67mlr3lvrka2rboeoudr0t5-19376809&sEventCode=SP1411IT&sDocument=Webinar1_Download. Accessed on 18 November 2014.
60. EPIA — European Photovoltaic Association, Retrospective measures at national level and their impact on the photovoltaic sector, Brussels (2013).
61. IEA — International Energy Agency, *High Penetration of PV in Local Distribution Grids* (IEA. Photovoltaic Power Systems Programme, Paris, 2014).
62. IEA — International Energy Agency, *Technology Roadmap. Solar Photovoltaic Energy* (IEA, Paris, 2014).
63. Photon International —*The Solar Power Magazine, Markets* (Photon Publishing GmbH, Germany, 2014).
64. IEA — International Energy Agency, *Africa Energy Outlook. A focus on Energy prospects in Sub-Saharan Africa* (IEA. World Energy Outlook Special Reports, Paris, 2014).

Chapter 5

Biomass

Anthony Turhollow
Environmental Sciences Division, Oak Ridge National Laboratory
P.O. Box 2008, Oak Ridge, TN, USA
turhollowaf@ornl.gov

Biomass resources and conversion technologies are diverse. Substantial biomass resources exist including woody crops, herbaceous perennials and annuals, forest resources, agricultural residues, and algae. Conversion processes available include fermentation, gasification, pyrolysis, anaerobic digestion, combustion, and transesterification. Bioderived products include liquid fuels (e.g. ethanol, biodiesel, and gasoline and diesel substitutes), gases, electricity, biochemical, and wood pellets. At present the major sources of biomass-derived liquid fuels are from first generation biofuels; ethanol from maize and sugar cane (89 billion liters in 2013) and biodiesel from vegetable oils and fats (24 billion liters in 2011). For other than traditional uses, policy in the forms of mandates, targets, subsidies, and greenhouse gas emission targets has largely been driving biomass utilization. Second generation biofuels have been slow to take off.

1 Introduction

Biomass is organic matter that can be directly utilized or transformed into other forms of energy. It includes organic material of agricultural origin (grains, crop residues, vegetable oils, sugar crops, manure and purpose grown cellulosic crops and woody material (residues and thinnings, pulping liquors, fuelwood and purpose grown short rotation woody crops). These resources are substantial, with more than 1 billion metric tons (1 megagram, Mg) estimated to be available in the United States alone.[1–3] Biomass makes up 10% of the world's primary energy consumption.[4]

There are multiple methods of converting biomass into other forms of energy. At present, most biomass is combusted in traditional fireplaces or stoves. Biomass is composed of carbohydrates (sugars and starch), cellulose, hemi-cellulose, lignin, lipids (vegetable oil), protein, ash and other minor constituents. It may also contain significant amounts of water, up to 80% (wet-weight basis) in fresh material. Starch from maize (corn) and other grains and sucrose from sugar cane are converted on large scales into ethanol by fermentation. The oils from oilseed crops and fats are converted into biodiesel by transesterification and hydrotreating. These are first generation biofuels. Second generation biofuels are just coming online.

2 Biomass Products

2.1 *Ethanol*

The primary liquid biofuel produced is ethanol, 88 billion liters (L) in 2013, with the United States (57%) and Brazil (27%) accounting for the majority. World ethanol production increased up until 2010 and then hit a plateau (Table 1).

Presently ethanol is produced primarily from first-generation feedstocks — maize and sugarcane. Current ethanol yields from maize and sugarcane in the United States and Brazil are 4000 and 7500 L per hectare (ha), respectively. Ethanol yields are projected to increase to 5100 and 9800 L per ha in the United States and Brazil, respectively (Fig. 1).

2.1.1 *Ethanol production in Brazil*

Brazil is the second largest producer of ethanol in the world, producing 24 billion L in 2013, utilizing sugar from sugarcane. Production and use of sugarcane-derived ethanol for fuel began in 1975 when Proálcool was started in response to high oil prices and a crisis in the sugar market. Use of ethanol as a fuel was stimulated by feedstock availability, a supportive

Table 1: Ethanol production (in billions of L)[5,6] (USDOE/EIA 2014, RFA 2014).

Region	2001	2003	2005	2007	2009	2011	2013
United States	6.68	10.61	14.78	24.68	41.40	52.72	50.34
Canada	0.23	0.23	0.26	0.8	1.16	1.74	1.98
Brazil	11.46	14.47	16.04	22.55	26.10	22.75	23.72
Europe	0.14	0.39	0.86	1.82	3.44	4.22	5.19
China	0	0.80	1.20	1.67	2.15	2.26	2.63
India	0.17	0.19	0.21	0.26	0.35	0.35	2.06
Rest of World	0.07	0.30	0.58	1.83	2.29	2.58	2.75
World	18.76	27.00	33.95	53.64	76.96	86.66	88.68

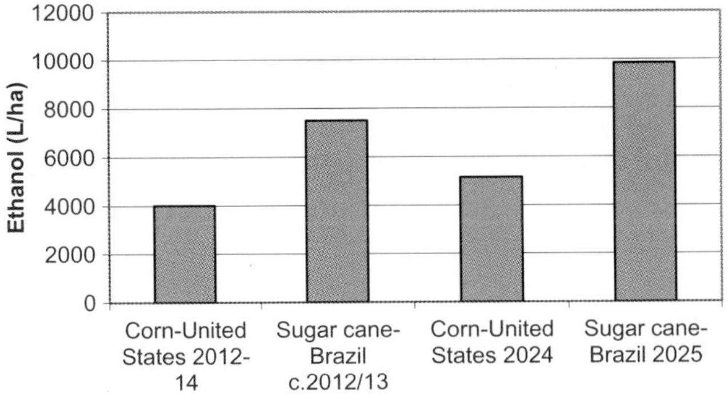

Fig. 1. Per ha ethanol production from maize and sugarcane.
Source: Sugarcane Brazil: sugarcane.org,[8] 2025 de Cerqueira Leite *et al.*[9]; Corn: United States, based on average 2012–2014 corn yield 9.55 Mg per ha[10,11] and 420 L per Mg,[12] Corn-United States-2014 based on USDA[13] corn yield 11.52 Mg per ha and assume 447 L per Mg.

government policy, and improvements in sugarcane production and ethanol conversion processes.[7]

Based on Fig. 5 in Ref. 7, sugarcane yields have increased from about 50 tons per ha in 1970–1975 to 90 tons per ha in 2010 and ethanol yield has increased from about 40 to 80 L per ton in the same timeframe.

[Note that the Valdes publication lists tons[a] and it is not clear whether this is metric tons or short tons.] Ethanol yield has increased from 2000 to 7200 L per ha from 1970–1975 to 2010. In São Paulo province, where the majority of ethanol is produced, yields are even higher.[14]

In the early 1980s, vehicles that could use hydrous ethanol (using neat-ethanol engines) became attractive to consumers because the government ensured that the price of hydrous ethanol was 64.5% that of gasoline. Sales of neat-ethanol vehicles and their market share increased to over 90% of vehicle sales. However, there was not a guaranteed supply of hydrous ethanol, and a shortage of ethanol in the early 1990s led to a crisis and gradual abandonment of neat-ethanol vehicles. In 2003, flex-fueled vehicles were introduced, which are capable of utilizing ethanol–gasoline blends ranging from 0 to 100% ethanol.[7,13]

[a]1 ton = 1 metric ton = 1000 kg = 10^6 g = 1 megagram = 1 Mg = 2205 lbs.
1 long ton = 2240 lbs = 1.016 ton; This long ton was the pre-metric British system ton.
1 short ton = 2000 lbs = 0.907 ton.

The area planted to sugarcane has been increasing, with 4.273, 4.805, and 9.191 million ha in 1990, 2000, and 2010, respectively, and is expected to reach 10.045 million ha in 2020 (Table 10 in Ref. 7). USDA/FAS projects sugar cane area at 9.9 million ha in 2014.[15] About 60% of the sugarcane produced is used for ethanol production and this is expected to remain about the same. Yields have been increasing over time and investments are being made in higher-yielding varieties and practices. Brazil's ethanol industry in 2014 is estimated to operate at 57% of installed capacity. The trajectory of Brazilian ethanol capacity has changed since the Valdes estimates, which he projected to reach 46.4 million L by 2018. Capacity reached a peak in 2011 at 41.4 million L, but has declined slightly to 39.7 million L in 2014 and the same is projected for 2015.[16]

2.1.2 *Ethanol production in the United States*

Maize is the dominant feedstock for first generation ethanol production in the United States. Ethanol production took off in the mid-2000s, increasing 8-fold between 2001 and 2011 (Table 1). Most ethanol is now produced in dry mills. Initially many facilities ground the corn, hydrolyzed the starch fraction and fermented that to produce ethanol, and then the residual, the distillers dried grain, was dried and sold as a by-product. These facilities have become more efficient and sophisticated over time, with ethanol yields increasing (to an average of 420 L per Mg in 2012) and many now removing [distillers] corn oil. Oil extracted averaged 4.3 and 20.9 kg/Mg in 2008 and 2012, respectively.[12] Depending on the processes and technologies utilized, oil extracted can range from 31.5 to 50.4 kg/Mg.[12] The corn oil is considered inedible and has a high free fatty acid content. It sells at a discount compared to soybean oil. It is utilized to produce biodiesel and for livestock feed. Biodiesel production is now being integrated with some ethanol facilities utilizing the corn oil.[17] Technology to convert the cellulosic portion in the corn to ethanol is beginning to be utilized. Quad County Corn Processors is a 132 million L corn ethanol facility that is adding 7.6 million L of ethanol production from the cellulosic portion, or approximately 24 L per Mg of corn.[18] ICM has a process called Generation 1.5 that also converts the fiber portion of corn grain to ethanol at existing facilities which increases overall ethanol yield by 7–10% over just converting the starch portion, or about an increase of 30 to 40 L per Mg.[19] These technologies have the potential to increase ethanol production from the currently utilized maize by 3.8 to 5.6 billion L without increasing maize input. The resulting distillers dried grain is a higher protein feed.[18,19]

2.1.3 *Second-generation Ethanol production*

A limited number of large-scale second generation ethanol facilities, based on cellulosic feedstocks have recently, or are about to, come online. These include four facilities in the United States (Abengoa's Hugoton, Kansas plant — 95 million L plus 18 MW electricity using corn stover, wheat straw, and sorghum stover); DuPont's Nevada, Iowa plant — 114 million L using corn stover; IneosBio's Vero Beach, Florida plant — 30 million L plus 6 MW electricity using vegetative matter and municipal solid waste (MSW), Poet-DSM's Project Liberty in Emmetsburg, Iowa — 95 million L using corn stover), 1 in Canada (Enerkem's Edmonton, Alberta plant — 38 million L using sorted MSW), 1 in Europe (Beta Renewables Crescento, Italy plant — 75 million L using primarily rice straw), and 1 in Brazil (GranBio São Miguel dos Campos, Alagoas plant — 82 million L using sugar cane bagasse and sugar cane trash).[20–22]

2.2 *Biodiesel*

The term biodiesel refers to diesel fuels and substitutes produced from biomass. The primary process for producing biodiesel is transesterification. Vegetable oils and animal fats (triglycerides) are reacted with an alcohol (typically methanol) and a chemical catalyst (typically lye) to produce an ester (typically methyl ester) [hereafter referred to as a fatty acid methyl ester (FAME)] and glycerin. Enzymatic transesterification is about to become commercial.[23] Advantages of enzymatic transesterification over chemical transesterification include: ability to convert high free fatty acid feedstocks (which are often lower in cost), lower energy consumption, production of higher quality glycerin, no caustic chemicals required, reduces final processing, little excess methanol, and lower capital cost.[24] The major disadvantage is the cost of enzymes. Another process that is now being utilized is the hydrotreating of vegetable oils and animal fats to produce hydrotreated vegetable oils (HVOs).

Biodiesel production increased rapidly from 2001 to 2009, increasing by almost 15 fold. Biodiesel production was 24 billion L in 2011, with the most produced in the European Union (45%), followed by the United States (16%), Argentina (12%) and Brazil (11%) (Table 2). Asian countries in the aggregate produce 11% of world biodiesel production, and this fraction is increasing as Indonesia, Thailand, and Malaysia increase production. The countries with the highest production are Germany, the United States, Brazil, Argentina, and Indonesia.[25] In the United States, soybean

Table 2: Biodiesel production (in billions of L).

Region	2001	2003	2005	2007	2009	2011	2013
Canada	0.00	0.00	0.01	0.10	0.12	0.16	0.47
United States	0.03	0.06	0.35	1.90	2.00	3.75	5.14
Argentina	0.01	0.01	0.01	0.21	1.37	2.81	2.26
Brazil	0.00	0.00	0.00	0.41	1.65	2.74	2.92
Colombia	0.00	0.00	0.00	0.01	0.34	0.53	0.53
European Union	1.11	1.94	3.70	7.31	10.52	10.71	10.89
China	0.01	0.01	0.05	0.12	0.36	0.46	0.97
Indonesia	0.00	0.00	0.01	0.02	0.36	1.19	2.45
Malaysia	0.00	0.00	0.00	0.07	0.27	0.06	0.39
South Korea	0.00	0.00	0.01	0.10	0.30	0.37	
Thailand	0.00	0.00	0.02	0.07	0.62	0.61	1.06
Rest of world	0.00	0.01	0.06	0.12	0.45	0.59	
World	1.16	2.03	4.23	10.62	18.36	23.98	

Note: Not all countries available for 2013 from USDA/GAIN reports.
Source: 2001–2011 — USDOE/EIA,[25] 2013 — USDA/FAS.[16,26–34]

oil accounts for slightly over half the feedstock, while in the EU rapeseed oil is the most widely-used feedstock.[26,27]

Hydrotreating triglycerides (vegetable oils and fats) to produce HVOs is an alternative to esterification. Hydrogen is used to remove oxygen from triglycerides. HVOs have qualities superior to FAMEs, including better cold temperature properties, higher cetane value [than either FAME or petroleum-based diesel (hereafter referred to as diesel)], has an energy density similar to diesel, and better storage properties.[35] Finland's Neste Oil has four facilities producing HVOs: 2 at Porvoo, Finland (capacity each 170,000 Mg/year of HVO), Rotterdam (800,000 Mg/year), and Singapore (800,000 Mg/year).[36] In Louisiana, USA, the Diamond Green Diesel facility produces more than 500 million L per year utilizing animal fats and used cooking oil.[37]

2.3 *Electricity*

In 2012, 370 TWh of electricity was produced from bioenergy, 1.5% of the world's electricity production, and IEA projects this will reach 560 TWh in 2018.[4] Electricity from biomass can be generated in a number of ways, direct combustion in power plants, utilizing biogas, using gasification, and co-firing with coal. In 2012, in the European Union (EU) and United States production of biobased electricity was 27.9 and 57.6 TWh, respectively, and increased in the United States to 59.9 TWh in 2013.[38,39]

2.4 *Biochemicals*

Many chemicals can be derived from biomass. In 2010, 50 million Mg of bio-based chemicals and polymers were produced.[40] The OECD (2009) estimated that bio-based chemicals contributed (in terms of value) 1.8% of chemical production and are expected to contribute between 12 to 20% of the value of the chemical market by 2015.[41] Nexant (2014) estimated the renewable chemical market potential in the United States for 2012, 2017, and 2022 (Table 3).[42]

The primary output of chemicals is dominated by a small number of chemicals: methanol, ethylene, propylene, butadiene, benzene, toluene, and xylene. These are then converted to polymers and plastics and many other specialty chemicals. Technically almost all petrochemicals could be replaced by biochemicals, but bio-based chemicals are often more expensive and they must prove equal in quality to their fossil-based counterparts. Potential platforms to produce biochemicals include: Syngas, biogas, C6/C5 sugars, plant-based oils, algae oil, organic solutions, lignin, and pyrolysis oils. Potential bioproducts can be characterized as compounds containing C_1, C_2, C_3, C_4, C_5, C_6, and C_n.[40]

A number of studies have tried to prioritize biochemicals from the myriad of choices. These include Besson *et al.*,[43] Gallezot,[44] Bozell and Petersen,[45] Holladay *et al.*,[46] and Elliot.[47]

Vegetable oils and animal fats (primarily triglycerides) are used to produce oleochemicals. Oleochemicals include lubricants, surfactants, paints, detergents, soaps, and cosmetics. In 2004, 17 million Mg of oils and fats were used for olechemical production (including fuels).[48] Note that in 2004 fuel use was relatively small, approximately 1.8 million Mg. One potential market for vegetable oils is their use as lubricants (including motor oils). World lubricant consumption was 34 million Mg in 2011 and biobased lubricants

Table 3: Biochemicals in the United States 2012–2022 (1000 Mg) (Nexant).[42]

Chemical type	2012	2017	2022
C2	0	0	1000
C3	40	100	300
C4	100	150	400
Aromatics	<10	100	500
Specialty oils	20	300	1000
Total	160	750	3200

accounted for less than 1% of this market. The advantages of biolubricants over petroleum-based lubricants include lubricity, high viscosity, high flash point, biodegradability, low aquatic toxicity, and low bioaccumulation. Disadvantages include poor thermal and oxidative stability.[49]

2.4.1 *Nitrogen fertilizer produced from biomass*

Nitrogen fertilizer is important for agricultural productivity and a major world chemical. It is also a large source of greenhouse gases. Natural gas is the primary feedstock for ammonia production, which can then be converted into other forms of nitrogen (e.g. urea, ammonium nitrate). Ahlgren *et al.* studied the environmental impacts of ammonium nitrate produced from willow (Salix) and wheat straw and found greenhouse gas emissions to be 22–30%, eutophication potential to be 50 to 150% higher and acidification potential to be similar to those of the fertilizer using natural gas as the feedstock.[50] Ahlgren *et al.* studied the environmental impacts of using biogas produced from maize and ley grass to produce ammonium nitrate, and found the greenhouse gas impacts to be between 47 and 60%, eutophication potential to be 10 to 12 times higher and acidification potential to be 2.5 to 4 times higher than those from using natural gas as the feedstock for ammonium nitrate production.[51] Andersson and Lungren modeled the production of ammonia from biomass and found that to achieve an internal rate of return of 10–20%, the selling price would have to be between 581–882 and 509–774 Euro per Mg (approximately $725–1100 and $635–970 per Mg assuming $1 = 0.80 Euro) for a standalone production facility and a facility integrated with a pulp mill, respectively.[52] The price of ammonia in April 2014 in the United States was $938 per Mg.[53] BioNitrogen is planning to build facilities to convert biomass into urea via gasification and catalytic conversion. Its planned facility in Taylor County, FL, USA has an expected capacity of 160,000 Mg of urea and a capital cost of between $100 and $220 million.[54,55]

2.5 *Wood Pellets*

The use of wood pellets has increased over the past few years and is expected to continue to do so in the near future. Pellets are not only used for home heating but increasingly for power generation. The European Union is the largest producer and consumer of wood pellets. The Food and Agriculture Organization of the United Nations (FAO) began collecting statistics for wood pellets starting with 2012 with world production of 19.8 and 21.6

Table 4: Wood pellet production 2012 and 2013 (million Mg).

Region	2012	2013
Belarus	354	354
Bosnia and Herzegovina	83	184
Canada	2400	2400
China	100	100
Russian Federation	791	810
Serbia	82	167
South Africa	115	115
Switzerland	180	180
Ukraine	210	210
United States	4100	5000
Vietnam	50	170
European Union	11010	11513
Rest of world	327	426
World	19801	21629

Source: FAOSTAT.[56]

million Mg in 2012 and 2013, respectively (FAOSTAT).[56] The EU produced about 55% (c. 11 million Mg), and consumed about 75–80% of wood pellets in 2012 and 2013. EU consumption is projected to be 21 million Mg in 2015, up from 17.5 million Mg in 2013.[27] Other large producers [and major exporters] are the United States (21–23%) and Canada (11–12% of world production) (Table 4).

Wood pellets are seen as a means of reducing net carbon emissions. In Canada, targeted proposals to reduce carbon emissions would require 14 million Mg of wood pellets to replace coal by 2020, which is far above estimated 2014 production of 3.2 million Mg[32] and is unlikely to be achieved. In the United Kingdom, Drax Power plans to convert three of its six boilers at the Drax Power Station from coal to biomass, which would require 7 million Mg of wood pellets, more than 25% of current world production. Drax is investing $375 million in the United States in two pellet production facilities (in Louisiana and Mississippi), with an annual combined capacity of 900,000 Mg and port facilities in Baton Rouge, Louisiana.[57] World wood pellet production capacity will have to increase significantly to meet the increased demands from power generation and home heating.

3 Thermal Processes

Thermal processes to convert biomass into biofuels and electricity include pyrolysis, gasification and combustion. In thermal processes, all the fractions of the biomass (excluding the ash) are converted into energy. In the case of wood, this allows for up to 99% of the biomass to potentially be utilized to make fuel products. A biological fermentation process might get up to 460 L of ethanol per dry Mg of biomass, a thermal process utilizing wood and fermentation of the syngas could get up to 530 L of ethanol per dry Mg.[58]

3.1 *Pyrolysis and Hydropyrolysis*

Pyrolysis is heating in the absence of oxygen, while controlling temperature, pressure, and residence time to determine the formation of gaseous, liquid and solid products (Table 5). Fast pyrolysis sets conditions (short residence time, lower temperature process (350–500°C), rapid quenching) such that mostly liquid fuel (referred to as pyrolysis- or bio-oil) (70 wt. %+) is produced. Pyrolysis takes place at atmospheric pressure and requires a low moisture (<10%) content feedstock (to minimize moisture in the bio-oil) and small particle size (c. 2–3 mm).[59,60]

There are issues associated with the acidity and stability of the bio-oil. Bio-oil has a low pH (2.5), high water content (15–30%), and a heating value of 16–19 MJ per kg (higher heating value).[61] Using a pyrolysis process, wood and herbaceous materials have the potential of producing 620 and 530 L per dry Mg of ethanol equivalent, respectively.[62]

Wood has a higher output of product because of its higher carbon (energy) content compared to herbaceous material. Wood is higher in lignin than herbaceous materials and thermal processes take advantage of this.

Bio-oil can be utilized for a number of products — liquid boiler fuel; petroleum-like fuels by direct utilization, upgrading via hydrotreating and hydrocracking (Fig. 2), gasification, fermentation, reform to hydrogen; and phenolics replacement. The solid produced (char) may be utilized as a

Table 5: Pyrolysis products (%) depending on conditions.[59]

Mode	Conditions	Liquid	Solid (char)	Gas
Fast	500°C, 1 s vapor residence	75	(12)	13
Intermediate	500°C, 10–20 s vapor residence	50	(25)	25
Gasification (slow)	400°C, long solids residence	30	(35)	35
Gasification	800°C, long vapor residence	5	10	85
Torrefaction (slow)	~290°C, 10–60 min solid residence	0	80	20

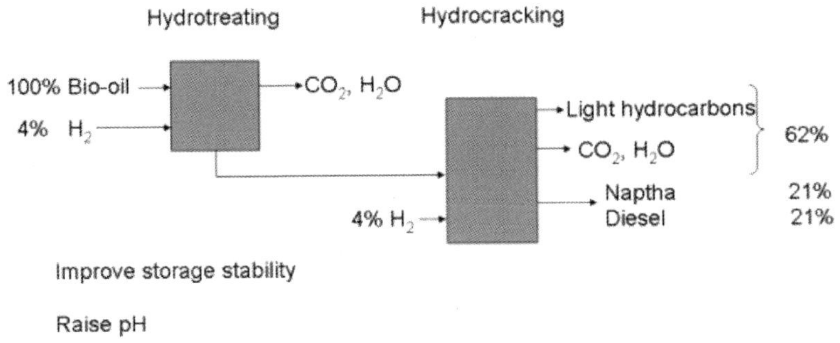

Fig. 2. Upgrading pyrolysis oil.[58]

soil amendment, activated carbon, and process heat or for energy for the pyrolysis process. The gaseous product can also provide energy for the pyrolysis process.

Jones *et al.* estimate a selling price for ethanol of \$0.539/L of "fuel product" (\$0.354/L ethanol equivalent) (2007\$) for a 2000 dry Mg per day facility yielding 417 L per dry Mg of hybrid poplar chips.[63] Fuel product consists of diesel and gasoline blendstocks. The process is: pretreatment (drying of hybrid poplar to less than 10% moisture and grinding to 2–7 mm particle size), fast pyrolysis, hydrotreating, hydrocracking and product (gasoline and diesel blendstock) separation. Natural gas is steam reformed to provide hydrogen for hydrotreating. Hybrid poplar chips (50% moisture) have a delivered price of \$64.50 per dry Mg. The capital cost of a stand-alone facility is \$303 million. For a facility collocated (integrated) with an existing petroleum refinery the capital cost is reduced to \$188 million and product price is reduced to \$0.460 per L (\$0.301 per L ethanol equivalent).[63]

Hydropyrolysis represents an alternative to pyrolysis. For use as transportation fuels, pyrolysis oils have to be upgraded. Some of the undesirable properties of pyrolysis oils include high acidity (high total acid number), lower energy density than diesel, high oxygen content (c. 40%), chemical instability, high water content, and meturgical incompatibility with metals used in conventional transport containers and oil refineries. Using catalytic hydropyrolysis or integrated hydropyrolysis, biomass can be directly converted into a hydrocarbon product which is compatible with petroleum refineries. In integrated hydropyrolysis and hydroconversion all the hydrogen needed for hydropyrolysis is provided by reforming the light gases produced within the process.[64]

Table 6: Two possible gasification to product pathways.

Step 1 (Biomass conversion)	Step 2 (Syngas conversion)	Step 3 (Product generation)
Gasification to syngas	Synthesis into mixed alcohols	Distill to pure alcohols
	Catalytic conversion into hydrocarbon fuels	Refining to diesel and naphtha (for jet fuel) fractions

3.2 *Gasification*

Gasification converts the carbon fraction in biomass into synthesis gas consisting primarily of carbon monoxide and hydrogen. Synthesis gas has the potential for a variety of uses: combustion, in gas engines, in gas turbines, in fuel cells, upgraded to methane and methanol, fermented to ethanol and utilized in Fischer–Tropsch processes to produce liquid fuels including diesel. Major research challenges include cost-effective hot gas cleaning and improved catalysts for liquid fuel production.[65] Two potential gasification conversion pathways are shown in Table 6.

Worley and Yale[66] estimate capital costs (2011$) for three gasification technologies for a 1,000 dry Mg per day facility utilizing wood chips and bark. Technology 1 consists of an oxygen-blown partial oxidation bubbling fluidized bed and a tar reformer (reactor vessel filled with solid catalyst block to crack tars. The cost is $70.6 million. Technology 2 uses an indirect heating circulating fluidized bed with heating of the bed occurring in a separate combustor utilizing char combustion air and the tar reformer with a bubbling fluidized bed design. The cost is $59.7 million. Technology 3 consists of an oxygen-blown partial oxidation bubbling fluidized bed with an unknown (not revealed by the vendor) technology used for tar reforming. Its cost is $70.7 million.

3.3 *Combustion*

Combustion of biomass is generally not energy efficient, with conversions in the 20–25% range for electricity generation (i.e. 20–25% of the energy in the biomass is converted into electricity). There are two primary reasons for this: (1) the scale of the electricity generation facility is usually small, less than 25 MW_e, which makes investments that can make the process more energy efficient uneconomic and (2) the moisture content of the biomass is usually high, which makes the conversion less efficient (Fig. 3). First

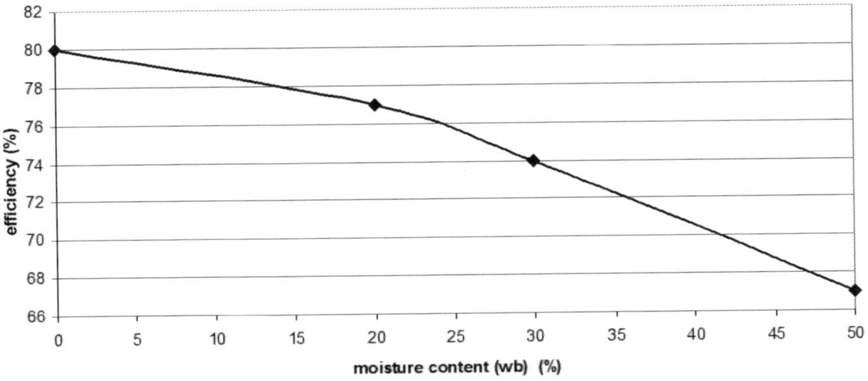

Fig. 3. The effect of moisture content on conversion efficiency.[58]

gasifying the biomass and then generating electricity, or utilizing pellets (c. 10% moisture) would be more energy efficient.

4 Biological Processes

4.1 *Anaerobic Digestion*

For anaerobic digestion, microorganisms break down biodegradable organic material in the absence of oxygen, producing a mixture of gases, primarily methane and carbon dioxide. Compared to natural gas (higher heating value of around 38,200 kJ per m^3) biogas has a medium heat value that ranges from 19,000 to 29,000 kJ per m.[3] However, biogas can be upgraded to natural gas quality. IEA Task 37 lists 347 upgrading facilities in the world, with 279 in Europe, 144 in Germany alone.[67]

Anaerobic digestion is used at sewage treatment and industrial facilities to treat organic wastes and on farms to treat animal wastes. India and China make widespread use of anaerobic digesters in rural areas to provide fuel. According to Burns, there were 37 million anaerobic digesters in China (of which only 16,000 were large scale) and 4 million in India in 2009.[68] There are many more small-scale than large-scale anaerobic digesters. The Anaerobic Digestion China Working Group estimated that at the end of 2012 there were over 20,000 large-scale (over 300 m^3) anaerobic digesters in China, over 30% of the world's total.[69]

In the United States, USDA/USEPA/USDOE estimate that at present there are 2,116 currently operating biogas facilities (livestock manure, land-fill gas, and water resource recovery facilities), with a potential of 13,008 facilities generating 370 PJ per year of biogas. Building the additional 11,000 systems would require a capital cost of $33 billion.[70]

While on-farm digesters are typically used to treat livestock manure, other organic feedstocks such as crop residues or off-farm food processing wastes can also be utilized. After processing, the remaining organic matter now has reduced odor pollution, reduced pathogens, and reduced biological oxygen demand, while retaining most of its nutrient value for possible use as a fertilizer. Most anaerobic digesters in the United States supply biogas to an engine that produces electricity.

The main on-farm digester designs are: Covered lagoons; plug-flow; and mixed or stirred.[71] Crenshaw[72] reports costs for two covered lagoons as, $778,586 for 496 cows and $1,265,194 for 1,600 cows, and Lazarus[71] reports regression equations for costs based on Crenshaw's analysis[72] (see Table 7).

At the Poet-DSM cellulosic ethanol facility in Emmetsburg, IA, USA, an anaerobic digester will supply the thermal needs of not only the cellulosic ethanol facility, but also the thermal load of the collocated corn-to-ethanol facility.[73]

4.2 *Fermentation*

Fermentation is the main process currently used to produce liquid transportation fuel in the form of ethanol from maize in the United States and sugarcane in Brazil. Fermentation converts carbohydrates into ethanol, although other products can be produced as well. The lignin fraction is not converted in fermentation processes and is a byproduct.

Table 7: Anaerobic digester costs.[71]

	Digester type		
	Plug flow		Mixed
	700 cow	2,800 cows	
Digester cost ($)	678,064 + 563/cow		354,866 + 61/cow
Ancillary items	13%		
Cost ($)	1.2 million	2.5 million	
Operating & maintenance[a]	5%		
Output (m^3/cow-day)	0.71–3.82		
Electrical output (kWh/cow-year)[b]	1000		
Breakeven electricity price ($/kWh)	0.22	0.12	

[a]Digester with electrical generation.
[b]1.98 m^3 per cow-day, 60% methane, 27% thermal conversion to electricity, 90% engine run time.

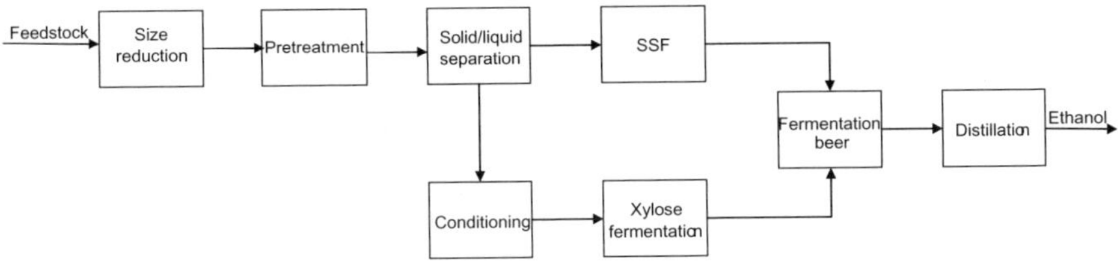

Fig. 4. One possible pathway for biological ethanol production.[58]

Cellulosic feedstocks can utilize many pathways to be converted into ethanol. Both grasses and hardwoods have xylose as their main hemi-cellulosic sugar, whereas softwoods have mannose as their main hemi-cellulosic sugar. The discussion that follows is geared toward xylose as the main hemi-cellulosic sugar.

The main steps are: (1) size reduction, (2) pretreatment to release xylose and cellulose, (3) solid (to simultaneous saccharification and fermentation)/liquid (liquor to conditioning) separation, (4) conditioning prior to xylose fermentation, (5) xylose fermentation, (6) simultaneous saccharification and fermentation of glucose (SSF), and (7) distillation of the ethanol product from the two fermentation processes (see Fig. 4). There are many possible variations on this process, such as using the stream (or part of it) from xylose fermentation to dilute the solids going into the SSF reactor.

There are a number of choices for the pretreatment step (e.g. dilute acid, concentrated acid, steam explosion, ammonia fiber expansion (AFEX), organosolvents).

Humbird *et al.* estimate a selling price for ethanol of \$2.15 (2007\$) for a 2000 dry Mg per day facility and yielding 330 L per dry Mg of corn stover.[74] The process is dilute-acid pretreatment, enzymatic hydrolysis and cofermentation of cellulose and hemicellulose. Lignin is combusted to provide steam and electricity. Enzymes are produced on site. The corn stover is assumed to arrive at the facility in a milled state and costs \$64.50 per dry Mg.[74]

5 Dedicated Energy Crops

Included in dedicated energy crops are herbaceous (e.g. switchgrass, sorghum, energy cane, Miscanthus) and short rotation woody crops (e.g. hybrid poplar, willow, eucalyptus, pine). The major advantages dedicated energy crops have over conventional food crops are the potential to produce higher yields per unit land area and to utilize land that food crops cannot use.

5.1 *Switchgrass*

Native to North America, switchgrass (*Panicum virgatum*) is a perennial, thin-stemmed warm-season (C4 plant) grass. [The two main photosynthetic pathways in plants are C3 and C4. C3 plants fix carbon dioxide through photorespiration and require stomatal opening to acquire CO_2. C4 plants acquire CO_2 from malate and do not require open stomata, thus providing higher water use efficiency and produce more biomass in hotter, drier climates than C3 plants. Under conditions of moderate temperatures and available soil water, C3 plants typically have an advantage in CO_2 fixation and thus overall growth.[75]] Switchgrass has been targeted for development as a potential energy crop.

Parrish and Fike[76] and Mitchell *et al.*[77] provide overviews of switchgrass production. It is well adapted to low rainfall areas and is efficient in nutrient use. It has two major ploidity levels (tetraploid and octaploid) and two ecotypes — upland and lowland. The upland variety has finer stems and a lower potential yield than lowland varieties, where lowland varieties are adapted. Yield is linked to temperature and the timing of reproductive growth and is connected to photoperiod. By moving lower latitude cultivars to higher latitudes, the length of the growing season can be extended and hence the potential yield increased. However, frost tolerance can be an issue. This is being addressed by breeding. Planting is done with seed. Weed control is necessary for successful stand establishment. Full yield is not expected until the third year of growth. For sustained optimum yield, nitrogen fertilization is needed. If harvesting is carried out after a killing frost, then 10 kg of nitrogen (N) per Mg of harvested yield is recommended, otherwise 20 kg N per Mg is recommended.[77] One harvest per year after senescence and translocation of nutrients allows stands to persist for an extended period. Two harvests per year can increase the yield in some areas, but additional nutrients will need to be applied to compensate for nutrients removed in the midseason harvest, and stand longevity may suffer. In addition to its potential use as a biofuel, switchgrass can also be utilized for soil erosion control and as a filter strip along streams to intercept soil, nutrients, and pesticides.

Most of the switchgrass currently grown is from varieties developed for forage. In the past few years, varieties specifically bred for bioenergy have

Table 8: Switchgrass hybrid yield.[78]

Strain	Yield (Mg per ha) relative to Kanlow
Kanlow	1.01
Sumner	0.94
Kanlow × *Sumner* F_1 *hybrid*	0.67

been released. F_1 hybrids of Kanlow (lowland variety) and Sumner (upland variety) produced higher yields than either parent at Mead, Nebraska (Table 8).[78] In 2014, a lowland cultivar 'Liberty' that is adapted to plant hardiness zones 4, 5, and 6[79] (most of the central and eastern United States north of latitude 37°N) was released. It produces higher yields than lowland or upland varieties and better winter survival than the lowland cultivars 'Kanlow' and 'Alamo' north of latitude 40°N. Yield averaged 18 dry Mg per ha for 2009 to 2011 at Mead, NE, USA.[80]

5.2 *Sorghum*

Sorghum (*Sorghum bicolor*) is a morphologically diverse C4 grass with grain, forage, sweet, and energy varieties. Grain sorghum is short stemmed and produces as much residue as grain. The grain from grain sorghum is currently used for ethanol production and the residue can be collected for use as a bioenergy feedstock. Forage sorghum is designed for use as livestock feed. Sweet and energy sorghums have potential application as bioenergy feedstocks.[81] Sorghums, because they are annual and not perennial, can be rotated with traditional food crops, allowing added flexibility to production systems.

Sweet sorghum has a sweet stalk containing sugar. It also produces a large amount of lignocellulosic biomass in the stalk and may also produce grain. The International Crops Research Institute for the Semi-Arid Tropics (ICRISAT) is developing sweet sorghums adapted to Asia, Africa, and the Americas. Two sorghum crops per annum are possible in areas with longer growing seasons. ICRISAT is developing sweet sorghums that not only produce sugar but also relatively large amounts of grains.[82,83] It has the potential to complement sugarcane by being harvested during the period of the year when sugarcane is not being harvested. While sweet sorghum sugar yields can equal those of sugarcane, its sugar is not used for refined sugar production because its high starch content interferes with sucrose crystallization and hastens the conversion of sucrose to glucose and fructose.[84]

Energy sorghums are bred for high biomass yields. Energy sorghums are photoperiod sensitive and will not flower in temperate climates, thereby utilizing the entire possible growing season to accumulate maximum biomass, primarily lignocellulose. Texas A&M University has a breeding program for energy sorghum, with a yield goal of 34 to 45 dry Mg per ha-year.[85]

5.3 *Miscanthus*

Miscanthus (*Miscanthus* × *giganteus*) is a tall thick-stemmed C4 grass related to sugarcane. It is a sterile triploid. It can be highly productive and has the advantage over other herbaceous crops of being able to be harvested in the spring after overwintering. The yield will be about 1/3 lower in the winter but the harvested plant matter will be drier than in the fall. A major disadvantage of Miscanthus is that it is vegetatively propagated (using rhizomes) and requires a specialized planter for mechanical planting.

Costs to establish Miscanthus were estimated by Aravindhakshan *et al.*,[86] Jain *et al.*[87] and Khanna *et al.*[88] at $705, $2,957, $613 per ha, respectively. Jain *et al.* have a higher cost because they assume a cost of $0.25 per rhizome and 10,000 rhizomes per ha, for a cost of $2,500 per ha. Aravindhakshan *et al.* base their rhizome cost on Khanna *et al.*, which is a planting density of 10,000 rhizomes per ha at a cost of $335 per ha. Two papers[86,88] compared the economics of switchgrass and Miscanthus production and come to different conclusions on the relative economics of the two crops because of different assumptions. When harvested in spring, Miscanthus has very low nutrient requirements because most of its nutrients have been translocated back into its roots. Christian *et al.*[89] reported no response to nitrogen in a 14-year study in the United Kingdom, while Ercoli *et al.*[90] reported response to nitrogen in Italy. Rather than looking at switchgrass and Miscanthus as competing crops, because of their differing harvest windows, they can be seen as complementary crops by providing fresh supplies of biomass at different times of the year and by diversifying biomass supply sources.

5.4 *Sugarcane and Energy Cane*

Sugarcane (*Saccharum spp.*) is a thick-stemmed tall C4 grass that is cultivated as a perennial crop, primarily for its storage of sucrose (a disaccharide consisting of glucose and fructose) in its stem. It is genetically complex from interspecific hybridization efforts, primarily from *Saccharum officinarum* and *Saccharum spontaneum*. Sugarcane is grown as a monoculture and replanted every four to five years in the United States. It is vegetatively propagated, with one acre of seedcane being able to plant 6 to 10 acres of sugarcane. Vegetative planting is expensive both in materials and labor.

Energy cane is bred for higher biomass yields (and has lower stalk sugar) than sugarcane. Stalk numbers and heights are higher than sugarcane, and as a result, one acre of energy cane can provide 13 acres of seed cane. Energy cane is also expected to have at least two more harvests than the four or five one gets from sugarcane. Annualized planting costs are expected to be about $143 per ha, versus $309 per ha for sugarcane. In Louisiana, United States, Salassi *et al.* (2014) report energy cane varieties [wet (dry) Mg per ha] have yielded 65 to 99 (14 to 20) in the first year (plant cane), 55 to 105 (14 to 20) in the second year (first stubble), 114 to 162 (29 to 37) in the third year (second stubble), and 61 to 111 (15 to 22) in the fourth year (third stubble) in a trial at St. Gabriel.[91] They assume that for the fifth year (fourth stubble), sixth year (fifth stubble), and seventh year (sixth stubble), that yields are 85, 82, and 79% of the average the yield of years one to four. For energy cane with 5, 6, or 7 years of harvest they report costs ($ per dry Mg), based on sugarcane practices, for variable costs of about 74, fixed costs of 21, overhead costs 4.50, land rent of 25, and total costs of 125.[91]

Because of the need to get energy cane off the field before the next growing season and its thick stalks with a waxy coating, energy cane, like sugarcane, will be harvested green and dewatered or, for energy cane, ensiled if it is to be stored for later use.

Sugarcane breeding begins with hybridization with the introgression of desirable traits from the wild relative of sugarcane, *Saccharum spontaneum.* Early generation progeny from these crosses exhibit high levels of hybrid vigor, including increased cold tolerance, better rationing ability (re-sprouting from the roots), increased tolerance of moisture extremes, increased disease and insect tolerance, and more efficient nutrient utilization.[92] Early generation hybrids are ideal candidates for energy cane for biomass.

5.5 *Hybrid Poplar*

Almost 30 species are included in the genus *Populus.* Hybrids of species within and among *Populus* are referred to as "hybrid poplars." Poplars are planted at relatively dense spacing (e.g. 1,680 trees per ha) using cuttings for energy purposes. Initially weed control is necessary, using herbicides and/or mechanical cultivation. Insects and diseases can also be issues. Harvest utilizes either standard pulpwood harvest systems or purpose designed machines that combine felling and chipping or bundling in one machine.

Yields of newly selected poplar genotypes in small plot experiments have exceeded 15 dry Mg per ha on good agricultural soils in southern Wisconsin and Iowa.[93,94] Yields of 10, 13, and 20 dry Mg per ha, and with appropriate research, yields possibly twice as high in the future, can be expected in the midwestern, southern, and northwestern United States, respectively. Production costs (to the forest landing) in the United States are estimated at between $28 and $66 per dry Mg.[2]

5.6 *Willow*

Interest in shrub willow (*Salix* spp.) as an energy crop is focused in Europe and North America, particularly in Sweden, the United States and Canada. Trials began in Sweden in the mid-1970s and in the United States in 1986. Characteristics that make willow a good candidate for biomass include: (1) ease of propagation from dormant hardwood cuttings; (2) a broad underutilized genetic base; (3) ease of breeding for multiple characteristics and (4) ability to re-sprout after multiple cuttings.[2]

To produce biomass from willow, genetically improved varieties are planted in prepared fields where weeds have been controlled. Weed control utilizes mechanical and chemical means and should begin in the fall before planting if perennial weeds are present. Planting of willows takes place as early in spring as feasible at 15,000 unrooted, dormant cuttings per ha using mechanized planters attached to tractors, capable of planting 0.8 ha per hour. After the first year of growth during the dormant season, the willows are cut back to just above ground level to encourage coppice regrowth of multiple stems. After 3 to 4 years of growth, the willows are mechanically harvested during the dormant season after the leaves have been dropped, utilizing forage harvesters with cutting heads specifically designed for willow coppice material. The forage harvester produces uniform, consistently sized chips that can be collected and delivered with no additional processing and then additional size reduction and/or drying can take place at an end-user facility.

Willow requires fertilizer, about 100 kg per ha, in the spring after each harvest. There are three to four years between harvests. Seven harvests may be possible before replanting is required because the willow stools have expanded too much to allow harvesting.

Fertilized, irrigated yields of unimproved willow varieties have exceeded 27 dry Mg per ha.[95,96] On non-irrigated research-scale trials in central New York (in the United States), yields ranged from 8.5 to 11.6 dry Mg per ha.[95,97,98] Second rotation yields of the five best producing unimproved varieties showed yield 18% to 62% higher than first rotation.[99] Breeding and selection of willow in the United States began in the mid-1990s. New varieties from the initial rounds of breeding in the late 1990s have produced yields up to 40% and 70% greater than first and second rotations using standard varieties used in the early yield trials.

The major costs of production are harvesting, establishment and land rent.[100] If the frequency of harvesting is reduced by increasing the rotation length from three to four years, harvesting costs are reduced by \$14 to \$16 per dry Mg. Planting stock accounts for about two-thirds of establishment costs. Reducing cost per cutting from \$0.12 to \$0.10, reduces establishment costs by \$262 per ha. Higher yields can also reduce costs.

5.7 *Eucalyptus*

There are over 700 species of eucalyptus, but fewer than 15 have commercial significance. They are the most widely planted species in the world, widely grown in the tropics and subtropics. They are bred for fast, uniform growth, self-pruning, regrowing from the stump (coppicing) after harvest. Eucalyptus are used for roundwood, pulp, and charcoal, and are a major source of low-cost fiber. Brazil and other South American countries are major producers. Eucalypts have great potential for use as an energy feedstock.

For conventional pulpwood production, stands are typically planted at 1,500–2,500 trees per ha and harvested every 6 to 10 years. For biomass production eucalypts in Florida, USA would be planted more densely at 8,400 trees per ha and harvested every 3 to 4 years.[101] *Eucalyptus* spp. yielded 17–32 dry Mg per ha-year after 3 to 5 years of growth in central Florida on a clay settling area, which is similar to the 20–31 dry Mg per ha-year estimated for Florida.[102]

5.8 *Oilseeds*

Oilseeds can provide feedstock for biodiesel production. Different feedstocks give very different per ha biodiesel yields (Fig. 5). There are some feedstocks that may potentially give high per ha biodiesel yields, including jatropha

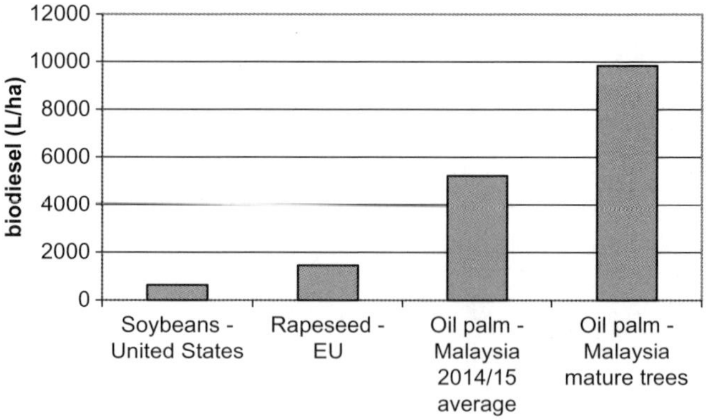

Fig. 5. Per ha biodiesel production.

Sources: Soybeans-United States based on soybean yield of 2.92 Mg per ha (3 year average 2012–2014)[10,11] and 19.42% oil; Rapeseed-EU rapeseed yield of 3.12 Mg per ha and 41.85% oil[104] Oil palm Malaysia 2014/15 forecast average and mature trees based on palm oil production of 20,350,000 Mg (100% oil) and palm kernel production of 4,790,000 Mg (45.68% oil) and 4,800,000 ha and 2,550,000 mature [tree] ha equivalents[105]; and assuming 0.899 kg oil per L of biodiesel.

Table 9: Biodiesel yield of current and potential sources.

Crop	Yield (L/ha)	Data source
Soybeans — US current	630	
Soybeans — US 2025[a]	870	
Rapeseed — EU current	1,450	
Jatropha — India estimate	1,890	Dar[103]
Oil palm — Malaysia mature current	9,830	Wahab[105]
Microalgae — projected	9,350–60,770	USDOE/EERE[106]

[a]United Soybean Board goal for 2025 is 60 bushels/acre (4.03 Mg/ha).

and algae (Table 9). Dar[103] reports that in India, the National Biofuel Centre of the Petroleum Conservation Research Association estimates jatropha seed yield of 1.5 Mg per ha, or an oil yield of about 540 L per ha, while the Centre of Excellence for Jatropha Biodiesel Promotion in Rajasthan estimates, for intensively managed plantations, a seed yield of 10 Mg per ha or an oil yield of about 3,400 L per ha.

Many different vegetable oils and fats are used for biodiesel production. In the United States, these include soybean, corn, canola, and palm oil, and yellow grease, white grease, tallow, poultry fat, and other recycled fats and feeds. In 2014, soybeans made up 53% of the biodiesel feedstock in the United States.[26]

5.9 *Algae*

Algae can potentially produce high amounts of biomass-derived fuels from small areas of land and can utilize saline waters. However, algae have harvest and handling challenges. Two methods of cultivating microalgae and cyanobacteria are open ponds and closed photobioreactors. Macroalgae (seaweed) has different requirements and are grown in coastal facilities or offshore and are not considered further.

Research on algae for fuels started in the 1950s with work on utilizing algae to produce methane via anaerobic digestion. Interest was revived after the oil crisis in the 1970s, and focused on methane and hydrogen. The U.S. Government sponsored research on microalgae was centered at the Solar Energy Research Institute in Colorado (later renamed the National Renewable Energy Laboratory) with the Aquatic Species Program (ASP), and the focus turned to lipid oils and diesel replacement. There was still work on hydrogen. The ASP started in 1978 and continued through 1996. Over 3,000 species were screened and 300 promising species selected primarily green algae (*Chlorophyceae*) and diatoms (*Bacillariophyceae*). Choosing the right starting species is important. While a foundation was laid, much work remains to be done. There is a need to understand biosynthetic pathways and regulatory mechanisms and to develop genetic strategies. Dewatering and lipid extraction are significant technical and cost challenges.[107,108]

Davis *et al.* estimate the cost of producing diesel-equivalent from algae of $4.92 per L.[109] [The current price of diesel (refiner price to end users, excluding taxes) is around $0.80 per L.[110]] Over half the cost is for ponds and pond liners. The result is sensitive to the lipid fraction of the algae and the algae growth rate.

6 Policies Affecting Biofuels

Much of biofuel use in the developed and developing world is driven by government policies. In the United States, much of first generation ethanol and biodiesel production was the result of government subsidies in the form of exemptions from taxes on diesel that are no longer available at the Federal level, but are still available in a number of states. These fuels have become established primarily as blends in the gasoline and diesel markets. As a result of the Renewable Fuel Standard (RFS) set by the Environmental

Protection Agency (EPA), first effective in 2005 and then revised in 2007, there are targets for ethanol (with a maximum ceiling of 56.8 billion L of corn from ethanol) and biodiesel use. For 2014 (2022) the RFS set targets of 54.5 (56.8), 6.6 (60.6), 14.2 (79.1), 3.8 (minimum of 3.8), and 68.7 (136.3) billion L of conventional biofuels (primarily corn ethanol), cellulosic biofuels, advanced biofuels (includes cellulosic and biodiesel), biodiesel, and total renewable fuels respectively.[111]

In the EU, the EU Energy and Climate Change Package (ECCP) was adopted in April 2013 by the European Council and included the Renewable Energy Directive (RED). The ECCP included goals for 2020 of: a 20% reduction in greenhouse gases compared to 1990, a 20% improvement in energy efficiency compared to the forecast made for 2020, and a 20% share for renewable energy in the energy mix, with a minimum target of 10% renewable energy in the transport sector. In January 2014, the European Commission published its Communication on 2030 climate and energy goals, which includes a 40% reduction in greenhouse gas emissions compared to 1990 and renewable energy of at least 27%. Second generation biofuels (made from lignocellulosic, non-food cellulosic, and waste and residue materials count double toward goals and renewable electricity used by cars counts 2.5, meaning that the actual portion of renewable fuels in the transport sector can be less than 10%.[27] The European Industrial Bioenergy Initiative was launched in October 2010, to promote and commercialize second-generation biofuels. The amount of conventional biofuels to be included in transport fuels is still fluid.[27] The EU also has a countervailing duty on the import of biodiesel from Argentina of 24.6%. Europe has imposed non-tariff trade barriers to the import of Indonesian biodiesel.

In Brazil the ethanol content of gasoline is 25% and there are tax incentives for ethanol flex fuel vehicles; and for biodiesel there was a 5% blend requirement until July 2014 that was scheduled to increase to 6% on July 1, 2014 and 7% in November 2014.[16] In Argentina, there is a 10% biodiesel mandate (including diesel used in power plants) and ethanol is being blended at 7.5%. Argentina has export taxes on biodiesel.[28]

In Indonesia, the bioethanol mandate is currently 7% in transportation and industry (3% for subsidized gasoline) and in 2015 this is scheduled to increase to 10% (5% for subsidized gasoline) with further increases scheduled for 2020 and 2025. The biodiesel mandate is 7% for transportation (5% for subsidized gasoline), and 10% for industry and electricity and in 2015 this is scheduled to increase to 10% (5% for subsidized gasoline) with further increases scheduled for 2020 and 2025.[29] In Malaysia, there is a biodiesel mandate of 5% as of July 2014, but the distribution capacity is not in place, particularly in East Malaysia.[30] In China, the mandatory blend rate for ethanol in gasoline is 10% and there is a subsidy for ethanol production which since 2012 stands at 6 cents per L ($79.4 per Mg). There is no mandate for biodiesel. In its 5-year plan, China says it will increase renewable energy use (which includes biofuels).[31]

Other countries have biofuels mandates in place. These were summarized at the end of December 2013 in Biofuels Digest.[112]

7 Conclusions and Outlook

Biofuels contribute 10% of the world's primary energy, much of it in the form of traditional uses of the fuels such as cooking. In the last 10 years the production of first generation liquid biofuels (primarily ethanol and biodiesel) has increased greatly. Ethanol production has increased from 19 to 89 billion L from 2001 to 2013. Biodiesel production increased from 1.2 to 24 billion L from 2001 to 2011. First generation biofuels come from food crops such as maize, sugar cane, soybeans, rapeseed (canola), and the oil palm. Second generation biofuels from cellulosic sources are just starting to come online in Europe, Brazil, and the United States, but their production is small at present. The use of biomass to generate electricity has steadily increased over time. Recently there has been a surge in the use of wood pellets, especially for electricity generation to meet carbon emission policies in the EU. In the OECD–FAO agricultural outlook, output of ethanol and biodiesel are expected to increase by 50% from 2013 to 2023.[113] Policy will continue to be important for the production and use of biomass and biofuels.

Both conversion technologies and second-generation feedstock production are advancing. There are many potential conversion technologies to produce solid, liquid, and gaseous fuels. First-generation biofuels are limited by food production. Even though there will continue to be incremental changes in first-generation production and conversion technologies (e.g. higher maize yields, higher conversion efficiencies), for much greater biofuels production higher yielding crops (e.g. sorghum, miscanthus, poplars) and crops more suitable for more marginal lands are needed, and the costs of conversion technologies must decline to produce fuels that are economically competitive with fossil-based sources. Policy can drive the production of second-generation biofuels, but the costs of those policies cannot be too high.

Acknowledgments

This work was supported by the Biomass Energy Technologies Office of the U.S. Department of Energy. Prepared by Oak Ridge National Laboratory, managed by UT-Battelle for the U.S. Department of energy under contract DE-AC05-00OR22725. I would like to thank Laurence Eaton for reviewing this chapter. This chapter draws heavily from Chapter 11 (pp. 225–243) in The World Scientific Handbook of Energy, Vol. 3 published in 2013 by World Scientific Publishing Company.

References

1. R. Perlack, L. Wright, A. Turhollow, R. Graham, B. Stokes and D. Erbach, *Biomass as Feedstock for a Bioenergy and Bioproducts Industry: The Technical Feasibility of a Billion-ton Annual Supply*, ORNL/TM-2005/66 (Oak Ridge National Laboratory, Oak Ridge, TN, USA, 2005).
2. USDOE, *Biomass as Feedstock for a Bioenergy and Bioproducts Industry: An Update to the Billion-ton Annual Supply*, R. Perlack, B. Stokes (Leads), ORNL/TM-2010/224 (Oak Ridge National Laboratory, Oak Ridge, TN, USA, 2011).
3. A. Turhollow, R. Perlack, L. Eaton, M. Langholtz, C. Brandt, M. Downing, L. Wright, K. Skog, C. Hellwinckel, B. Stokes and P. Lebow, The updated billion-ton resource assessment, *Biomass Bioenerg* **70** (2014), pp. 149–164.
4. IEA, Renewables, International Energy Agency (2014). Available at: http://www.iea.org/topics/renewables/subtopics/bioenergy/. Accessed 3 November 2014.
5. USDOE/EIA, *International Energy Statistics, Renewable, Biofuels: Fuel Ethanol Production* (U.S. Department of Energy, Energy Information Administration Washington, D.C., 2014). Available at: http://www.eia.gov/cfapps/ipdbproject/iedindex3.cfm?tid=79&pid=80&aid=1&cid=regions&syid=2000&eyid=2011&unit=TBPD, Accessed 12 September 2014.
6. Renewable Fuels Association (RFA), Statistics (2011). Available at: http://www.ethanolrfa.org/pages/World-Fuel-Ethanol-Production. Accessed 6 August 2014.
7. C. Valdes, *Brazil's Ethanol Industry: Looking Forward, BIO-02* (U.S. Department of Agriculture, Economic Research Service, 2011).
8. sugarcane.org, Answers to frequently asked questions, question 9 (2014). Available at: http://sugarcane.org/media-center/faqs. Accessed 14 October 2014.
9. R. de Cerqueira Leite, M. Verde Leal, L. Barbosa Cortez, W. Griffin and M. Gaya Scandiffio, Can Brazil replace 5% of the 2025 gasoline world demand with ethanol? *Energy* **34** (2009), pp. 655–661.
10. USDA/NASS, *Crop Production 2013 Summary* (U.S. Department of Agriculture, National Agricultural Statistics Service, Washington, D.C., 2014).

11. USDA/NASS, *Crop Production*, 10 October 2014, (U.S. Department of Agriculture, National Agricultural Statistics Service, Washington, D.C., 2014).
12. S. Mueller and J. Kwik, *2012 Corn Ethanol Emerging Plant Energy and Environmental Technologies* (University of Illinois at Chicago, Energy Resources Center, 2013).
13. USDA/OCE-WAOB, *USDA Agricultural Projections to 2023, Long-term Projections Report OCE-2014-1* (U.S. Department of Agriculture, Office of the Chief Economist and World Agricultural Outlook Board, Washington, D.C., 2014).
14. J. Goldemberg, The Brazilian biofuels industry, *Biotechnology for Biofuels* **1** (2008), p. 6.
15. S. Barros, *Sugar Annual*, GAIN Report BR14001 (USDA Foreign Agricultural Service, Sao Paulo, Brazil, 2014).
16. S. Barros, Brazil biofuels annual, GAIN Report BR14004 (USDA Foreign Agricultural Service, Sao Paulo, Brazil, 2014).
17. Anon., Kansas ethanol plant to co-locate renewable diesel facility on-site, *Biodiesel Magazine*, 13 (2014).
18. Anon., Quad County produces cellulosic ethanol, *Ethanol Producer Magazine*, 24 (2014).
19. J. Javers, ICM generation 1.5 cellulose to ethanol, Presentation at International Fuel Ethanol Workshop & Expo 11 June (2013).
20. R. Janssen, A. Turhollow, D. Rutz and R. Mergner, Production facilities for second-generation biofuels in the USA and the EU — current status and future perspectives, *Biofuels Bioprod. Bioref.* **7** (2013), pp. 647–665.
21. Beta Renewables, Crescentino's biorefinery grand opening, press release, 9 October (2013). Available at: http://www.betarenewables.com/press-release-detail/2/crescentinos-biorefinery-grand-opening, accessed 22 October 2014.
22. Gran Bio, Gran Bio begins cellulosic ethanol production in Brazil, *Biomass Magazine*, 24 September (2014). Available at: http://www.biomassmagazine.com/articles/10991/granbio-begins-cellulosic-ethanol-production-in-brazil. Accessed 26 September 2014.
23. J. Lane, Biodiesel 2.0-the next chapter for America's favorite advanced biofuel, *Biofuels Digest* 20 January (2014). Available at: biofuelsdigest.com/bdigest/2014/01/20/biodiesel-2-0-the-next-chapter-for-americas-favoriteadvanced-biofuel/. Accessed 27 August 2014.
24. R. Holden, Commercializing enzymatic biodiesel production, *Biodiesel Magazine*, 3 January (2014). Available at: http://www.biodieselmagazine.com/articles/9481/commercializing-enzymatic-biodiesel-production. Accessed 23 October 2014.
25. USDOE/EIA, International energy statistics, renewable, biofuels: Biodiesel production (U.S. Department of Energy, Energy Information Administration Washington D.C., 2014). Available at: http://www.eia.gov/cfapps/ipdb project/iedindex3.cfm?tid=79&pid=81&aid=1&cid=regions&syid=2000. Accessed 12 September 2014.

26. DOE/EIA, Monthly biodiesel production report, July (U.S. Department of Energy, Energy Information Administration Washington D.C., 2014).
27. B. Flach, K. Bendz and S. Lieberz, EU biofuels annual, GAIN Report No. NL4025 (USDA Foreign Agricultural Service, The Hague, The Netherlands).
28. K. Joseph, Argentina biofuels annual, GAIN Report (USDA Foreign Agricultural Service, Buenos Aires, Argentina, 2014).
29. R. Wright and I Wiyono, Indonesia biofuels annual, GAIN Report No. ID1420 (USDA Foreign Agricultural Service, Jakarta, Indonesia, 2014).
30. A. Wahab, Malaysia biofuels annual, GAIN Report No. MY4100 (USDA Foreign Agricultural Service, Kuala Lampur, Malaysia, 2014).
31. R. Scott and J. Junyang, China — Peoples Republic of biofuels annual, GAIN Report 13040 (USDA Foreign Agricultural Service, Beijing, China, 2014).
32. B. Evans, Canada biofuels annual, GAIN Report CA13034 (USDA Foreign Agricultural Service, Ottawa, Canada, 2014).
33. A. Gilbert and L. Pinzon, Colombia biofuels annual, GAIN Report (USDA Foreign Agricultural Service, Bogota, Colombia, 2014).
34. S. Preechajarn and P. Prasertsri, Thailand biofuels annual, GAIN Report TH4057 (USDA Foreign Agricultural Service, Bangkok, Thailand, 2014).
35. H. Aatola, M. Larmi and T. Sarjovaara, Hydrotreated vegetable oil (HVO) as a renewable diesel fuel: Trade-off between NOx, particulate emission, and fuel consumption of a heavy duty engine, SAE international Paper No. 2008-01-2500 (2008).
36. H. Aatola, M. Larmi, T. Sarjovaara and S. Mikkonen, Hydrotreated vegetable oil (HVO) as a renewable diesel fuel: Trade-off between NO_x, particulate emission, and fuel consumption of a heavy duty engine, presentation (2008). Available at: http://www.google.com/url?sa=t&rct=j&q=&esrc=s&source=web&cd=2&cad=rja&uact=8&sqi=2&ved=0CCcQFjAB&url=http%3A%2F%2Fwww.nesteoil.com%2Fbinary.asp%3FGUID%3D585DB098-061A-4460-B43B-BC7BE431049D&ei=_zBJVObKCoiZyATQiYHwDQ&usg=AFQjCNGh86kJ2qBndBPKuhhPjhwrhe0xjg&sig2=Ri1ZX0e6x6-c77hescpCug. Accessed 24 September 2014.
37. Diamond Green Diesel, Home. Available at: http://www.diamondgreendiesel.com/Pages/default.aspx. Accessed 23 October 2014.
38. Eurostat, Source data for tables and figures: renewable energy: Tables and figures. Available at: http://epp.eurostat.ec.europa.eu/statistics_explained/images/6/61/Renewable_energy_YB2014.xlsx. Accessed 3 November 2014.
39. DOE/EIA, Monthly energy review October 2014 (U.S. Department of Energy, Energy Information Administration, Washington, D.C., 2014).
40. E. de Jong, A. Higson, P. Walsh and M. Wallisch, Biobased chemicals: Value added from biorefineries (IEA Bioenergy, Task 42 Biorefineies, 2011).
41. OECD, The bioeconomy to 2030, designing a policy agenda, main findings and policy conclusions. Available at: http://www.oecd.org/futures/longtermtechnologicalsocietalchallenges/42837897.pdf. Accessed 17 October 2014.

42. Nexant, Renewable chemicals & materials opportunity assessment: Major job creation and agricultural sector engine. Available at: http://www.usda.gov/oce/reports/energy/USDA_RenewChems_Jan2014.pdf. Accessed 20 June 2014.
43. M. Besson, P. Gallezot and C. Pinel, Conversion of biomass into chemicals over metal catalysts, *Chem. Rev.* **114** (2014), pp. 1827–1870.
44. P. Gallezot, Conversion of biomass to slected chemical products, *Chem. Soc. Rev.* **41** (2012), pp. 1538–1558.
45. J. Bozell and G. Petersen, Technology development for the production of biobased products from biorefinery carbohydrates — the US Department of energy's "top 10," *Green Chem* **12**(4) (2010), pp. 539–554.
46. J. Holladay, J. Bozell, J. White and D. Johnson, Top value added chemicals from biomass, Volume II — results of screening for potential candidates from biorefinery lignin, PNNL-16983 (Pacific Northwest National Laboratory, Richland, WA, USA, 2007).
47. D. Elliott, Chemicals from biomass, in *Encyclopedia of Energy*, Vol. 1, ed. C. Cleland (Elsevier, Oxford, United Kingdom, 2004), pp. 163–174.
48. J. C. Mol, Catalytic metathesis of unsaturated fatty acid esters and oils, *Top. Catal.* **27**(1–4) (2004), pp. 97–104.
49. B. Vijayendran, Biobased motor oils are ready for primetime, *Ind. Biotechnol.* **10**(2) (2014), pp. 64–68.
50. S. Ahlgren, A. Baky, S. Bernesson, Å. Nordberg, O. Norén and P.-A. Hansson, Ammonium nitrate fertilizer production based on biomass — environmental effects from a life cycle perspective, *Biores. Technol.* **99** (2008), pp. 8034–8041.
51. S. Ahlgren, S. Bernesson, Å. Nordberg and P.-A. Hansson, Nitrogen fertilizer production based on biogas — energy input, environmental impact and land use, *Biores. Technol.* **101** (2010), pp. 7181–7184.
52. J. Andersson and J. Lundgren, Techno-economic analysis of ammonia production via integrated biomass gasification, *Appl. Energ.* **130** (2014), pp. 484–490.
53. USDA/NASS, Agricultural prices, April 2014. (U.S. Department of Agriculture, National Agricultural Statistics Service, Washington, D.C., 2014).
54. BioNitrogen, Investor fact sheet Q3 2014. Available at: http://www.bionitrogen.com/wp-content/uploads/2014/04/BioNitrogen-Investor-Fact-Sheet. pdf. Accessed 3 November 2014.
55. Taco Times, Taylor County, FL|BioNitrogen closes on 55-acre land deal. Available at: http://www.bionitrogen.com/plant_construction_updates/taylor-county-fl-bionitrogen-closes-on-55-acre-land-deal/. Accessed 3 November 2014.
56. FAOSTAT, Forestry, forestry production and trade, wood pellets, Food and Agriculture Organization of the United Nations — Statistics Division. Available at: http://faostat3.fao.org/download/F/FO/E. Accessed 22 October 2014.

57. T. Portz, Investing upstream, *Pellet Mill Magazine*, **Q4 2014** (2014), pp. 17–21.
58. M. Downing and A. F. Turhollow, Jr., Biomass, *The World Scientific Handbook of Energy — Materials and Energy*, Vol. 3, ed. G.M. Crawley (World Scientific, Singapore, 2013), pp. 225–243.
59. A. V. Bridgwater, Review of fast pyrolysis of biomass and product upgrading, *Biomass Bioenerg.* **38** (2012), pp. 68–94.
60. R. C. Brown and J. Holmgren, Fast pyrolysis and bio-oil upgrading (2011). Available at: http://www.ars.usda.gov/sp2UserFiles/Program/307/biomasstoDiesel/RobertBrown&JenniferHolmgrenpresentationslides.pdf. Accessed 24 October 2014.
61. E. Taylor, Coproducts and byproducts of woody biorefinery processing, in *Proceedings of the Role of Extension in Energy*, eds. B. C. English, J. Menard and K. Jensen, 30 June–1 July (Little Rock, AR, USA, 2009), pp. 62–63. Available at: http://www.farmfoundation.org/news/articlefiles/1704-Final% 20proceedings%209-14-09.pdf. Accessed 24 October 2014.
62. J. Scahill, Pyrolysis — a promising new path to liquid fuels, Presented at *Can forests meet our energy needs? The future of forest biomass*, in Colorado, 21 February (Colorado State University, Fort Collins, CO, USA, 2008). Available at: http://digitool.library.colostate.edu///exlibris/dtl/d3_1/apache_media/L2V4bGlicmlzL2R0bC9kM18xL2FwYWNoZV9tZWRpYS80OTI2.pdf. Accessed 24 October 2014.
63. S. Jones, J. Holladay, C. Valkenburg, D. Stevens, C. Walton, C. Kinchin, D. Elliot, and S. Czernak, Production of gasoline and diesel from biomass via fast pyrolysis, hydrotreating and hydrocracking: A Design Case, PNNL-18284 (Pacific Northwest National Laboratory, Richland, WA, USA, 2009).
64. T. L. Marker, L. G. Felix, M. B. Linck and M. J. Roberts, Integrated hydropyrolysis and hydroconversion (IH^2) for the direct production of gasoline and diesel fuels or blending components from biomass, part 1: Proof of principle testing, *Environ. Prog. Sust. Energ.* **31**(2) (2012), pp. 191–199.
65. USDOE/EERE, Using heat and chemistry to make fuels and power: Thermochemical conversion, U.S. Department of Energy, Energy Efficiency & Renewable Energy, Biomass Program (2010). Available at: http://www1.eere.energy.gov/library/asset_handler.aspx?src=http://www1.eere.energy.gov/biomass/pdfs/thermochemical_four_pager.pdf&id=4374. Accessed 24 October 2014.
66. M. Worley and J. Yale, Biomass gasification technology assessment: Consolidation report, NREL/SR-5100-57085 (National Renewable Energy Laboratory, Golden CO, USA, 2012).
67. IEA Task 37, Up-grading plant list. Available at: http://www.iea-biogas.net/plant-list.html?file=files/daten-redaktion/download/Up-grading_Plant_List.xls. Accessed 14 October 2014.
68. R. Burns, Current state of manure anaerobic digestion in the U.S. and beyond, presentation at: Energy production from anaerobic digestion of dairy manure, Madison, WI, USA 28–29 September 2009 as cited in: J. Arogo Ogejo, J. Ignosh, S. Lansing, G. Felton, Anaerobic digesters an overview.

Available at: http://www.epa.gov/agstar/documents/meeting_120210_lansing.pdf. Accessed 10 November 2014.

69. Anaerobic Digestion China Working Group, International Water Association, Background. Available at: http://www.iwahq.org/213/communities/specialist-groups/list-of-groups/anaerobic-digestion/ad-china-wg.html. Accessed 14 October 2014.
70. USDA, USEPA, USDOE, Biogas opportunities roadmap (2014).
71. W. F. Lazarus, Anaerobic digester technology. In *Proceedings of the Role of Extension in Energy*, eds. B. C. English, J. Menard, and K. Jensen, 30 June–1 July (Little Rock, AR, USA, 2009), pp. 42–45. Available at: http://www.farmfoundation.org/news/articlefiles/1704-Final%20proceedings%209-14-09.pdf. Accessed 10 November 2014.
72. J. Crenshaw, What's a digester cost these days? Presented at the 2009 AgSTAR National Conference, 24–25 February 2009, Baltimore, MD, USA. Available at: http://www.epa.gov/agstar/documents/conf09/crenshaw_digester_cost.pdf. Accessed 10 November 2014.
73. K. Fletcher, NABCE 2014 comes to an end with Poet-DSM fantasy fuel plant tour, Ethanol Producer Magazine 16 October (2014). Available at: http://www.ethanolproducer.com/articles/11554/nabce-2014-comes-to-an-end-with-poet-dsm-fantasy-fuel-plant-tour. Accessed 4 November 2014.
74. D. Humbird, R. Davis, L. Tao, C. Kinchin, A. Aden, P. Schoen, J. Lukas, B. Olthof and M. Worley, Process design and economics for biochemical conversion of lignocellulosic biomass to ethanol: Dilute-acid pretreatment and enzymatic hydrolysis of corn stover, NREL/TP-5100-47764 (National Renewable Energy Laboratory, Golden CO, USA, 2011).
75. C. S. Byrt, C. P. L Grof and R. T. Furbank, C_4 plants as biofuel feedstocks: Optimising biomass production and feedstock quality from a lignocellulosic perspective, *J. Integrative Plant Biol.* **53**(2) (2011), pp. 120–135.
76. D. J. Parrish and J. H. Fike, The biology and agronomy of switchgrass for biofuels, *Crit. Rev. Plant Sci.* **24** (2005), pp. 423–459.
77. R. Mitchell, K. Vogel and M Schmer, Growing switchgrass for biofuels (Iowa State University Extension and Outreach, Ames, IA, USA, 2013).
78. K. P. Vogel and R. B. Mitchell, Heterosis in switchgrass: Biomass yield in swards, *Crop Science* **48**(2008) (1983), pp. 2159–21641.
79. USDA/ARS, USDA plant hardiness zone map. Available at: http://planthardiness.ars.usda.gov/PHZMWeb/. Accessed 15 October 2014.
80. K. Vogel, R. Mitchell, M. Casler and G. Sarath, Registration of 'Liberty' switchgrass, *J. Plant Registrations* **8** (2014), pp. 242–247.
81. A. F. Turhollow, E. G. Webb and M. E. Downing, Review of sorghum production practices: Applications for bioenergy, (ORNL/TM-2010/7, Oak Ridge National Laboratory, Oak Ridge, TN, USA, 2010).
82. V. S. Belum Reddy, S. Ramesh, P. Sanjana Reddy, B. Ramaiah, P. M. Salimath and R. Kachapur, Sweet sorghum — A potential alternative raw material for bio-ethanol and bio-energy, *Sorghum Millets Newslett.* **46** (2005), pp. 79–96.

83. V. S. Belum Reddy, Willaim D. Dar, P. Parthasarathy Rao, P. Srinivasa Rao, A. Ashok Kumar and P. Sanjana Reddy, Sweet sorghum as a bioethanol feedstock: Challenges and opportunities, International Conference on Sorghum for Biofuel, 19–22 August (Houston, TX, USA, 2008). Available at: http://www.ars.usda.gov/meetings/Sorghum/presentations/Reddy.pdf. Accessed 2 May 2011.
84. T. L. Tew, R. M. Cobill and E. P. Richard, Jr., Evaluation of sweet sorghum and *sorghum x sudangrass* hybrids as feedstocks for ethanol production, *Bioenergy Res.* **1** (2008), pp. 147–152.
85. S. R. Schill, Tall Texas sorghum, *Ethanol Producer Magazine*, November 2007. Available at: http://www.ethanolproducer.com/articles/3424/tall-texas-sorghum. Accessed 4 November 2014.
86. S. C. Aravindhakshan, F. M. Epplin and C. M. Taliafarro, Economics of switchgrass and Miscanthus relative to coal as feedstock for generating electricity, *Biomass Bioenerg.* **35**(1) (2011), pp. 308–319.
87. A. K. Jain, M. Khanna, M. Erickson and H. Huang, An integrated biogeochemical and economic analysis of bioenergy crops in the Midwestern United States, *Global Change Biol. Bioenerg.* **2**(5) (2010), pp. 217–234.
88. M. Khanna, B. Dhungana and J. Clifton-Brown, Costs of producing Miscanthus and switchgrass for bioenergy in Illinois, *Biomass Bioenerg.* **32** (2008), pp. 482–493.
89. D. G. Christian, A. B. Richie and N. E. Yates, Growth, yield and mineral content of *Miscanthus x giganteus* grown as a biofuel for 14 successive harvests, *Ind. Crop. Prod.* **28** (2008), pp. 320–327.
90. L. Ercoli, M. Mariotti, A. Masoni and E. Bonari, Effect of irrigation and nitrogen fertilization on biomass yield and efficiency of energy use in crop production of Miscanthus, *Field Crops Res.* **63** (1999), pp. 3–11.
91. M. Salassi, K. Brown, B. Hilburn, M. Deliberto, K. Gravois, T. Mark and L. Falconer, Farm-scale cost of producing perennial energy cane, *Bioenerg. Res.* **7** (2014), pp. 609–619.
92. B. L. Legendre and D. M. Burner, Biomass production of sugarcane cultivars and early-generation hybrids, *Biomass Bioenerg.* **8**(2) (1995), pp. 55–61.
93. D. E. Riemenschneider, H. E. Stelzer and G. S. Foster, Quantitative genetics of poplars and poplar hybrids, in *Biology of Populus and its Implications for Management and Conservation*, eds. R. F. Stettler, H. D. Bradshaw, Jr, P. E. Heilamn and T. M. Hinkley, Part I, Chapter 7 (NRC Research Press, National Research Council of Canada, Ottawa, ON, Canada, 1996), pp. 159–181.
94. R. S. Zalesny Jr., R. B. Hall, J. A. Zalesny, B. G. McMahon, W. E. Berguson and G. R. Stanosz, Biomass and genotype x environment interactions of Populus energy crops in the Midwestern United States, *Bioenergy Res.* **2** (2009), pp. 106–122.
95. H. G. Adegbidi, T. A. Volk, E. H. White, R. D. Briggs, L. P. Abrahamson and D. H. Bickelhaupt, Biomass and nutrient export by willow clones in experimental bioenergy plantations in New York, *Biomass Bioenerg.* **20** (2001), pp. 389–398.

96. M. Labrecque and T. I. Teodorescu, High biomass yield achieved by Salix clones in SRIC following two 3-year coppice rotations on abandoned farmland in southern Quebec, Canada, *Biomass Bioenerg.* **25** (2003), pp. 135–146.
97. H. G. Adegbidi, R. D. Briggs, T. A. Volk, E. H. White and L. P. Abrahamson, Effect of organic amendments and slow-release nitrogen fertilizer on willow biomass production and soil chemical characteristics, *Biomass Bioenerg.* **25** (2003), pp. 389–398.
98. T. A. Volk, L. P. Abrahamson, C. A. Nowak, L. B. Smart, P. J. Tharakan and E. H. White, The development of short-rotation willow in the northeastern United States for bioenergy and bioproducts, agroforestry and phytoremediation, *Biomass Bioenerg.* **30** (2006), pp. 715–727.
99. T. A. Volk, B. D. Kiernan, R. Kopp and L. P. Abrahamson, First- and second-rotation yields of willow clones at two sites in New York State, Fifth Biomass Conference of the Americas, 17–21 September (Orlando, FL, USA, 2001). Available at: http://www.brdisolutions.com/pdfs/bcota/abstracts/22/z146.pdf.
100. T. Bucholz and T. A Volk, Improving the profitability of willow biomass crops — identifying opportunities with a crop budget model, *Bioenergy Res.* **4**(2) (2011), pp. 85–95.
101. M. Langholtz, D. R. Carter, D. L. Rockwood and J. R. R. Alavalapati. The economic feasibility of reclaiming phosphate mined lands with short-rotation woody crops in Florida, *J. Forest Economics* **12**(4) (2007), pp. 237–249.
102. M. Rahmani, A. W. Hodges, J. Stricker and C. F. Kiker, Economic analysis of biomass crop production in Florida, in Making a business from biomass in energy, environment, chemical, fibers and materials, eds. R. P. Overend and E. Chornet, Third Biomass Conference of the Americas, 24–29 August (Elsevier Science, New York, NY, USA), pp. 91–99.
103. W. Dar, Research needed to cut risk to biofuel farmers, *Science and Development Network* (2007). Available at: http://www.scidev.net/en/opinions/research-needed-to-cut-risks-to-biofuel-farmers.html. Accessed 4 November 2014.
104. R. Krautgartner, X. Audran, L. E. Rehder, M. Boshnkova, M. Dobrescu, B. Flach, J. Wilson, O. Bettini, M. Guerrero and K. Bendz, EU-28 oilseeds and products annual, GAIN Report Number: AU1407 (USDA, Foreign Agricultural Service, Vienna, Austria, 2014).
105. A. G. Wahab, Malaysia oilseeds and products annual, GAIN Report Number: MY4006 (USDA Foreign Agricultural Service, Kuala Lumpur, Malaysia, 2014).
106. USDOE/EERE, National algal biofuels technology roadmap, U.S. Department of Energy, Energy Efficiency & Renewable Energy, Biomass Program, 2010. Available at: http://www1.eere.energy.gov/biomass/pdfs/algal_ biofuels_roadmap.pdf. Accessed 4 November 2014.
107. E. D. Jarvis, Aquatic species program: Lessons learned, Presentation at the AFSOR Workshop on Algal Oil for Jet Fuel Production, 19–21

February (Washington, DC, USA, 2008). Available at: http://www.nrel.gov/biomass/pdfs/43232.pdf. Accessed 29 April 2011.

108. J. Sheehan, T. Dunahay, J. Benemann and P. Roessler, A look back at the U.S. Department of Energy's Aquatic Species Program — Biodiesel from algae, NREL/TP-580-24190 (National Renewable Energy Laboratory, Golden, CO, USA, 1998).

109. R. Davis, D. Fishman, E. Frank, M. Wigmosta, A. Aden, A. Coleman, P. Pienkos, R. Skaggs, E. Venteris and M. Wang, Renewable diesel from algal lipids: An integrated baseline for cost, emisions, and resource potential from a harmonized model, ANL/ESD/12-4, NREL/TP-5100-55431, PNNL-21437 (Argonne National Laboratory, Argonne, IL, USA; National Renewable Energy Laboratory, Golden CO, USA; Pacific Northwest National Laboratory, Richland, WA, USA, 2012).

110. USDOE/EIA, Monthly energy review October, DOE/EIA-0035(2014/10) (U.S. Department of Energy, Energy Information Administration Washington D.C., 2014).

111. U.S. Congress, Energy Independence and Security Act of 2007, 110th Congress, 1st session HR 6.4, January (2007).

112. J. Lane, Biofuels mandates around the world, Biofuels digest 31 Dec. (2013). Available at: http://www.biofuelsdigest.com/bdigest/2013/12/31/biofuels-mandates-around-the-world-2014/. Accessed 16 October 2014.

113. OECD-FAO, OECD-FAO Agricultural outlook: Biofuels, see all data for biofuels, (2014). Available at: http://stats.oecd.org/viewhtml.aspx?QueryId=58648&vh=0000&vf=0&l&il=&lang=en. Accessed 18 November 2014.

Chapter 6

Artificial Photosynthesis

Nathan Skillen* and Peter K. J. Robertson
Centre for the Theory and Application of Catalysis (CenTACat), School of Chemistry and Chemical Engineering, Queen's University Belfast, David Keir Building, Stranmillis Road, Belfast, BT9 5AG, UK
**n.skillen@qub.ac.uk*

Photosynthesis is a natural mechanism which is capable of not only converting solar energy to a potential food stock but also reducing atmospheric CO_2 levels. The process of artificial photosynthesis attempts to mimic this mechanism through the development of materials and technology that can harvest solar energy for high energy conversion to renewable fuel. The potential of using a hugely sustainable source of energy such as solar with H_2O and CO_2, as the only feedstock's, is very desirable in the current global energy situation. A number of approaches have been developed in an attempt to achieve this including; dye sensitized solar cells (DSSCs), photocatalytic H_2O oxidation and CO_2 reduction and nanobiocatalytic complexes. Due to the complexity and intricacy of the interaction between the compounds and structures involved in photosynthesis, replicating the process is faced with a number of challenges; harvesting solar light whilst maintaining a high yield of product, electron transfer and charge separation, carbon fixation and high turnover number. Moreover, photosynthesis is capable of proceeding under a range of varying environmental conditions such as diffuse light conditions. While significant challenges are faced, impressive breakthroughs and advancements have been made in the past decade which contributes towards the development of an artificial photosynthetic system. Discussed herein is the research that has been conducted in artificial photosynthesis covering a range of applications, emerging technologies and the work of dedicated Artificial Photosynthesis System (APS) institutes.

1 Introduction

The chemical process of photosynthesis, in which CO_2 is converted to carbohydrates while H_2O is oxidized to O_2 and H^+, is a desirable mechanism to be replicated in view of the current global energy concerns. The process of photosynthesis has provided the majority of stored energy available on the planet. The overuse of stored energy, primarily through the burning of fossil fuels, has resulted in the depletion of fuel stocks to the point that natural resources are unable to replenish that which has been used. Furthermore, the levels of CO_2 emissions have long surpassed the level at which nature is capable of keeping a safe balance. Therefore, a significant challenge remains in not only developing renewable fuels that do not further elevate CO_2 emissions but also the capping and potential decreasing of current anthropogenic CO_2 levels.

The concept of utilizing solar light for not only CO_2 sequestration but solar fuel generation is the basis for artificial photosynthesis systems (APS) currently under development. The potential of artificial photosynthesis continues to grow with significant advances being made during the past decade. In recent years, the development of novel heterogeneous catalysts for solar light harvesting, H_2O oxidation and CO_2 reduction along with advances in photoelectrochemical cells have contributed to the growth of the field. Furthermore, increasing atmospheric CO_2 concentrations coupled with the demand for renewable energy and the staggering potential of solar energy (4.3×10^{20} J hr^{-1} reaching earth's surface) have been the primary driving forces. Currently there are a broad range of approaches that are being developed for APS including chemical and photocatalytic,[1–4] bimetallic[5] and biomimetic[6,7] and nanobiocatalytic.[8]

The chemistry involved in the process of photosynthesis has been well studied and the mechanisms are understood at a precise and detailed level. With the discovery of the Photosystem II (PS II) crystal structure in 2004 by Ferriera *et al.*, the photo-induced electron transfer chain was published which gave an explanation for the high yield of photosynthesis.[9] This cascade effect, in conjunction with efficient light harvesting compounds, are examples of both chemical and mechanistic challenges that must be overcome by APS. Kalyanasundaram and Graetzel[10] stated that artificial photosynthesis is an '*approach wherein the important structural elements and reaction features of natural photosynthesis are used in conceptually simpler systems to achieve results of natural photosynthesis*'. The chemical and structural complexity of photosynthesis is challenging to replicate in any artificial system and it is a noteworthy point that the natural process is one which is utilized for plant growth and as such the rate of photosynthesis varies greatly depending on the species of plant. Despite this, solar energy conversion efficiency in natural photosynthesis is only ~2–3%. APS will also aim to improve upon this.

This chapter will present an insight into artificial photosynthesis, covering the areas of current applications including heterogeneous and homogenous systems, emerging technologies and novel materials. A comprehensive, although not exhaustive, review has been compiled to provide the reader with an understanding of the fundamentals of this field of research. As a useful starting point, the natural process is reviewed to identify the key areas that need to be replicated including light harvesting, electron transfer and carbon fixation. The chapter will then discuss approaches deployed for artificial photosynthesis, encompassing photocatalytic systems, H_2O oxidation, CO_2 reduction and conversion to fuels, solar energy conversion and dye sensitized solar cells (DSSCs). Focus shall also be given to key dedicated institutes that have been developed, specifically focusing on artificial photosynthesis, such as the work conducted by the Joint Centre for Artificial Photosynthesis (JCAP) and the Swedish consortium for artificial photosynthesis.

2 Natural Photosynthesis

When discussing the topic of artificial photosynthesis it is important to review the natural process and highlight the areas which can be reproduced. This section will provide the reader with an overview of the natural process including the chemistry involved and the mechanistic pathways which generate high value products. Key areas of the process will be identified along with areas which are capable of being mimicked and these areas will be further discussed in subsequent sections. Photosynthesis can be subdivided into a number of key areas which require consideration. Generally, the mechanism consists of two processes referred to as the '*light*' and '*dark*' reactions. Despite the names, these systems are linked and regulated by a series of enzymes to ensure the operation of an energy efficient system (see Fig. 1).

The light reactions are those dependent on the presence of light as the source of energy in the photolysis of H_2O to molecular O_2 and H^+ (see Eq. (1)).

$$2H_2O \xrightarrow[light]{} 4H^+ + O_2 + 4e^-. \tag{1}$$

The dark reaction, or Calvin cycle, operates independently of illumination and results in the fixation of CO_2 to produce carbohydrates. On a more specific level, it is possible to split photosynthesis into four key processes; energy absorption, electron transfer, ATP generation and carbon fixation.

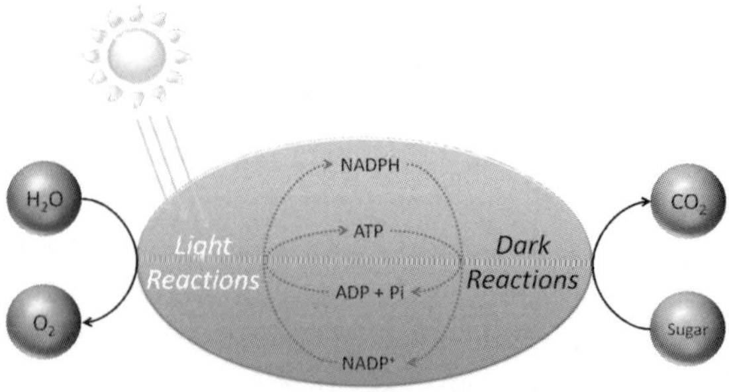

Fig. 1. Schematic representation of photosynthesis.

This section will discuss photosynthesis under these four headings with a view towards their replication in artificial systems.

Firstly, it should be acknowledged that the arrangement of chemicals and organelle structures at the microscopic level within the plant, is critical to the efficiency of the natural process. The chloroplast is the 'working' part of the plant cell in relation to photosynthesis. Within the chloroplast an array of 'antennas' and 'reaction centers' can be found organized into flattened disks known as thylakoids. An assembly of these flattened disks are referred to as grana. The thylakoid is surrounded by the thylakoid membrane and contains a single continuous compartment known as the thylakoid lumen. Specifically, it is the thylakoid membrane within the chloroplast that is the site of photosynthesis. Figure 2 shows the basic structure of the thylakoids.

2.1 *Key Features*

2.1.1 *Energy absorption*

The absorption of light energy is the initial stage of the photosynthetic process. The light energy absorbed is utilized to remove electrons from unwilling donors, which in plants is H_2O, in order to form O_2. The energy supply is then used for the transfer of electrons to a primary electron acceptor. Numerous molecules are located in the thylakoid structure, which are capable of absorbing light from the visible portion of the solar spectrum (see Fig. 3). Chlorophyll (Chl) a is the primary molecule for photon capture in green plants, absorbing light in the blue–violet and orange–red regions of the spectrum. Also present are a number of accessory pigments, including xanthophylls, carotenoids and Chl b, which are capable of absorbing light within the yellow–green–orange section of the spectrum. The arrangement

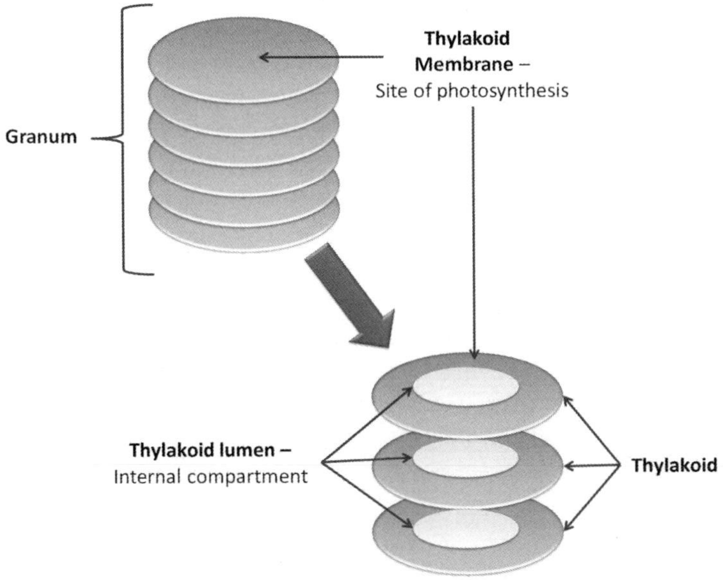

Fig. 2. Stacked thylakoids as a granum.

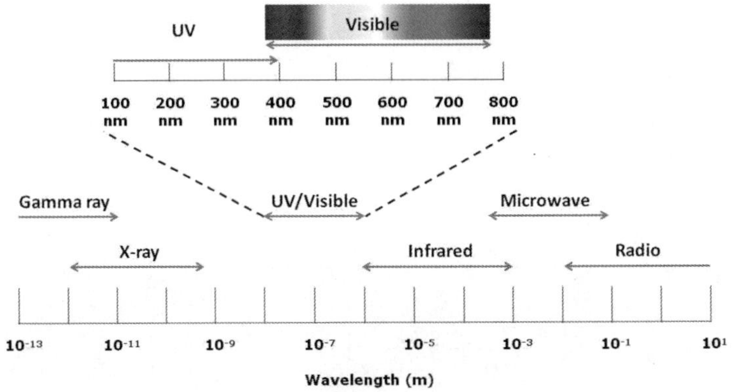

Fig. 3. Electromagnetic spectrum.

of Chl and additional light absorbing molecules provides the plant with an antenna system which is the first key area artificial systems must mimic. The antenna is not only capable of capturing photons, it also relays them onto reaction centers as part of electron-transfer processes. The highly significant feature of this process is the ability of the plant to carry this out under varying light intensities, thus photosynthesis efficiencies are comparable under direct and diffuse light conditions.

2.1.2 *Electron transfer and ATP generation*

Following energy absorption, electron transfer occurs through a series of reaction centers. After the initial electron acceptor, the transfer process occurs through a chain of electron transport molecules in the thylakoid membrane until the final acceptor is reached. Typically the final acceptor in natural photosynthesis is $NADP^+$, which subsequently becomes NADPH. In plants, the transport of electrons is coupled to the movement of protons from the stroma to the thylakoid lumen, forming a pH gradient across the thylakoid membrane (pH lumen <pH stroma).

Figure 4 shows an overview of the light dependant stage in the natural process, while Eq. (2) shows the overall equation for light absorption and

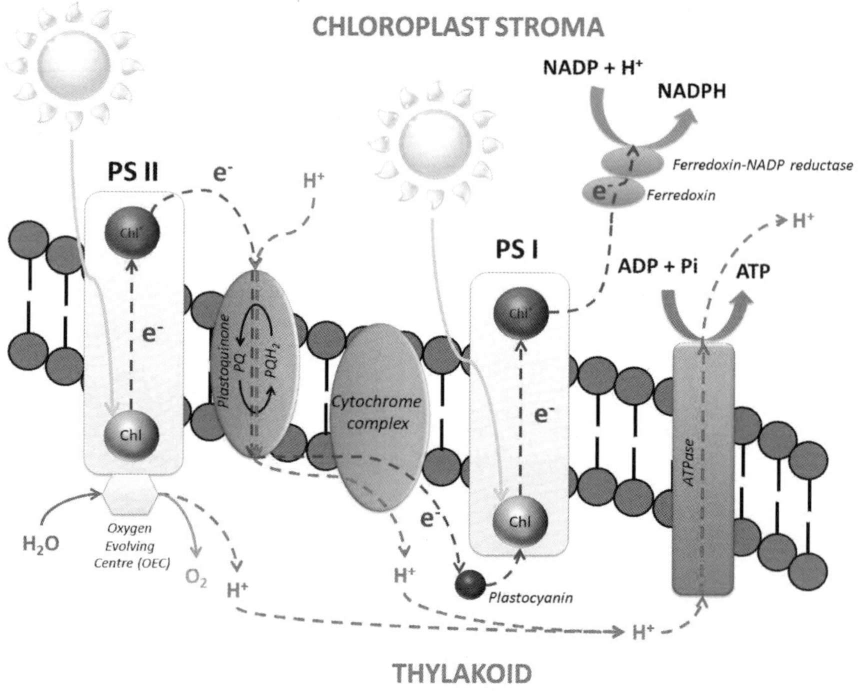

Fig. 4. Schematic representation of the 'light' reaction in photosynthesis.

electron transfer.

$$2H_2O + 2NADP^+ \xrightarrow[light]{} 2H^+ + 2NADPH + O_2. \qquad (2)$$

The mechanism in Fig. 4 shows the capturing of light by Chl in PS II resulting in a series of electron transfer reactions through the thylakoid membrane. Upon absorption of photons, excitation of Chl *a* (P680) occurs which initiates the photo-induced electron transfer process. As shown in Fig. 4, the process of 'Chl' becoming 'Chl*' is the driving force for the oxidation of H_2O to occur. Electrons are extracted from H_2O at the calcium-manganese center to form O_2 and H^+. The electron transfer mechanism occurs through a chain of acceptors including plastoquinone, cyctochrome complex and plastocyanin before being delivered to PS I.

Similar to PS II, PS I excited electrons (P700) are transferred to an acceptor, ferrodoxin, before being utilised for the reduction of $NADP^+$ to NADPH, a reaction which is facilitated by the enzyme ferredoxin-NADP reductase. In conjunction with the transfer of electrons, protons are transported through the system also. Protons from the oxidation of H_2O at the calcium-manganese center along with those that travel across the membrane via plastoquinone are used for the conversion of ATP from ADP and Pi via ATPase. The final products of this mechanism, NADPH and ATP are transported to the Calvin cycle where they are used in the fixation of carbon. This mechanism was one of the most important discoveries in research into natural photosynthesis. The mechanism involving the movement of electrons through a number of acceptors allows for very efficient charge separation which explains the high turnover number observed in photosynthesis. The replication of this process is fundamental to the development of an energy efficient APS.

2.1.3 *Carbon fixation*

The second stage of photosynthesis (Calvin cycle) involves the fixation of CO_2 to facilitate sugar production. The ATP and NADPH generated in the light reaction supply the energy and electrons which drive the Calvin cycle. The cycle can be considered as three phases; carbon fixation, reduction and ribulose-1,5-bisphosphate (RuBP) regeneration. A schematic representation of the cycle is summarized in Fig. 5. The initial stage of the cycle is the incorporation of CO_2 into RuBP to 3-phosphoglycerate (3-PGA), a carboxylation reaction catalyzed by the enzyme RuBisCo. Following this, 3-PGA is phosphorylzed to 1,3-bisphosphoglycerate by ATP and the catalyst

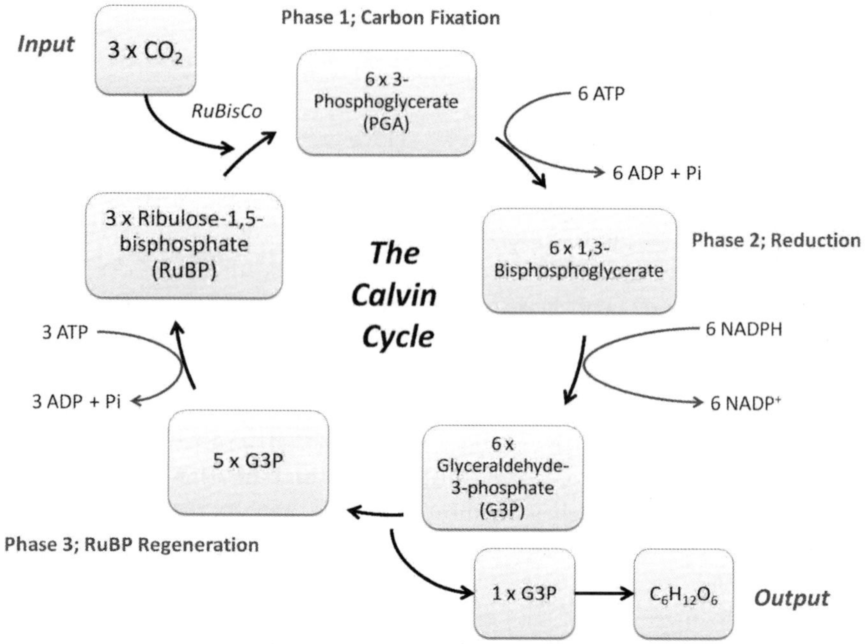

Fig. 5. Flow diagram of the Calvin cycle.

phosphoglycerate kinase, which in turn produces ADP and Pi. During the carbon fixation phase, 1 molecule of CO_2, which enters the cycle, will yield 2, 3-PGA molecules. The 'energy cost' of this fixation stage is two molecules of ATP per CO_2 molecule fixed. During phase 2, 1,3-bisphosphoglycerate is reduced to G3P by G3P dehydrogenase and NADPH, which itself undergoes oxidation to $NADP^+$. As was observed with ATP, two molecules of NADPH are required per CO_2 molecule fixed. The final phase is the completion of the cycle in order to regenerate RuBP. This phase involves a series of enzyme controlled reactions and a number of intermediates during which five molecules of G3P are converted into three molecules of RuBP at a cost of three molecules of ATP. As only five of the six molecules of G3P generated from three molecules of CO_2 are required, there is a net gain of one molecule of G3P. The one molecule of G3P will contribute towards the production of the carbohydrate products of photosynthesis which requires two molecules of G3P in order to yield one molecule of $C_6H_{12}O_6$. The overall equation for the process is shown in Eq. (3).

$$6CO_2 + 18ATP + 12NADPH + 12H_2O \rightarrow C_6H_{12}O_6 + 18ADP + 18Pi + 12NADP^+ + 6H^+. \quad (3)$$

The stages discussed here are those that are key to the natural process of photosynthesis and those that must be mimicked in an artificial system. Replicating the complexity of the natural process is not a feasible approach. Moreover, direct replication of photosynthesis may yield a system with an equally low solar energy conversion efficiency (~2–3%). There is, therefore, a great need to develop systems which can replicate the core chemical process occurring in natural photosynthesis while increasing the yield of product from solar photonic activation. Light absorption and electron transfer are the first stages which should be considered in an artificial system. The light absorbing pigments, antennae and reaction complexes of the natural system are highly efficient and proceed under varying environmental conditions such as light intensity. The development of materials that are capable of absorbing light over a broad range of wavelengths and efficiently transferring electrons to excited states while preventing any form of recombination is a significant challenge. The generation of ATP in the natural system is one which perhaps does not have to be precisely replicated. However, the process by which ATP is generated is one of interest. The photolysis of H_2O to yield H_2/H^+ and O_2 has received significant attention in the field of photocatalysis and photoelectrochemistry with a view towards potential solar energy conversion and renewable fuel generation. Finally, carbon fixation in the Calvin cycle is perhaps the most appealing and significant area where attempts are being made to replicate the process. The utilization and potential sequestration of atmospheric CO_2 in a system which can also generate energy has become a global goal.

3 Artificial Photosynthesis

This area of research is becoming densely populated with published articles that investigate new and alternative approaches to achieving artificial photosynthesis. Generally, the aim of APS is the capturing and converting of light energy into fuels. Therefore, this chapter shall attempt to discuss the most significant approaches that are currently being developed. The reader is also directed to a number of excellent review papers which will give a broad appreciation of the scope of work being conducted in this area.[1–4,11]

3.1 *Light Harvesting*

A key stage in both natural and artificial photosynthesis is light harvesting and the process of 'capturing photons'. Nature achieves this by using light harvesting complexes within an organized array of pigments located within the chloroplasts. A range of pigments allows for a broad spectrum of light to be absorbed. The structure and chemical properties of the pigments and chloroplasts ensure both the movement of photons to the correct location and avoid any issues of overloading at high light intensities. In order to mimic this system, it is essential to first look at the nature of the solar electromagnetic spectrum. While the proportion of solar energy which reaches the surface of the Earth is sufficient for the photosynthesis process, there are advantages in concentrating the direct portion of the sunlight that falls on the earth to use in artificial photosynthesis. Diffuse sunlight is also available both for natural photosynthesis and APS.

A range of chemical compounds can be utilized for photon capture including modified photosynthetic pigments, dye molecules along with metal oxide and nitride semiconductors. This section shall focus on the use of photosensitizers that can be applied to APS including DSSC's and photocatalysts.

3.1.1 *Photosensitizers*

3.1.1.1 Dye sensitized solar cells

The dye sensitized solar cell comprises a photoelectrode constructed usually from glass coated with a conducting layer such as indium tin oxide (ITO) onto which a semiconductor photocatalyst film is deposited, most commonly titanium dioxide. A photosensitizing dye is adsorbed onto the semiconductor photocatalyst and the counter electrode, such as platinum, is separated from the photoelectrode by an electrolyte containing suitable redox system such as I_3^-/I^- system. When irradiated with light of a suitable wavelength, the photoexcited dye injects an electron into the semiconductor material which is then transported around the circuit to the counter electrode. At the counter electrode the tri iodide is reduced by two electrons to iodide. The iodide in the electrode subsequently donates an electron to the oxidised dye, regenerating the parent dye molecule (scheme 1 and Fig. 6).

An excellent review on photosensitizers for dye sensitised solar cells has been prepared by Basheer *et al.*[12]

$$\mathrm{Dye} + h\nu \rightarrow \mathrm{Dye}^* \quad (1)$$

$$\mathrm{Dye}^* + \mathrm{PC} \rightarrow \mathrm{PC}^- + \mathrm{Dye}^+ \quad (2)$$

$$\mathrm{I}_3^- + 2\mathrm{e}^- \rightarrow 3\mathrm{I}^- \quad (3)$$

$$\mathrm{Dye}^+ + 3/2\mathrm{I}^- \rightarrow \mathrm{Dye} + 1/2\mathrm{I}_3^- \quad (4)$$

Scheme 1

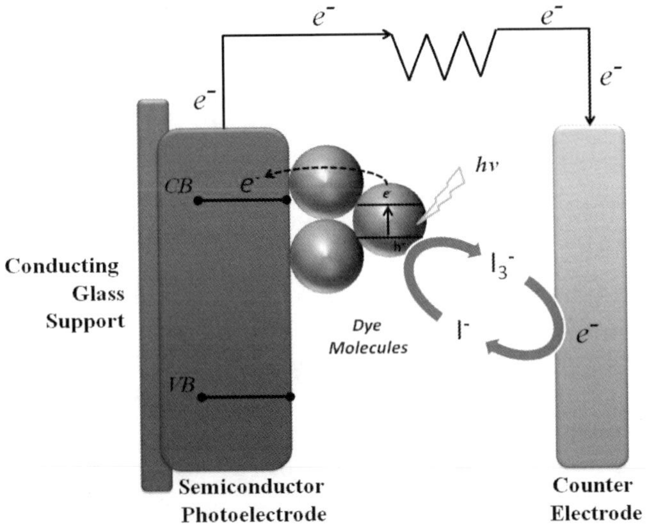

Fig. 6. Dye sensitized solar cell.

O'Regan and Graetzel[13] reported the development of the first DSSC in their seminal Nature paper in 1991. They reported the use of a ruthenium bipyridyl dye adsorbed to a 10-μm-thick, optically transparent film of titanium dioxide coated ITO glass photoelectrode. A conversion yield of light to electric current of 12% was obtained under diffuse daylight. Furthermore, the turnover number of the unit was reported to be over 5 million demonstrating the robustness of the overall system.

Since Graetzel's first report, research on ruthenium based dyes has probably been one of the most active and fruitful areas of research into photosensitizers for DSSCs. The most effective materials are the N3, N749, and N719 dyes which are all ruthenium organometallic materials based around the $RuL_xL_ySCN_2$. The ligands L and L′ are polypyridyl compounds.[12] The efficiencies of the N3, N719 and N749 dyes are reported as 8.2, 10.1, and 10.4 respectively under AM 1.5 irradiation.[12]

The use of the Chl derivatives and natural porphyrins, as photosensitizers on colloidal TiO_2 photoelectrodes was subsequently reported by Kay *et al.*[14] Energy conversions of 2.6% were obtained for copper Chl under simulated solar irradiation. In a study of different Chl derivatives it was found that free carboxyl groups were necessary for both adsorption and sensitization on TiO_2 photoelectrodes.

Campbell *et al.*[15] investigated the use of six porphyrin dyes as photosensitizers in DSSCs. Unfortunately, these proved to be less effective compared to ruthenium bipyridyl based dyes with overall efficiencies of less than 5% under AM1.5 solar conditions.

The development of four novel ruthenium based heteroleptic photosensitizers was reported by Lagref *et al.*[16] The most effective dye was the $[Ru(H_2dcbpy)(bhcbpy)(NCS)_2]$ derivative which demonstrated a conversion efficiency of 74% under AM1.5 solar irradiation.

Twenty natural dyes extracted from materials including leaves, fruits and traditional Chinese medicines, were assessed as photosensitizers for dye sensitised solar cells by Zhou *et al.*[17] The overall performance however was rather disappointing with incident photon to current conversion efficiencies (IPCE) ranging between 0.3 and just over 1% efficiencies. The most effective material was extracted from the rind of the mangosteen fruit but only demonstrated an IPCE of 1.17%.

The efficiency of natural dyes was also assessed by Hug *et al.*[18] They also concluded that the dyes extracted from natural materials had a significantly poorer performance than that obtained for synthetic dyes. While the best performing dyes were Chl derivatives extracted from spinach, they still only achieved an IPCE of around 4% under AM1.5 solar irradiation.

Park and Han described the application of a mixed TiO_2/Nb_2O_3 layer coated on the surface of nanoporous TiO_2 photoelectrodes using Ruthenizer 535-bisTBA (N719 dye) as a photosensitizer.[19] A conversion efficiency of 10.7% was obtained for this electrode compared to 7.9% which was obtained for the dye sensitized electrode which contained no over-layer. The enhanced efficiency was proposed to a number of factors including reduced recombination, enhanced light harvesting and electron collection efficiency.

Despite the extensive research on a broad range of dye sensitisers, to date the ruthenium based dyes remain the most efficient materials with conversion efficiencies of just over 10%.[12]

3.1.1.2 Photocatalysts

Photocatalysis is one of the primary methods adopted for artificial photosynthesis, especially as photosynthesis can be considered as oxidation and reduction half reactions; the initial photooxidation of H_2O to O_2 and H^+, followed by the reduction of CO_2 via carbon fixation. Researchers therefore have attempted to develop catalysts and reactor technology that can

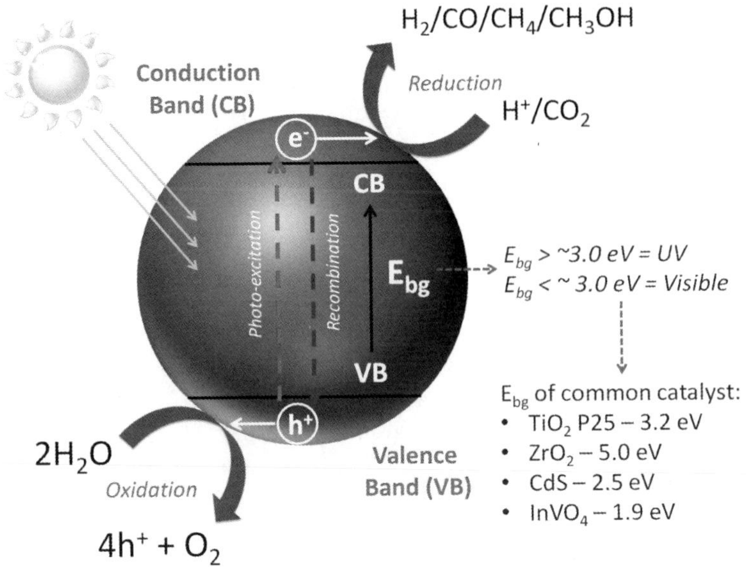

Fig. 7. Schematic representation of photocatalysis for artificial photosynthesis.

either split H_2O or reduce CO_2 along with a combined system which can completely replicate photosynthesis.

The chemical process of photocatalysis is shown in Fig. 7. The process of photocatalysis is initiated by the absorption of light energy and thus dictated by the electronic structure of the material. As the schematic shows, upon absorption, excitation occurs resulting in the formation of electron-hole pairs and subsequently redox reactions with adsorbed species on the surface. The E_{bg} indicates the level of energy required to initiate excitation with a wide band gap ($>\sim 3.0\,\text{eV}$) requiring UV light and a narrow band gap ($<\sim 3.0\,\text{eV}$) suited to the visible portion of the spectrum.

The challenge faced by researchers is maintaining the level of efficiency and stability observed in UV based catalysts, when using a visible catalyst where typically photocorrosion (CdS) and redox potentials are a significant issue. The typical catalyst engineering approaches adopted to narrow the band gap and increase solar photon absorption include; adjustment and modification to the conduction band (CB) and/or the valence band (VB) along with modulation of the CB/VB.

Despite high levels of H_2O oxidation achieved by a range of catalysts, a major limitation is often the requirement of UV photoactivation. The requirements for a visible light catalyst capable of stoichiometric H_2O oxidation are stringent and as such there are limited examples. Whilst there are examples of H_2 production under visible irradiation, the presence of a sacrificial electron donor (SED) is often required. This section shall discuss some promising catalysts that possess characteristics suitable for efficient photon capture in an APS.

While a great deal of focus has always been on metal oxides with a d^0 electron configuration, potentially of more interest for APS is the recent development of metal nitrides with a d^{10} electron configuration. The VBs of metal nitrides are composed of N2p, which has a higher energy level than that of O2p in metal oxides. Therefore the energy band gap (E_{bg}) is narrower, presuming the CB remains the same, which allows for excitation by photons with a wavelength >400 nm. Examples of metal nitrides being used include β-Ge_3N_4,[20] doped GaN and $(Ga_{1-x}Zn)(N_{1-x}O_x)$.[20–24] Maeda *et al.*[22] reported their samples of $(Ga_{1-x}Zn)(N_{1-x}O_x)$ to have an E_{bg} in the range of 2.6–2.8 eV, which is lower than that of the individual catalysts ZnO (3.2 eV) and GaN (3.4 eV) and low enough to permit excitation by absorbed visible photons. More recently, Lee *et al.*[24] have further reduced the E_{bg} to 2.2 eV by increasing the value of x to 0.87 to produce the catalyst $(Ga_{0.13}Zn)(N_{0.13}O_{0.87})$, which they concluded can significantly increase the absorption of solar photons. The catalyst was found to be active when used with both RuO_2 and Rh doped with lanthanides, producing stoichiometric quantities of H_2 and O_2 under visible illumination.[22,23]

3.2 *Electron Separation and Transfer*

Focus has been placed on single catalysts and Z-scheme mechanisms which are capable of converting light energy, particularly the solar spectrum, to chemical energy that can be harvested as fuel. Furthermore, this section will discuss materials that are deployed for H_2O oxidation, CO_2 reduction and combined systems.

As discussed previously, photocatalysts can be deployed as light capturing units. However upon excitation, charge separation becomes the primary consideration. Discussed here is work conducted towards increasing charged species mobility and electron transfer with a view to APS.

3.2.1 *Single oxide photocatalysts*

Since Fujishima and Honda's findings on the H_2O splitting capability of TiO_2 in 1972,[25] there has been a surge of publications on not only TiO_2

based materials but a range of photocatalysts. The breakthroughs in photocatalytic materials for APS have predominantly been achieved for the oxidation of H_2O, however more recently this has stimulated development also into CO_2 reduction. While TiO_2 remains a starting material for many novel catalyst materials, a range of oxide and nitride materials have been developed that show activity in both H_2O splitting and CO_2 reduction along with activation by solar photons.

Beginning with the work conducted into H_2O oxidation, a significant amount of catalyst engineering is focused on achieving the key features of natural photosynthesis; solar absorption and electron transfer through charge separation. While there are excellent examples of engineered materials capable of increasing efficiency through suppressed recombination, increased charge separation and the presence of separated active site, there remain a limited number of single catalysts capable of achieving this as a result of solar excitation.

In the field of photocatalytic H_2O splitting, compounds with a vacant *d* orbital (d^0) and filled *d* orbital (d^{10}) have been shown to be effective.[26–31] Specifically titantes, tantalates, niobates, vanadates and tungstanates with d^0 electronic configurations have been extensively researched along with metal oxides and nitrides with a d^{10} configuration.[26,27] The majority of these catalysts have shown increased efficiency as a result of crystal structure and bandgap engineering. Efficient catalyst structures for H_2O oxidation have included anatase, perovskite, layered structure, tunnel structure and cubic pyrochlore and within this a varied range of novel catalyst materials have been published. Despite the difference in crystal phase and chemical composition, the efficiency of these catalysts is often dependent on the presence of a metal oxide corner unit in the structure.[32] The incorporation of an MO_6 and MO_4 (M = transition/typical metal) octahedral and tetrahedral unit and moreover, the distortion, metal cation and the ratio of corner units present, lead to increased mobility and separation of charged particles.[33–35] The result is increased activity towards stoichiometric evolution of H_2 and O_2. The increased mobility is a result of the catalysts high electrochemical conductivity, allowing charged particles to reach an active site located on the catalyst surface.[36]

Research published by Inoue[27] and Sato *et al.*[33] also showed that the dipole moment could indicate the level of M–O–M bond distortion and found that non-active catalysts had a dipole moment of 0 and were distortion free. Inoue[27] suggested that the length of metal-oxygen bonds impacts photoexcitation. The poor symmetry formed as a result of distorted

Table 1: Increased H_2O splitting over catalysts with a distorted structure and metal cation corner unit.

Catalyst	Corner unit	Distortion: M–O–M angle (°C)	Distortion: Dipole moment (D)	Activity (μmol hr^{-1}): H_2	Activity (μmol hr^{-1}): O_2	Refs.
$MgGa_2O_4$	GaO_6	—	0	~0.8	0	27
$SrGa_2O_4$	GaO_4	—	0.8, 1.2	~7.5	~3.5	27
$BaGa_2O_4$	GaO_4	—	1.7, 1.1, 2.58	~3.5	~2	27
$SbNbO_4$	NbO_6	150	—	2.5	—	33
$SbTaO_4$	TaO_6	130	—	5.8	—	33
Y_2TiO_2	TiO_6	—	—	850	420	26
$Gd_2Ti_2O_7$	TiO_6	—	—	400	198	26
La_2TiO_7	TiO_6	—	—	307	152	26
La_3TaO_7	TaO_6	—	—	4	0	26
Gd_3TaO_7	TaO_6/GdO_6	—	—	2	0	26
Y_3TaO_7	TaO_8	—	—	Trace	0	26

octahedral and tetrahedral structures results in isolated orbitals, which gives rise to different photoexcitation efficiencies. Inoue also suggested that the local internal fields of the compounds, as a result of the dipole moment, could increase electron–hole separation and suppress recombination.

For the convenience of the reader, Table 1 shows a selection of catalysts that highlight the increased activity resulting from the presence of metal cation corner units and structure distortion.

While discussing photocatalysis, it is noteworthy to cover catalysts which are capable of both H_2O oxidation and CO_2 reduction with a view towards solar excitation. The development of such a catalyst would be a significant step towards the aims of APS. While a number of H_2O oxidation and CO_2 reduction catalysts have been reported separately, there are very few examples of a single catalyst capable of facilitating both reactions. This is primarily due to the significant challenges faced by a single photocatalyst, namely suitable band positioning, reaction competition and CO_2 solubility. Moreover, there is the added challenge of conducting this reaction under solar irradiation for use in APS. To date only two catalysts have been reported to achieve this under UV irradiation. Sayama and Arakawa in 1993[37] showed that with ZrO_2, H_2, O_2, and CO were generated at rates of 19.5, 10.8, and 2.5 μmol h^{-1} respectively while $BaLa_4TiO_{15}$ produced H_2, O_2, CO, and HCOOH at rates of 10, 7, 4.3, and 0.3 μmol h^{-1}. Sayama and Arakawa[37] suggested that the highly negative flat-band potential of ZrO_2 coupled with the wide band gap increased the activity. There are obvious limitations to both these systems beyond the low yield of products, including large band gaps between 4 and 5 eV which are not suitable for solar excitation and the presence of $NaHCO_3$ additives.

In 2011, a communication in the Journal of the American Chemical Society (JACS) from Cao *et al.*[38] detailed the potential of carbon nanoparticles for CO_2 reduction and H_2 generation. Functionalized carbon nanoparticles coated with either Ag or Pt showed absorption within the solar spectrum with a distinct peak at 550 nm. Using Ag-coated nano particles and under visible irradiation (425–720 nm), CO_2 reduction was achieved forming formic acid with a quantum yield of $\sim$0.3%. In a separate experiment (with no CO_2 present), Pt coated nanoparticles were found to be active for H_2 generation, although no H_2 quantification had been done at the time of the publication.

Recently one of the most promising single catalysts to be developed for potential applications in APS is $KTaO_3$ 'nanoflakes'.[39] The authors reported both the oxidation of H_2O and reduction of CO_2 simultaneously, with the corresponding products of O_2, H_2, and CO detected in a system that contained no electron donor (other than H_2O) and no electrical bias. The authors found that particle size and morphology were critical to photoactivity and specifically product selectivity. These factors were controlled through three synthesis routes including the solid-state reaction (SS-KTO), solvothermal methods in hexane–water (Hex-KTO) and pure ethanol (Eth-KTO). This catalyst, while still primarily requiring UV excitation, addresses a number of the key challenges faced in this field. Firstly, potassium tantalate possesses desirable band potentials for the reactions with a bottom CB edge at -0.7 eV and top VB edge at 2.9 eV versus NHE at pH 0, giving a band gap of 3.6 eV. This provides a balance between utilising a minor proportion of solar photons, while maintaining the potential required for H_2O oxidation and CO_2 reduction in the absence of a sacrificial electron donor. Secondly, proton reduction is the major competing reaction to CO_2/CO conversion. Proton reduction, while proceeding in a similar reaction mechanism to CO_2 reduction, is a more favorable reaction due to the redox energies involved; -0.41 V versus NHE at pH 7 for H_2/H_2O, while -0.51 V versus NHE at pH 7 for CO/CO_2. Moreover, the localized concentration of H_2O molecules is high in comparison to CO_2 as a result of low CO_2 solubility. Previously, Handoko and Tang[40] demonstrated that altering the facet of Cu_2O modified product selectivity. The low index facet $\{100\}$ was found to exhibit higher CO_2 reduction activity over that of high index facets $\{111\}$ due to the presence of active sites that favor CO_2

reduction. In their more recent work, Li and colleagues[39] found loading a co-catalyst (Ag) introduced CO_2 favorable reduction sites, while preventing recombination. The results indicated that in the presence of Ag-$KTaO_3$, CO_2 reduction doubled with a decrease in H_2 also noted.

3.2.2 *Z-scheme photocatalysts*

The alternative approach to a single catalyst mechanism is the Z-scheme system which is capable of mimicking photosynthesis more closely,[41–44] Fig. 8. The presence of both an oxidizing and reducing catalyst means the challenging requirements faced by a single catalyst are alleviated. This dual catalyst approach allows the utilization of a wider range of visible light materials as the change in Gibbs free energy required to drive each photocatalyst is reduced when compared to the one step system.[31] In order for this system to operate, an electron mediator is required. Typical mediators that are reported include IO_3^-/I^-[45] and Fe^{3+}/Fe^{2+}[46,47] which act as redox shuttles to facilitate the reactions.

A number of Z-scheme systems have been reported in relation to H_2O oxidation to generate the production of O_2 and H_2 under irradiation by

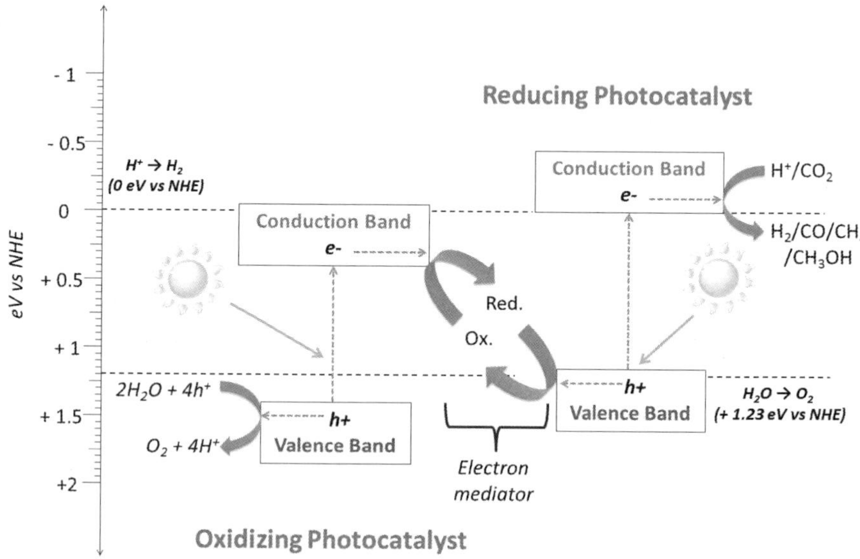

Fig. 8. Schematic representation of Z-scheme photocatalysis for artificial photosynthesis (reproduced from Ref. 31, with kind permission of Springer).

UV and visible light; Sayama and colleagues reported the production of O_2 and H_2 over WO_3 and Cr/Ta-$SrTiO_3$ respectively using an iodate/iodide redox couple under visible illumination.[43,44] Using the electron mediator Fe^{3+}/Fe^{2+}, the splitting of H_2O was also achieved under visible irradiation using $BiVO_4$ and Rh-$SrTiO_3$ for O_2 and H_2 generation respectively.[46] In addition to solely utilizing Z-scheme for water splitting, the combined reduction of CO_2 with this approach is now also being reported.[48] Of particular interest is the deployment of a 'twin' reactor that shows great potential for APS.[48–50] The twin reactor utilizes a modified Z-scheme approach, where an oxidizing and reducing catalyst are located in two reactor compartments separated by a nafion membrane (see Fig. 9). A number of papers have been published which utilise this reactor for H_2O splitting due to the fact that it allows for separate evolution of both O_2 and H_2.[49] For a more detailed insight into membrane reactors, a number of designs for H_2 production have recently been reviewed by Molinari *et al.*[48] More recently, a twin membrane reactor was deployed for H_2O oxidation and CO_2 reduction.[50]

In 'compartment 1', the oxidizing catalyst photocatalytically splits H_2O to generate O_2 and H^+ in the presence of an electron mediator. The O_2 can be safely removed from the compartment while the H^+ travels across the

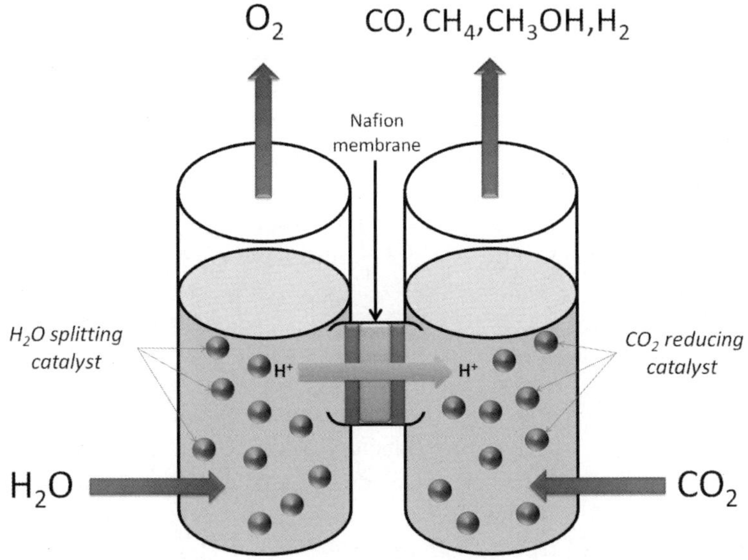

Fig. 9. Schematic representation of 'Z-scheme' reactor for H_2O splitting and CO_2 reduction (reproduced from Ref. 50, with kind permission of Elsevier).

membrane to 'compartment 2' to act as a source of H_2 for the hydrogenation of CO_2 in the presence of a CO_2 reducing catalyst. In compartment 2 products such as CO, CH_4, CH_3OH, and H_2 can be generated. This Z-scheme approach and reactor design have a number of advantages; separate evolution streams of O_2 and H_2 are generated preventing any back reactions, no 'extra' H_2 is required to be injected into the system as H_2 produced from H_2O splitting can be used directly in CO_2 hydrogenation and the removal of O_2 from the systems prevents any oxidation of hydrocarbon products. Lee *et al.* recently found that the combination of Pt-$CuGaAlO_4$ and Pt-$SrTiO_3$:Rh for photo reduction along with WO_3 for photo oxidation was effective. The efficiency of the system was a result of increased H_2 evolution from Pt-$SrTiO_3$ which further increased the photo reduction of CO_2 over Pt-$CuGaAlO_4$. The photocatalytic quantum efficiency (PQE) of the system was calculated to be 0.0051%, which is still low in comparison to natural photosynthesis. However, the authors concluded that with optimization of the system, the PQE could be significantly increased.

3.2.3 *Nanobiocatalytic assemblies*

An interesting area of artificial photosynthesis is the application of nanobiocatalytic assemblies to achieve an efficient system. While previous approaches discussed are focused on a single key process e.g. photocatalysis, nanobiocatalytic assemblies can be considered as a hybrid approach to APS which often combines biocatalysis and photocatalysis. Such systems have recently been reviewed by Kim *et al.*,[8] in which a number of approaches combine a range of biological and chemical processes. The key parameter to be achieved in such a system is sufficient charge separation between key components. In view of this, the following section briefly discusses key publications of nanobiocatalytic assemblies that have been developed to mimic PS I and II.

3.2.3.1 Mimicking PS I

Primarily, both PS I and II are reaction centers which allow efficient electron transfer along an electron transport chain. Upon excitation in PS I, photoexcited P700 will transfer electrons to ferredoxin which eventually is picked up by NADP/NADPH. This process is conducted through a number of electron acceptors to enhance the rate of charge separation. The fundamental step here is the initial excitation of P700 which allows for rapid electron transfer. In order to mimic this, APS have investigated the coupling of sensitisers to electron acceptors to encourage rapid transfer of charged species. This has been achieved through a number of designs published in the literature including; graphene based systems, CdS quantum dot-TiO_2 nanotube arrays, synthetic woods and diphenylalanine self-assemblies.[8]

CdS quantum dot sensitized TiO_2 nanotubes were investigated in 2011 by Ryu and colleagues who presented a number of interesting findings.[51] Various scenarios were investigated including the use of Al_2O_3 as nanotube arrays and nanoparticle in place of nanotubes. Ryu *et al.* found that a high electron–hole recombination rate resulted in CdS–Al_2O_3 showing minimal activity towards NADH regeneration. In contrast, under optimum conditions CdS–TiO_2 nanotubes achieved ~50% NADH regeneration which Ryu *et al.* proposed was due to rapid charge separation at the CdS–TiO_2 interface. Upon electron injection from CdS to TiO_2, electrons were transferred to mediators, which acted as a redox cycle, resulting in NADH regeneration. In the CdS–Al_2O_3 system, however, the injection of electrons into the CB of Al_2O_3 was suppressed due to the unfavorable thermodynamics and thus recombination increased. Using a nanoparticle film in replace of nanotubes showed a four-fold reduction in NADH regeneration. It was found that the morphology and structure of nanotubes over nanoparticles is optimal for electron collection and transfer. The rate of reaction species diffusion through the nanotube assemblies was also proposed to be greater than that of the nanoparticles, which would increase NADH regeneration.

Another study of particular interest is the one conducted by Yadav *et al.*[52] which deployed a graphene based photocatalyst coupled with an enzyme in order to mimic photosynthesis. The electrochemical properties of graphene are well known. However, there are two key features which make it an excellent material for APS; a large surface area and high carrier mobility, which permits multiple electrons to be transported. Graphene, a single-layered sheet of sp^2-bonded carbon atoms, was used in the form of graphene oxide and coupled to multi-anthraquinone substituted porphyrin to demonstrate excellent charge separation and mobility along with CO_2 reduction for the production of solar fuel. A major challenge in any photocatalyst–biocatalyst system is the efficient harvesting of visible light that can trigger electron transfer, resulting in the desired enzymatic reaction taking place. Yadav *et al.* showed that graphene oxide is an excellent visible light catalyst which allows for NADH regeneration, thus initiating the enzymatic reduction of CO_2 to formic acid. The performance of the catalyst was measured in comparison to a previous photocatalyst used by the same group, namely $W_2Fe_4Ta_2O_{17}$. In relation to both yield of NADH and formic acid the graphene-based system outperformed the previous catalyst; 45.54% NADH yield and 110.55 μmol of formic acid in comparison to 14.50% NADH yield and 14.25 μmol of formic acid for the $W_2Fe_4Ta_2O_{17}$ system. The properties of graphene allowed for increased electron movement along the photoexcited chain which accelerated the NADH regeneration process. The authors also concluded that the graphene acted as an electron reservoir, which aided in the transport of multiple electrons.

3.2.3.2 Mimicking PS II

PS II is the initial stage in the electron transfer process in photosynthesis and ultimately provides the electrons for the process to occur. The excitation of electrons from light absorption initiates the transfer of electrons to the first acceptors in the transport chain. Coupled with PS II is the oxygen evolving center (OEC), which replenishes the loss of excited electrons through the splitting of H_2O. As discussed previously the replication of an OEC has been a focus point for many areas in an attempt to replicate photosynthesis and generate H_2 as a solar fuel. Efficient electron transfer in this aspect of artificial photosynthesis is a key parameter as often low quantum yields are reported as a result of charge species recombination occurring at a faster rate than the catalytic four-electron oxidation of H_2O. This issue was addressed by Zhao and colleagues in their study in 2012,[53] focusing on DSSCs coupled with a biomimetic electron mediator. The electron transport assembly constructed by Zhao *et al.* included the anchoring of benzimidazole-phenol (BIP) to IrO_2 nanoparticles by 2-carboxyethylphosponic acid (CEPA), which was then linked to a TiO_2 photoelectrode. The presence of the 3p-Ru dye linked to the TiO_2 electrode acted as the sensitiser. The purpose of BIP in this assembly was to mimic the tyrosine–histidine pairing in the natural process. The PS II system has two redox active amino acids, tyrosine and histidine, which aid the movement of electrons and protons from the $CaMn_4O_5$ catalyst in the OEC to the photo excited electron P680. This process occurs through the deprotonation and protonation of the phenyl and imidazole groups of tyrosine and histidine. In this instance, the BIP-IrO_2 assembly accelerated the rate of electron injection by increasing electron transfer between Ir (IV) and Ru (III). The results showed that in the absence of the BIP mediator, the quantum yield was less than half that of the BIP based system. Under optimum conditions the system produced an internal quantum efficiency of ~2.3% under blue light.

An interesting approach to light harvesting and charge separation was presented by Nam *et al.* in 2010[54] using a virus template assembly of porphyrins. The use of biological material as a template or 'scaffold' is advantageous due to pre-existing chemical bonds, electrostatic interactions and precise nanoscale organization. Of particular interest as a scaffold are viruses which posses a highly organized protein coat structure. Nam *et al.* found that the M13 viruses were suitable for use in an APS, as the proteins present in the coat are easily modified chemically. The research group chose the porphyrins Zn(II) deuteroporphyrin IX 2,4-bis(ethylene glycol) (ZnD-PEG) as the model pigment for light absorption and were conjugated to the M13 virus coat via a carbodimide coupling reaction. Using inductively coupled plasma-atomic emission spectrometry, it was found the between ~1564 and 2900 porphyrins were conjugated per virus sample. The results indicated that the photons absorbed by the pigments were transported through long range dipole-dipole interactions and were quenched in trap sites that were generated as a result of the pigment binding to the virus coat. Furthermore, the group hypothesized that the insertion or deletion of amino acids on the virus coat would alter the distance of interactions and could further increase electron transport and thus energy transfer.

In another paper Nam *et al.*[55] applied the virus templated system to H_2O splitting, by the addition of a co-assembly on the M13 coat of zinc pyrophyrins and IrO_2 nanoparticle clusters. During their work, the group found the turnover number increased to ~796 for H_2O oxidation under visible light. The incorporation of an immobilization matrix improved the structural integrity of the nanoscale construct, producing not only a robust light harvesting unit but a catalyst capable of being recycled.

3.3 *Carbon Fixation*

The 'dark' reactions that occur in photosynthesis result in the fixation of CO_2 from the atmosphere to produce a usable source of energy, $C_6H_{12}O_6$. The driving force for this aspect of the process arises from H^+ carried by NADPH and an energy source produced by ATP via ATPase in the 'light' reactions. The reduction of CO_2 can be broadly divided into heterogeneous and homogenous systems. In a similar approach to H_2O oxidation, heterogeneous systems in APS are frequently photocatalytic based. However, despite significant advances in H_2O oxidation, the reduction of CO_2 over single or dual system catalysts remains a great challenge in artificial photosynthesis. CO_2 reduction faces a number of challenges which limit the overall conversion efficiency; the inability of photocatalyst materials to absorb the visible portions of the solar spectrum, insufficient charge separation, low solubility of CO_2 in H_2O (~33 μmol in 1 mL of H_2O at 100 kPa and room temperature),[56] CO_2 back reactions and competitive H_2 production from H_2O. Furthermore, as a range of products are generated from CO_2 reduction, with a number of multi-electron steps, the kinetic problem is significant.

In addition to the use of heterogeneous photocatalysts such as semiconductor materials, the photocatalytic reduction of carbon dioxide has also been achieved with a range of homogeneous photocatalysts. Some of the most extensively researched materials derive from ruthenium or rhenium bipyridyl complexes which act as either direct or indirect photosensitizers for the photocatalytic reduction of carbon dioxide.

3.3.1 *Heterogenous catalysts*

Since Inoue *et al.*[57] demonstrated the photoreduction of CO_2 with various catalysts using an electrical bias in 1979, a range of metal and mixed metal oxides have been tried as catalysts.[56–59] The mechanism of CO_2 reduction by heterogenous catalysis was discussed by Eggins *et al.* in 1998[60] and is summarized in Figs. 10 and 11. The one electron step (Fig. 10) involves the formation of a CO_2 radical anion that is capable of undergoing disproportionation, protonation and dimerization to produce CO, carbonate and oxalate respectively.

The 2 electron mechanism (Fig. 11) demonstrates the additional products formed from further reduction of oxalate. These products include glyoxylate, glycolate, tartrate and malate.

The three crystal phases of Ti-based materials, anatase, brookite and rutile, have received considerable interest. Work conducted by Liu *et al.*[61]

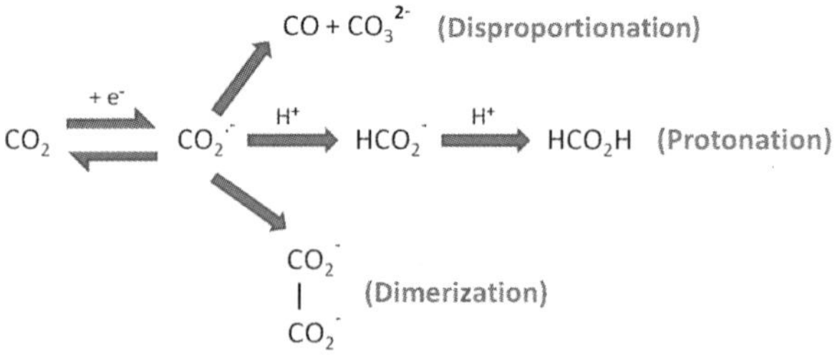

Fig. 10. Scheme showing the 1 electron CO_2 reduction mechanism (reproduced from Ref. 60, with kind permission of Elsevier).

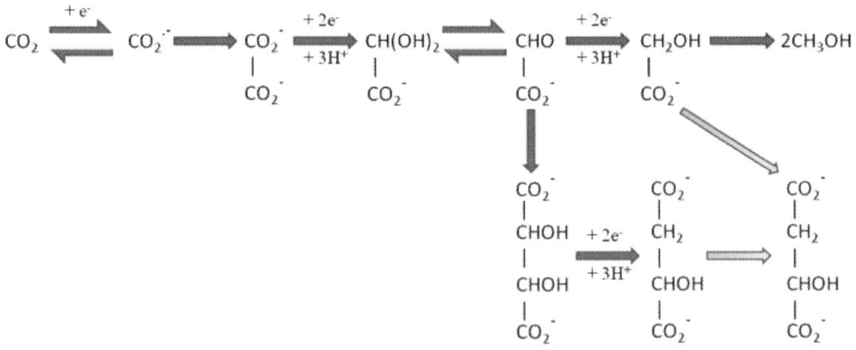

Fig. 11. Scheme showing the 2 electron CO_2 reduction mechanism (reproduced from Ref. 60, with kind permission of Elsevier).

concluded that the brookite phase of TiO_2 showed the highest activity of CO_2 reduction to CO and CH_4. In comparison, anatase showed minimal activity. However, anatase possessed a far more suitable band gap for visible light excitation. The most promising material was the mixed-phase TiO_2 which showed desirable properties for artificial photosynthesis; namely increased visible light harvesting and CO_2 reduction potential along with increased charge separation as a result of the rutile-anatase junction.

The deposition of a co-catalyst on a Ti-based catalyst has also showed both increased levels of CO_2 reduction along with product selectivity. A range of co-catalysts have been reported including Ag, Au, and Pt,[56,62,63] which are primarily deployed as a method for suppressing electron-hole recombination. Zhang *et al.* along with Vijayan and colleagues both demonstrated the impact of Pt on titania nanotubes (TNT) for use within visible light photocatalysts. Recently, Collado *et al.*[64] described the synergetic effect of Ag nanoparticles deposited onto TiO_2 for CO_2 reduction. The system used water vapour as an electron donor and found that upon the bare catalyst surface only CO and H_2 were produced, while CO, H_2 and a series of short chain hydrocarbons were formed by the addition of Ag nanoparticles. Under optimum conditions CH_4, C_2H_6, C_3H_8, and C_3H_6 were formed as a result of photo-excited electrons being injected into Ag particles on the catalyst surface. The increased hydrocarbon production was the result of the scavenging properties of Ag which prevented electron–hole recombination. Furthermore, the authors found that the coupling of TiO_2 and Ag shifts the apparent Fermi level, making the catalyst more reductive. The catalyst also showed a favored approach to the multi-electron mechanism for hydrocarbon formation, which is likely a result of the initially produced CO and H_2 forming $-CH_2-$ species that become building blocks for the further hydrocarbon chain formation. A similar observation was discussed by Ohno *et al.*[65] when using a hybrid composition of g–C_3N_4 and WO_3 that also used Ag as a co-catalyst.

A favorable approach to Ti based catalysts is the construction of TNT, due to the large surface area, reduced grain boundaries and facile charge transport paths of 1D nanomaterials.[66,67] The use of TNT can also increase the adsorption of CO_2 which is a major limiting factor in CO_2 reduction. Pt loaded TNT was found to increase CO_2 adsorption ability as a result of the synergetic effect of tubular morphology and mixed-valence states from the Pt.[68] TNT has also shown activity towards visible light harvesting as a result of increasing annealing temperature. Upon increasing the annealing temperature from 550 to 680°C, the rate of CH_4 production increased from 0.26 to 0.79 μmol m^{-2} hr^{-1} respectively. Varghese *et al.*[69] also showed visible light activation of TNTs under solar irradiation, as a result of N-doping. The results of the reduction, which produced a range of hydrocarbons, concluded that the presence of both Pt and Cu on the TNT surface was required to produce the optimum rate of 111 ppm cm^{-2} h^{-1}. N-doped mesoporous TiO_2 was also found to be active under visible light when loaded with noble metals such as Pt, Au and Ag. The higher work function of Pt, which facilitated the transfer of electrons from TiO_2 to Pt particles, generated a CH_4 yield of 2.9 μmol m^{-2} h^{-1}.

The search for a photocatalyst that posses a suitable band gap for visible and solar light harvesting is challenging, as the majority of catalysts which meet this criteria are unstable. Examples include CdS (2.5 eV) and Cu_2O (2.14 eV), both of which are subject to photocorrosion as a result of oxidation and reduction by the photogenerated charge carriers. The production of hydrocarbons from these photocatalysts has been found to reach maximum concentrations after a short time period and then decline due to the decreasing stability of the catalyst. Despite attempts to introduce a protective heterostructure layer to these catalysts using Al-doped ZnO, TiO_2 and RuO_2, the practical applications of using such catalysts in APS are limited.

A small number of non-Ti based visible light activated catalysts have also been reported to be photoactive; namely $BiVO_4$, $LaCoO_3$, $InTaO_4$, and N-doped Ta_2O_5. Jia *et al.*[70] reported a maximum yield of 19 and 128 μmol g^{-1} h^{-1} for formaldehyde and formic acid respectively over C and Fe co-doped $LaCoO_3$, under irradiation from a 125 W Xe lamp equipped with a 400 nm cut off filter. While a number of authors report the formation of CO and hydrocarbon products using these catalysts, it is often in the presence of an electron donor, which in itself presents a potential issue. It is favorable to use alcohols as hole scavengers, due to their ability to extract the photoinduced holes. Subsequently, however, the scavengers may oxidise themselves into CO_2, CO and other hydrocarbon products. Therefore, the formation and origin of photocatalytic products in these reactions is debatable.

An interesting bio-photocatalytic approach to CO_2 photoreduction is the coupling of dye sensitised TO_2 particles to a CO_2 reducing enzyme.[71] The CO_2 reducing enzyme CODH I from the anaerobic microbe *Carboxydothermus hydrogenoformans* (*Ch*) was selected due to its unique Ni_4Fe–4S active site which is capable of catalysing CO_2 in both directions (oxidation and reduction). The RuP sensitized TiO_2 nanoparticles functionalized with *Ch* CODH I were capable of generating CO from CO_2 using a two electron mechanism which is more thermodynamically favorable (E = −0.46 V versus SHE at pH 6) than the one electron radical mechanism (E = −1.90 V versus SHE). Under visible light (cut off filter at 420 nm) the system produced ∼5 μmol of CO after 4 h irradiation. Experiments conducted under solar irradiation also generated promising results, with CO production reaching ∼60% of the system under artificial visible light. The movement and injection of electrons from one component to another in this system suppressed recombination and ensured sufficient charge separation which is fundamental for CO_2 reduction to occur.

Previously mentioned in this chapter was the growing focus on graphene-based compounds for use in artificial photosynthesis assemblies. The attractive properties of graphene such as accelerated electron transfer and suitability for solar absorption make it a viable option for photocatalytic CO_2 reduction. The increased electron mobility increases charge separation, which increases the chances of trapping electrons at active sites allowing for maximum interaction with CO_2. Robust hollow spheres with alternating ultrathin sheets of titania ($Ti_{0.91}O_2$) and graphene were developed by Tu *et al.*[72] In comparison to TiO_2 and P25, a nine-fold improvement in CO_2 reduction (transformation to CO and CH_4) was recorded for G-$Ti_{0.91}O_2$ which is due to a number of properties. These include the rapid movement of charge carriers across the nanosheet structures along with injection into the graphene to increase charged species lifetime which in turn facilitates their participation in the CO_2 reduction mechanism. Furthermore, the authors also stated that the hollow spheres may act as 'photon trap-wells' that encourages the multiple-scattering of incident light to increase absorption. Graphene-TiO_2 thin films were synthesised for the reduction of CO_2 to CH_4 in H_2O by Liang *et al.*,[73] and showed a seven-fold improvement over bare TiO_2 films under visible light illumination.

3.3.2 *Homogenous catalysts*

There has been an extensive amount of research on the application of a number of homogeneous photocatalysts for conversion of CO_2 to potential fuel products. These fall into two main categories; the type I catalysts in which a dye such as a ruthenium bipyridine compound absorbs light and accepts an electron from a SED such as triethanolamine (Scheme 1). The reduced dye transfers an electron to a catalyst material that then reduces the CO_2 to subsequent products including carbon monoxide, methanol, formaldehyde and methane. With the type II mechanism on photoexcitation the sensitiser again reacts with a SED but this time the reduced photosensitiser directly transfers the electron to CO_2 (Scheme 2).[74]

$$\mathrm{PS} + \mathrm{h}\nu \rightarrow \mathrm{PS}^*$$
$$\mathrm{PS}^* + \mathrm{SED} \rightarrow \mathrm{PS}^-$$
$$\mathrm{PS}^- + \mathrm{Cat} \rightarrow \mathrm{PS} + \mathrm{Cat}^-$$
$$\mathrm{Cat}^- + \mathrm{CO_2} \rightarrow \mathrm{Cat} + \mathrm{CO_2^{\cdot -}} \rightarrow \mathrm{Products}$$

Scheme 1 — Type I Mechanism

$$\mathrm{PS} + \mathrm{h}\nu \rightarrow \mathrm{PS}^*$$
$$\mathrm{PS}^* + \mathrm{SED} \rightarrow \mathrm{PS}^-$$
$$\mathrm{PS}^- + \mathrm{CO_2} \rightarrow \mathrm{CO_2^{\cdot -}} \rightarrow \mathrm{Products}$$

Scheme 2 — Type II Mechanism

The use of Rh bipyridyl complex as a photosensitizers and C0 (II) ions as catalysts for the reduction of CO_2 to CO under visible light irradiation in acetonitrile was reported by Hawecker *et al.* in 1983.[75] They also reported the use of a Re (bipy)$(CO)_3$X (where X = halide) systems as a so called "single component" photosensitizer and catalyst system for the reduction of CO_2 to CO. The latter system was demonstrated to have a high level of selectivity for CO production over H_2 generation, with virtually no hydrogen being detected.

Matsuoka *et al.*[76] reported the reduction of CO_2 to CO, HCO_2^- and H_2 using electron mediation of cobalt macro cycle complexes, such as Cyclam, with p-Terphenyl photocatalysts in methanolic acetonitrile in the presence of tertiary amine SED. Under UV irradiation at 313 nm a maximum yield of 24 μmol of CO was obtained after 1 h reaction time using triethylamine as the SED. The effect of seven different amines as SEDs on the product yield was investigated. It was found that different amines as SEDs had a marked effect on the CO yield with 63 μmol of CO being generated when Tri-2-propanolamine was employed compared to 21 μmol of CO being generated in the presence of a Tri-n-butylamine electron donor.[76]

The use of cobalt-cyclam complexes as catalysts for carbon dioxide reduction has also been reported by Ogata *et al.*[77] They described the use of this complex in combination with a phenazine photosenstizers and also generated CO, HCO_2^- and H_2 as photoreduction products. In each case HCO_2^- was the predominant photoproduct with up to 28 μmol being generated compared to 1.6 and 0.15 μmol of CO and H_2 respectively being produced using a 2-ethylphenazine photocatalyst under UV irradiation at 313 nm.[77]

p-Terphenyl has also been used as a photosensitizer with cobalt corrin catalysts for CO_2 reduction. In the presence of a triethylamine SED, cobinamide produced an initial rate of formation of 0.48 and 0.30 μmol L^{-1} h^{-1} of CO and H_2 respectively in a 9/1 MeCN/MeOH solvent. In this case the rate of CO production was over 10 times higher than that obtained when cobalt tetra-m-tolylporphyrin was used as a catalyst.[78]

The photocatalytic reduction of CO_2 to CO in a high pressure system has been reported by Hori *et al.*[79] using a Rhenium bipyridyl complex. When CO_2 was irradiated for 24 h in the presence of [*fac*-Re (bpy)$(CO)_3P$ $(O^iPr)_3]^+$ at 1.36 MPa in a DMF/triethanolamine solvent, 348.3 μmol of CO were generated compared to 90.5 μmol CO at a pressure of 0.1 MPa. It was proposed that at higher pressures there was a greater dissolution of CO_2 into the solvent which increased the likelihood of an interaction with the catalyst.[79]

The utilization of supramolecular photocatalysts based on Ru/Re binuclear complexes combined with a 1,3-bis(4′-methyl-[2,2′]bipyridinyl-4-yl0-propan-2-ol (or bpyC3bpy for short) bridging ligand were reported to demonstrate a high level of efficiency for the reduction of CO_2 under visible light >500 nm. Using a $[Ru\text{-}ReP(OEt)_3]^{3+}$ photocatalyst a quantum yield of 0.21 and turnover number of 232 was obtained for the photoreduction of CO_2 to CO.[80]

4 Artificial Photosynthesis Institutes

The drive to establish an artificial photosynthetic system has led to the development of dedicated institutes which conduct research into various aspects of APS. Two such institutes highlighted in this chapter are the Swedish consortium for artificial photosynthesis and the JCAP.

4.1 *Swedish Consortium for Artificial Photosynthesis*

The Swedish Consortium for Artificial Photosynthesis was established in 1994 and over the past 20 years has conducted research into a broad range of fields including; molecular biology, biophysics, biochemistry along with organometallic and physical chemistry. The objective of the multi-disciplinary institute is the development of alternative fuels (H_2 and hydrocarbon based fuels) from solar energy via two approaches; artificial photosynthesis and photobiological systems.

Of particular interest within this consortium is the work conducted into charge separation through the development of an accumulative electron transfer[81] and a proton-coupled system.[82] Karlsson and colleagues[81] addressed the significant challenge of combining the initial photo adsorption process that generates a single electron–hole pair with the multi-electron steps involved with H_2O splitting and fuel production. Their method of charge accumulation was based around the coupling of a dye to TiO_2 nano particles; oligo(triarylamine)ruthenium(II) polypyridine anchored to TiO_2 (OTARu-TiO_2). The regenerative molecular system was capable of absorbing two photons for the production of two electron hole pairs. This attribute allows for a high yield accumulation of redox equivalents in a single linked system.

4.2 *Joint Centre for Artificial Photosynthesis (JCAP)*

JCAP is a dedicated research institute with the goal of developing cost-effective technology for solar fuel generation. The institute was established in 2010 as an Energy Innovation Hub under the U.S. Department of Energy (DOE) with a view towards utilizing only natural light, H_2O and CO_2 to produce a sustainable fuel. The group has dedicated research into light capture and conversion, heterogenous and molecular catalysis, molecular and nanoscale interfaces along with membrane and mesoscale assembly. Furthermore, while the majority of work conducted in this field is on a small lab scale, JCAP are also actively researching the potential scale up and prototyping of APS.

The volume of research and published output by JCAP is extensive and covers a broad range within artificial photosynthesis. The work conducted by Nate Lewis and his group at CalTech continues to be at the forefront of APS technology. Recent noteworthy publications include the use of cobalt complexes for electrocatalytic CO_2 reduction[83] stabilization of semiconductor photoanodes for H_2O oxidation[84] and the catalytic engineering of $ZnSn_xGe_{1-x}N_2$ from 2.0 to 3.1 eV.[85] Nitride based catalysts have received a great deal of interest in both H_2O splitting and APS due to the desirable band structure they posses which is favorable towards solar photon absorption.[20–24]

Of particular interest was the use of Si, GaAs and GaP as photoanodes by Hu *et al.*[84] These catalysts typically have a more favorable E_{bg} and VB potential than metal oxides for solar activated H_2O oxidation. However, they are susceptible to photo corrosion which limits their use. Hu and colleagues grew TiO_2 coatings by atomic layer deposition that not only stabilized these compounds but created electronic defects that increased hole conduction thus preventing recombination.

5 Future Outlook and Challenges

This chapter has attempted to provide a concise overview of artificial photosynthesis and the recent developments made in the field. Upon review, it is clear that while significant progress has been made in key areas such as light harvesting, charge separation and nanoscale system construction, there remain a number of crucial challenges yet to be overcome.

Challenges faced by research groups in this field include the development of robust light harvesting constructions, synthesis of efficient H_2O splitting catalysts and electron mediators along with improved CO_2 reducing catalysts. The efficiency of photosynthesis is credited to a number of processes but the photoexcited chain which carries electrons is key and must not be overlooked in artificial systems. Improving H_2O splitting to ensure a supply of electrons into PS II along with preventing recombination through efficient electron mediators and acceptors will drastically improve the overall performance of APS.

In the view of the authors an area which shows high potential is that of nanobiocatalytic research and the combination of biological and chemical processes. The requirements to achieve a reproducible photosynthetic system over a single compound are too demanding and therefore hybrid systems present a potentially more energy efficient system. This is mirrored by the work from institutes such as JCAP and the Swedish Consortium for

Artificial Photosynthesis where the coupling of regenerative photosensitizers to catalysts that can increase charge separation can be achieved.

References

1. A. C. Benniston and A. Harriman, Artificial photosynthesis, *Materials Today* **11**(12) (2008), pp. 26–34.
2. T. J. Meyer, Chemical approaches to artificial photosynthesis, *Acc. Chem. Res.* **22**(5) (1989), pp. 163–170.
3. D. Gust, T. A. Moore and A. L. Moore, Solar fuels via artificial photosynthesis, *Acc. Chem. Res.* **42** (2009), pp. 1890–1898.
4. N. S. Lewis and N. D. Nocera, Powering the planet: Chemical challenges in solar Energy utilization, *Proc. Natl. Acad. Sci. U.S.A.* **103** (2006), pp. 15729–15735.
5. K. Arifin, E. H. Majlan, W. R. W. Daud and M. B. Kassim, Bimetallic complexes in artificial photosynthesis for hydrogen production: A review, *Int. J. Hydrogen Energ.* **37** (2012), pp. 3066–3087.
6. M. Lee, J. H. Kim, S. H. Lee, S. H. Lee and C. B. Park, Biomimetic artificial Photosynthesis by light-harvesting synthetic wood, *ChemSusChem.* **4** (2011), pp. 581–586.
7. A. Magnuson, M. Anderlund, O. Johansson, P. Lindblad, R. Lomoth, T. Polivka, S. Ott, K. Stensjo, S. Styring, V. Sundstrom and L. Hammarstrom, Biomimetic and Microbial approaches to solar fuel generation, *Acc. Chem. Res.* **42**(12) (2009), pp. 1899–1909.
8. J. H. Kim, D. H. Nam and C. B. Park, Nanobiocatalytic assemblies for artificial photosynthesis, *Curr. Opin. Biotechnol.* **28** (2014), pp. 1–9.
9. K. N. Ferreira, T. M. Iverson, K. Maghlaoui, J. Barber and S. Iwata, Architecture of the photosynthetic oxygen-evolving center, *Science* **303**(5665) (2004), pp. 1831–1838.
10. K. Kalyanasundaram and M. Graetzel, Artificial photosynthesis: Biomimetic Approaches to solar energy conversion and storage, *Curr. Opin. Biotechnol.* **21**(3) (2010), pp. 298–310.
11. D. G. Nocera, The artificial leaf, *Acc. Chem. Res.* **45**(5) (2012), pp. 767–776.
12. B. Basheer, D. Mathew, B. K. George and C. P. Reghunadhan Nair, An overview on the spectrum of sensitizers: The heart of dye sensitized solar cells, *Solar Energ.* **108** (2014), pp. 479–507.
13. B. O'regan and M. Gratzel, A low-cost, high-efficiency solar cell based on Dye-sensitized colloidal TiO_2 films, *Nature* **353**(6346) (1991), pp. 737–740.
14. A. Kay and M. Graetzel, Artificial photosynthesis. 1. Photosensitization of Titania solar cells with chlorophyll derivatives and related natural porphyrins, *J. Phys. Chem.* **97**(23) (1993), pp. 6272–6277.
15. W. M. Campbell, K. W. Jolley, P. Wagner, K. Wagner, P. J. Walsh, K. C. Gordon, L. Schmidt-Mende, M. K. Nazeeruddin, Q. Wang, M. Grätzel and D. L. Officer, Highly efficient porphyrin sensitizers for dye-sensitized solar cells, *J. Phys. Chem. C* **111**(32) (2007), pp. 11760–11762.

16. J. Lagref, M. K. Nazeeruddin and M. Grätzel, Artificial photosynthesis based on dye-sensitized nanocrystalline TiO_2 solar cells, *Inorg. Chim. Acta* **361**(3) (2008), pp. 735–745.
17. H. Zhou, L. Wu, Y. Gao and T. Ma, Dye-sensitized solar cells using 20 Natural dyes as sensitizers, *J. Photochem. Photobio. Chem.* **219**(2–3) (2011), pp. 188–194.
18. H. Hug, M. Bader, P. Mair and T. Glatzel, Biophotovoltaics: Natural pigments in dye-sensitized solar cells, *Appl. Energ.* **115** (2014), pp. 216–225.
19. S. K. Park and Y. S. Han, Efficient dye-sensitized solar cells with Surface-modified photoelectrodes, *Solar Energ.* **110** (2014), pp. 260–267.
20. K. Maeda, K. Teramura, N. Saito, Y. Inoue and K. Domen, Improvement of Photocatalytic activity of (Ga1$-x$Znx)(N1$-x$Ox) solid solution for overall water Splitting by co-loading CR and another transition metal, *J. Catalysis* **243**(2) (2006), pp. 303–308.
21. K. Maeda, T. Takata, M. Hara, N. Saito, Y. Inoue, H. Kobayashi and K. Domen, GaN:ZnO solid solution as a photocatalyst for visible light driven overall water splittin, *J. Am. Chem. Soc.* **127** (2005), pp. 8286–8287.
22. K. Maeda, K. Teramura, T. Takata, M. Hara, N. Saito, K. Toda, Y. Inoue, H. Kobayashi and K. Domen, Overall water splitting on (Ga1-xZnx)(N1-xOx) solid Solution photocatalyst: Relationship between physical properties and Photocatalytic activity, *J. Phys. Chem. B* **109**(43) (2005), pp. 20504–20510.
23. M. Yoshida, K. Maeda, D. Lu, J. Kubota and K. Domen, Lanthanoid oxide layers on rhodium-loaded (Ga1-xZnx)(N1-xOx) photocatalyst as a modifier for overall water splitting under visible-light irradiation, *J. Phys. Chem. C* **17** (2013), pp. 14000–14006.
24. K. Lee, B. Tienes, M. Wilker, K. Schnitzebaumer and G. Dukovic, (Ga1-xZnx)(N1-xOx) nanocrystals: Visible absorbers with tunable composition and Absorption spectra, *Am. Chem. Soc. Nano Lett.* **12** (2012), pp. 3268–3272.
25. A. Fujishima and K. Honda, Lectrochemical photolysis of water at a semiconductor electrode, *Nature* **238** (1972), pp. 37–38.
26. A. Kudo and Y. Miseki, Heterogeneous photocatalyst materials for water splitting, *Roy. Soc. Chem.* **38** (2009), pp. 253–278.
27. Y. Inoue, Photocatalytic water splitting by RuO2-loaded metal oxides and nitrides with d0- and d10-related electronic configurations, *Energy Environmental Science* **2** (2009), pp. 364–386.
28. Y. Moriya, T. Takata and K. Domen, Recent progress in the development of (oxy)nitride photocatalysts for water splitting under visible-light irradiation, *Coord. Chem. Rev.* **257**(13–14) (2013), pp. 1957–1969.
29. K. Maeda, Photocatalytic water splitting using semiconductor particles: History and recent developments, *J. Photochem. Photobio. C* **12**(4) (2011), pp. 237–268.
30. R. Abe, Recent progress on photocatalytic and photoelectrochemical water Splitting under visible light irradiation, *J. Photochem. Photobio. C* **11**(4) (2010), pp. 179–209.

31. N. Skillen, C. McCullagh and M. Adams, *Photocatalytic Splitting of Water* (Springer, Berlin-Heidelberg, 2014), pp. 1–42.
32. R. Abe, M. Higashi, K. Sayama, Y. Abe and H. Sugihara, Photocatalytic activity of R3MO7 and R2Ti2O7 (R = Y, Gd, La; M = Nb, Ta) for water splitting into H2 and O2, *J. Phys. Chem. B* **110**(5) (2006), pp. 2219–2226.
33. J. Sato, N. Saito, H. Nishiyama and Y. Inoue, Photocatalytic water decomposition by RuO2-loaded antimonates, M2Sb2O7 (M = Ca, Sr), CaSb2O6 and NaSbO3, with d10 configuration, *J. Photochem. Photobio.* **148**(1–3) (2002), pp. 85–89.
34. S. H. Kim, S. Park, C. W. Lee, B. S. Han, S. W. Seo, J. S. Kim, I. S. Cho and K. S. Hong, Photophysical and photocatalytic water splitting performance of Stibiotantalite type-structure compounds, SbMO4 (M = NB, Ta), *Int. J. Hydrog. Energ.* **37**(22) (2012), pp. 16895–16902.
35. H. Kato and A. Kudo, Photocatalytic water splitting into H2 and O2 over various tantalate photocatalysts, *Catalysis Today* **78** (2003), pp. 561–569.
36. C. Zheng and A. West, Compound and solid-solution formation, phase equilibria and electrical properties in the ceramic system ZrO2-La2O3-Ta2O5, *J. Mater. Chem.* **1**(2) (1991), pp. 163–167.
37. K. Sayama and H. Arakawa, Photocatalytic decomposition of water and photocatalytic reduction of carbon dioxide over zirconia catalyst, *J. Phys. Chem.* **97**(3) (1993), pp. 531–533.
38. L. Cao, S. Sahu, P. Anil Kumar, C. E. Bunker, J. Xu, K. A. S. Fernando, P. Wang, E. A. Guliants, K. N. Tackett and Y. Sun, Carbon nanoparticles as visible-light photocatalysts for efficient CO2 conversion and beyond, *J. Am. Chem. Soc.* **133**(13) (2011), pp. 4754–4757.
39. K. Li, A. D. Handoko, M. Khraisheh and J. Tang, Photocatalytic reduction of CO2 and protons using water as an electron donor over potassium tantalate nanoflakes, *Nanoscale* **6**(16) (2014), pp. 9767–9773.
40. A. D. Handoko and J. Tang, Controllable proton and CO2 photoreduction over Cu2O with various morphologies, *Int. J. Hydrog. Energ.* **38**(29) (2013), pp. 13017–13022.
41. A. J. Bard, Photoelectrochemistry and heterogenous photocatalysis at Semiconductors, *J. Photochem. Photobio.* **11** (1979), pp. 59–75.
42. R. Abe, K. Sayama, K. Domen and H. Arakawa, A new type of water splitting system composed of two different TiO_2 photocatalysts (anatase, rutile) and a IO_3^-/I^- shuttle redox mediator, *Chem. Phys. Lett.* **344**(3) (2001), pp. 339–344.
43. K. Sayama, K. Mukasa, R. Abe, Y. Abe and H. Arakawa, A new photocatalytic water splitting system under visible light irradiation mimicking a *Z*-scheme Mechanism in photosynthesis, *J. Photochem. Photobio.* **148**(1–3) (2002), pp. 71–77.
44. K. Sayama, K. Mukasa, R. Abe, Y. Abe and H. Arakawa, Stoichiometric water Splitting into H2 and O2 using a mixture of two different photocatalysts and an IO3-/I- shuttle redox mediator under visible light irradiation, *Chem. Commun.* **7**(23) (2001), pp. 2416–2417.

45. R. Abe, T. Takata, H. Sugihara and K. Domen, Photocatalytic overallwater splitting under visible light by taon and WO3 with an IO3-/I- shuttle redox mediator, *Chem. Commun.* (2005), pp. 3829–3831.
46. H. Kato, Y. Sasaki, A. Iwase and A. Kudo, Role of iron ion electron Mediator on photocatalytic overall water splitting under visible light irradiation Using Z-scheme systems, *Chem. Soc. J.* **80**(12) (2007), pp. 2457–2464.
47. Y. Sakata, Y. Matsuda, T. Yanagida, K. Hirata, H. Imamura and K. Teramura, Effect of metal ion addition in a Ni supported Ga2O3 photocatalyst on the photocatalytic overall splitting of H2O, *Catal. Lett.* **125** (2008), pp. 22–26.
48. R. Molinari, T. Marino and P. Argurio, Photocatalytic membrane reactors for hydrogen production from water, *Int. J. Hydrog. Energ.* **39** (2014), pp. 7247–7261.
49. S.-C. Yu and C.-W. Huang, C.-H. Liao, J. C. S. Wu, S.-T. Chang, K.-H. Chen, A novel membrane reactor for separating hydrogen and oxygen in photocatalytic water splitting, *J. Membrane Sci.* **382** (2011), pp. 291–299.
50. W. Lee, C. Liao, M. Tsai, C. Huang and J. C. S. Wu, A novel twin reactor For CO2 photoreduction to mimic artificial photosynthesis, *Appl. Catal. B* **132–133** (2013), pp. 445–451.
51. J. Ryu, S. H. Lee, D. H. Nam and C. B. Park, Rational design and engineering of Quantum-dot-sensitized TiO2 nanotube arrays for artificial photosynthesis, *Adv. Mater.* **12** (2011), pp. 1883–1888.
52. R. K. Yadav, J. O. Baeg, G. H. Oh, N. J. Park, K. J. Kong, J. Kim, D. W. Hwang, S. K. Biswas, A photocatalyst-enzyme coupled artificial photosynthesis system for solar energy in production of formic acid from CO2, *J. Am. Chem. Soc.* **134** (2012), pp. 11455–11461.
53. Y. Zhao, J. R. Swierk, J. R. J. D. Megiatto, B. Sherman, W. J. Youngblood, D. Qin, D. M. Lentz, A. L. Moore, T. A. Moore, D. Gust, T. E. Mallouk, Improving the efficiency of water splitting in dye-sensitised solar cells by using biometric electron transfer mediator, *PNAS* **109**(39) (2012), pp. 15612–15616.
54. Y. S. Nam, T. Shin, H. Park, A. P. Magyar, K. Choi, G. Fantner, K. A. Nelson and A. M. Belcher, Virus-templated assembly of porphyrins into light-harvesting Nanoantennae, *J. Am. Chem. Soc.* **132** (2010), pp. 1462–1463.
55. Y. S. Nam, A. P. Magyar, D. Lee, J.-W. Kim, D. S. Yun, H. Park, T. S. Pollom, D. A. Weitz and A. M. Belcher, Biologically templated photocatalytic nanostructures for sustained light-driven water oxidation, *Nat. Nanotechnol.* **5** (2010), pp. 340–344.
56. K. Li, X. An, K. H. Park, M. Khraisheh and J. Tang, A critical review of CO2 photoconversion: Catalysts and reactors, *Catalysis Today* **224** (2014), pp. 3–12.
57. T. Inoue, A. Fujishima, S. Konishi and K. Honda, Photoelectrocatalytic reduction of carbon dioxide in aqueous suspensions of Semiconductor powders, *Nature* **277** (1979), pp. 637–638.
58. S. Das and W. M. A. Wan Daud, Photocatalytic CO2 transformation into fuel: A review on advances in photocatalyst and photoreactor, *Renew. Sust. Energ. Rev.* **39** (2014), pp. 765–805.

59. M. Anpo, Photocatalytic reduction of CO2 with H2O on highly dispersed Ti-oxide catalysts as a model of artificial photosynthesis, *J. CO2 Utilization* **1** (2013), pp. 8–17.
60. B. R. Eggins, P. K. J. Robertson, E. P. Murphy, E. Woods and J. T. S. Irvine, Factors affecting the photoelectrochemical fixation of carbon dioxide with semiconductor colloids, *J. Photochem. Photobio.* **118**(1) (1998), pp. 31–40.
61. L. Liu, H. Zhao, J. M. Andino and Y. Li, Photocatalytic CO2 reduction with H2O on TiO2 nanocrystals: Comparison of anatase, rutile and brookite polymorphs and exploration of surface chemistry, *ACS Catalysis* **2**(8) (2012), pp. 1817–1828.
62. B. K. Vijayan, N. M. Dimitrijevic, J. Wu and K. A. Gray, The effects of Pt Doping on the structure and visible light photoactivity of titania nanotubes, *J. Phys. Chem. C* **114**(49) (2010), pp. 21262–21269.
63. Q. Zhang, W. Han, Y. Hong and J. Yu, Photocatalytic reduction of CO2 with H2O on Pt-loaded TiO2 catalyst, *Catalysis Today* **148**(3–4) (2009), pp. 335–340.
64. L. Collado, P. Jana, B. Sierra, J. M. Coronado, P. Pizarro, D. P. Serrano and V. A. De La Pena O'Shea, Enhancement of hydrocarbon production via artificial photosynthesis due to synergetic effect of Ag supported on TiO2 and ZnO semiconductors, *Chem. Eng. J.* **224** (2013), pp. 128–135.
65. T. Ohno, N. Murakami, T. Koyanagi and Y. Yang, Photocatalytic reduction of CO2 over a hybrid photocatalyst composed of WO3 and graphic carbon nitride (g-C3N4) under visible light, *J. CO2 Utilisation* **6** (2014), pp. 17–25.
66. K. L. Schulte, P. A. Desario and K. A. Gray, Effect of crystal phase composition on the reductive and oxidative abilities of TiO2 nanotubes under UV and visible light, *Appl. Catalysis B* **97**(3–4) (2010), pp. 354–360.
67. R. S. Devan, R. A. Patil, J. Lin and Y. Ma, One-dimensional metal-oxide nanostructures: Recent developments in synthesis, characterization and applications, *Adv. Funct. Mater.* **22**(16) (2012), pp. 3326–3370.
68. K. Yu, W. Yu, M. Kuo, Y. Liou and S. Chien, Pt/titania-nanotube: A potential catalyst for CO2 adsorption and hydrogenation, *Appl. Catalysis B* **84**(1–2) (2008), pp. 112–118.
69. O. K. Varghese, M. Paulose, T. J. Latempa and C. A. Grimes, High-rate solar photocatalytic conversion of CO2 and water vapour to hydrocarbon fuels, *Nano Lett.* **9**(2) (2009), pp. 731–737.
70. L. Jia, J. Li and W. Fang, Enhanced visible-light active C and Fe co-doped LacoO3 for reduction of carbon dioxide, *Catalysis Commun.* **11**(2) (2009), pp. 87–90.
71. T. W. Woolerton, S. Sheard, E. Relsner, E. Pierce, S. W. Ragsdale, F. A. Armstrong, Efficient and clean photoreduction of CO2 to CO by enzyme-modified TiO2 nanoparticles using visible light, *J. Am. Chem. Soc. Commun.* **132** (2010), pp. 2132–2133.
72. W. Tu, Y. Zhou, Q. Liu, Z. Tian, J. Gao, X. Chen, H. Zhang, J. Liu and Z. Zou, Robust Hollow spheres consisting of alternating titania nanosheets

and graphene nanosheets with high photocatalytic activity for CO2 conversion into renewable fuels, *Adv. Funct. Mater.* **22**(6) (2012), pp. 1215–1221.
73. Y. T. Liang, B. K. Vijayan, K. A. Gray and M. Hersam, Minimizing graphene defects enhances titania nanocomposite-based photocatalytic reduction of CO2 for improved solar fuel production, *Nanoletters* **11**(7) (2011), pp. 2865–2870.
74. A. J. Morris, G. J. Meyer and E. Fujita, Molecular approaches to the Photocatalytic reduction of carbon dioxide for solar fuels, *Acc. Chem. Res.* **42**(12) (2009), pp. 1983–1994.
75. J. Hawecker, J. Lehn and R. Ziessel, Efficient photochemical reduction of CO2 to CO by visible light irradiation of systems containing Re(bipy)(CO)3X or Ru(bipy)32+-Co2+ combinations as homogeneous catalysts, *J. Chem. Soc. Chem. Commun.* **9** (1983), pp. 536–538.
76. S. Matsuoka, K. Yamamoto, T. Ogata, M. Kusaba, N. Nakashima, E. Fujita and S. Yanagida, Efficient and selective electron mediation of cobalt complexes with cyclam and related macrocycles in the p-terphenyl-catalyzed photoreduction of Carbon dioxide, *J. Am. Chem. Soc.* **115**(2) (1993), pp. 601–609.
77. T. Ogata, Y. Yamamoto, Y. Wada, K. Murakoshi, M. Kusaba, N. Nakashima, A. Ishida, S. Takamuku and S. Yanagida, Phenazine-photosensitized reduction of CO2 mediated by a cobalt-cyclam complex through electron and hydrogen transfer, *J. Phys. Chem.* **99**(31) (1995), pp. 11916–11922.
78. J. Grodkowski and P. Neta, Cobalt Corrin catalyzed photoreduction of CO2, *J. Phys. Chem. A* **104**(9) (2000), pp. 1848–1853.
79. H. Hori, K. Koike, Y. Suzuki, M. Ishizuka, J. Tanaka, K. Takeuchi and Y. Sasaki, High-pressure photocatalytic reduction of carbon dioxide using [fac-Re(bpy)(CO)3p(OiPr)3]+ (bpy = 2,2′-bipyridine), *J. Mol. Catalysis A* **179**(1–2) (2002), pp. 1–9.
80. S. Sato, K. Koike, H. Inoue and O. Ishitani, Highly efficient supramolecular photocatalysts for CO2 reduction using visible light, *Photochem. Photobiol. Sci.* **6**(4) (2007), pp. 454–461.
81. S. Karlsson, J. Boixel, Y. Pellegrin, E. Blart, H. Becker, F. Odobel and M. L. Hammarstrã, Accumulative charge separation inspired by photosynthesis, *J. Am. Chem. Soc.* **132**(51) (2010), pp. 17977–17979.
82. M. Zhang and M. L. Hammarstrã, Proton-coupled electron transfer from tryptophan: A concerted mechanism with water as proton acceptor, *J. Am. Chem. Soc.* **133**(23) (2011), pp. 8806–8809.
83. D. C. Lacy, C. C. L. McCrory and J. C. Peters, Studies of Cobalt-mediated electrocatalytic CO2 reduction using a redox-active ligand, *Inorg. Chem.* **53**(10) (2014), pp. 4980–4988.
84. S. Hu, M. R. Shaner, J. A. Beardslee, M. Lichterman, B. S. Brunschwig and N. S. Lewis, Amorphous TiO2 coatings stabilize Si, GaAs and GaP photoanodes for efficient water oxidation, *Science* **344**(6187) (2014), pp. 1005–1009.
85. P. Narang, S. Chen, N. C. Coronel, S. Gul, J. Yano, L. Wang, N. S. Lewis and H. A. Atwater, Bandgap tunability in Zn(Sn,Ge)N2 semiconductor alloys, *Adv. Mater.* **26**(8) (2014), pp. 1235–1241.

Chapter 7

Small-Scale PV Applications in Home and Business

Estefanía Caamaño-Martín, Miguel Ángel Egido
and Jorge Solórzano
Instituto de Energía Solar, Universidad Politécnica de Madrid
(Solar Energy Institute, Technical University of Madrid)
ETSI de Telecomunicación, Avda. Complutense 30, 28040 Madrid (Spain)

This chapter presents the main technical and non-technical aspects of distributed photovoltaic systems, both grid-connected and stand-alone. In particular, after an introduction where the main typologies and configurations are described, fundamentals on solar resource and operational parameters are presented. Photovoltaic (PV) applications in built areas and stand-alone PV applications will be then covered, including design, sizing, operation and maintenance, business models, policies and international experience.

1 Introduction

Photovoltaics (PV) is by far the world's fastest-growing electricity generation technology over the last century. With terrestrial use starting in the mid 1970's in water-pumping applications,[1] nowadays PV provides clean electricity for a diversity of uses ranging from small portable devices (e.g. mobile phones, portable computers) to buildings, public urban spaces and rural communities. As a consequence, PV has become the paradigm for "Distributed Generation" (DG), a term for which no consistent definition exists although it is generally considered as "electric power generation units connected directly to the distribution network or connected to the network on the customer site of the meter".[2] The inherent modular nature of PV technology, combined with the wide availability of its primary energy resource (solar radiation) allows expanding the DG concept to areas isolated from national electricity grids, where stand-alone PV systems (SAPVS) can be used to provide reliable electricity services to rural populations.

In this chapter, the main technical and non-technical aspects of PV systems designed for and installed next to consumers will be covered. They are grouped under the heading of "small-scale PV systems" to differentiate from the so-called "large-scale PV power plants" that are typically designed for maximizing the electricity production and installed relatively far from consumers (see Chapter 4). In particular, two main types of PV applications will be considered, the main characteristics of which are described next.

1.1 *PV Applications in the Built Environment*

With a PV generator with nominal power typically in the range of kW's (although examples of up to a few MW can also be found), PV applications are typically characterized by the following:

- The PV modules are integrated, in one way or another, in buildings and urban infrastructures (street equipment, barriers, shelters, aerial structures for lighting and advertising, etc.).
- The PV system outputs are connected to low-voltage electricity distribution grids or to the buildings' own electrical installations. (Note: an exception exists if, due to local grid codes, the PV system must be connected to a medium-voltage electricity distribution grid, in which case a conventional transformer substation is required).

The closeness between generation and consumption entails power and energy exchanges with the distribution grid, as shown in the following examples. Consider a single family house with a PV system installed on the roof that is connected to the low voltage electricity grid independently from the house electrical installation, both connections being physically very close. A block diagram of both electrical installations is shown in Fig. 1(a), where the PV generator and power conditioning equipment (DC/AC converter or inverter) are represented by the conventional symbols presented previously in this book (see Chapter 2). Figure 1(b) shows the daily development of the PV power fed into the distribution grid and the electrical power taken from such a grid on a sunny day, where the demand profile reflects the behavior of a user who consumes most electricity at the evenings. Note that there are two areas under the PV power and house power demand curves with striped diagonal patterns representing, respectively, the daily PV generation and house electrical consumption that would be registered by the corresponding electricity meters (E_{PV} and E_L). In addition, due to the physical proximity between both installations connection points to the distribution grid, whenever local generation and consumption coincide, the

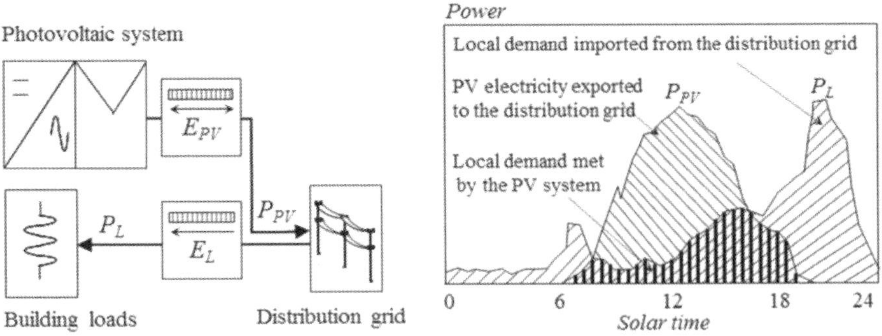

(a) Electrical block diagram

(b) Development of power flows in a sunny day

Fig. 1. Example of a PV system in a single family house connected to the distribution grid independently from the consumer's electrical installation.

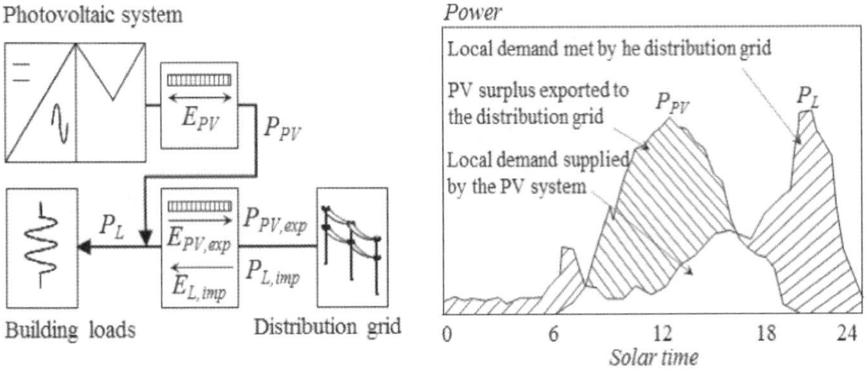

(a) Electrical block diagram

(b) Development of power flows in a sunny day

Fig. 2. Example of a PV system in a single family house connected to the consumer's electrical installation.

PV system will supply the local demand, shown in the figure by the area filled with a striped vertical pattern.

Consider now a different situation where the PV system is connected to the house electrical installation before the consumption meter, in the so-called "self-consumption" mode (see Fig. 2(a)). Note that in this case the PV system supplies the same amount of electrical demand as in the previous example, but now only the generation surplus (excess over the local demand) is fed into the distribution grid. This is shown in Fig. 2(b), where the white area represents the daily self-consumed PV generation.

The implications of both examples are different both in technical and economic terms: whereas from a technical point of view the demand experienced by the distribution grid is the same, in the second example the house would reduce the kWh going through the utility meter and, consequently its expenses in commercial electricity. Another important aspect is the value assigned to the PV electricity surplus that is fed into the grid, which is defined by the local electrical market and/or energy policies. These aspects will be covered in Sec. 4.4 of this chapter.

1.2 *Stand-Alone PV Applications*

Terrestrial PV applications started in the 1970's for SAPVS, mainly for water pumping and health centers managed by UNICEF. In fact, low power SAPVS are apparently very easy to install and operate: a PV module with 36 series-connected crystalline solar cells, plugged in parallel with a car battery and an electrical generator can be built for direct current charges (in further sections the low reliability of such a configuration will be explained). As a practical exercise, consider that we have to supply electricity for a service which requires some kWh per day in a location several kilometers away from the electrical grid, what is the best engineering solution? In comparison with diesel engines, SAPVS do not require fossil fuel, (its "fuel" is solar radiation) is well distributed at latitudes lower than 50° and predictable with enough confidence. (The wind resource, in comparison, is less extensively distributed and less predictable). In solar power generation there are no elements in motion, unlike other technologies (combustion engines, wind generators or hydropower), which improves the reliability and decreases maintenance costs. Also, time for implementation of a SAPVS is very low.

Despite these advantages, the main disadvantage of PV technology has been the cost, but this is changing. Thanks to the PV grid connected market evolution, which has exponentially increased in the last decade, PV module prices have dramatically decreased. For instance, in the budget share for a recent hybrid diesel PV stand-alone system of 3.6 kWp[a] in the Bolivian Amazon (December 2014), the PV generator cost is 26.1% (33.9%, if the support structure cost is also considered), electrical cables and protections and installation costs are 28.2% and transport cost is 6.9%. Definitively,

[a]The definition of Wp is in Chapter 2, Sec. 4.3.1.

the cost of isolated electrical generation is a consequence of this isolation. The cost per Wp is also interesting viz. 8.1 $/Wp.

A SAPVS is composed of a PV generator connected to an energy storage device; it requires always electrical control and power conditioning and can supply DC and/or AC electricity. PV technology has an additional property: the range of power generation could be from mW to MW with good cost efficiency.[b] For this reason the potential applications for SAPVS are huge: parking meters, safety signals, street lights, kiosks, etc., in urban environments; but in isolated locations is where the possibilities are even bigger: solar lanterns, solar home systems, solar villages, health centers, electricity supply for telecommunications, water pumping, and all kind of productive applications such as grain mill, drip irrigation, internet tele center, electricity shops for mobile phones, milk refrigeration, etc. All these and many more are part of the current landscape of PV applications. Even, more experimental engines such as solar cars or solar airplanes can be fully supplied by SAPVS, although are not yet economically viable.

While PV grid connected systems are designed to feed green electricity to local loads and/or the grid, SAPVS design focuses on supplying electricity to local loads autonomously. Consequently, the objective is not to maximize the electricity production on a yearly basis, but to guarantee the fulfillment of the electrical service, i.e. to minimize the loss of load probability. This reliability is not only related to PV modules and Balance of System (BOS) components failures, the variability of solar irradiation and the daily load are key factors. The design of a SAPVS requires sizing the peak power of PV generator and the storage capacity to supply a load in a location characterized for its solar irradiation and temperature. This task has to include general rules for electrical engineering: safety protections, cabling design, efficiencies, etc.

2 Solar Resource Availability

2.1 *Meteorological Resources and Modeling*

Meteorological resources data (solar radiation, ambient temperature) required for the design of small-scale PV applications depends on the accuracy of design required:

- In SAPVS, a guarantee of electrical service in the worst operational conditions is usually the main design goal. Therefore, monthly averages

[b]In 2014, a SAPVS of 5 MWp was inaugurated in a Chilean mine without subsidies.

of daily global solar radiation over the horizontal plane are required; monthly averages of daily maximum and minimum ambient temperature are also recommended, which allows calculating thermal losses of the PV generator (see Sec. 3.1).

- In grid-connected PV systems that are connected to public distribution grids, maximization of annual generation is usually the main goal. Consequently, the input data from the previous example is also commonly used, from which hourly values can be derived and used in the subsequent modeling phase.
- In grid-connected PV systems connected in the "self-consumption" mode where the interaction with the grid is a relevant aspect, hourly data that incorporate the natural variability of local solar radiation are recommended. The same applies to PV systems where the PV modules are architecturally integrated so that they influence the building's thermal behavior. In these cases, the preferable data are the so-called Typical Meteorological Years (TMYs), which are annual datasets that include natural diurnal and seasonal variations, representing typical climatic conditions for a given location over a long period of time (typically 30 years[3]). They have been frequently used in building simulation, in order to assess the heating and cooling demand. Specific variables depend on the activity domain and include solar radiation, air temperatures (mean, maximum, minimum), relative humidity, precipitation, etc.

Concerning the availability of solar radiation data, an increasing number of resources exist, ranging from historical datasets of national networks of weather stations to satellite-derived data.[4–6] Meteorological input data must then be processed in order to calculate the solar resource available on the PV module's plane:

- When monthly averages of daily horizontal global solar radiation are used, the best approach is to synthesize a full year of daily values from which hourly values can be estimated, following for example the Aguiar and Collares-Pereira methodologies.[7,8] TMYs already provide hourly solar radiation data.
- From the hourly values of horizontal global solar radiation the corresponding direct and diffuse components can be calculated using diffuse fraction or quasi-physically based models.[9,10]
- From the previous results the direct, diffuse and reflected components of hourly solar radiation can be derived. Particularly relevant is modeling of the diffuse component, for which the different approaches vary

from an isotropic model where solar radiation is considered to be evenly distributed throughout the sky hemisphere to more complex models that include anisotropic effects.[11,12] Direct and reflected components are more straightforward to calculate.[13]

- From these hourly components of global solar radiation, daily, monthly and/or annual values can be easily obtained.

Besides solar radiation, other meteorological variables that influence the operation of PV generators are; surroundings, ambient temperature and, at a lesser extent, wind speed. Particularly, when the ventilation of PV modules cannot be guaranteed (for example, when they constitute part or all of highly insulated envelopes), high solar cells' operating temperatures can occur leading to thermal losses (more than 85°C have been measured in central Europe[14]). Also to be considered is the PV modules composition: where the solar cells are laminated between two sheets of glass (glass–glass PV modules), their operation temperature is typically 2–3°C higher than in conventional glass-film PV modules where the solar cells are laminated between glass and the insulation film (usually PVF-polyvinyl fluoride). This is reflected in higher values of the "Nominal Operating Cell Temperature" (NOCT) parameter that characterizes the thermal performance of PV modules under natural convection[c]: it is typically around 45°C for glass-film PV modules and closer to 50°C for glass–glass PV modules.

2.2 *On the Positioning of PV Modules*

While it might be thought that the ultimate design goal of any grid connected PV system is to produce the maximum amount of electricity, this is not always the case, particularly in building integration where available surfaces for the PV modules may differ from ideal ones. But first, what is the optimal position for a PV module at a given location, in terms of azimuth and tilt angles? Several recommendations can be found in the literature, which can be summarized as:

- Grid-connected PV applications. Since a very common design objective of these PV systems is to maximize generation, the "optimal" position usually corresponds to maximum annual solar gains for fixed surfaces. To this aim, the optimal orientation is towards the Equator (to the south

[c]Measured with normal in-plane irradiance of 800 $W{\cdot}m^{-2}$, ambient temperature of 20°C and wind speed 1 $m{\cdot}s^{-1}$.

for locations in the northern hemisphere, to the north otherwise), and the optimal tilt angle lies relatively close to the value of the local latitude (ϕ).[13,15,16]

- Stand-alone PV applications. The previous recommendations also apply if the solar resource and loads are (more or less) evenly distributed throughout the year, which is in fact quite unusual. In general, to optimize the tilt angle is equivalent to maximize the ratio between mean daily solar irradiation and mean daily load on a monthly basis. However, under the assumption of constant electricity demand, the optimal position is the one that maximizes solar gains in the "worst month" (the one with minimum solar irradiation). For locations at mid-latitudes and with constant demand, this usually leads to tilt angles around $|\phi| + 10°$ facing to the equator. If this methodology is applied to locations in the inter-tropical belt, the result could be that the best tilt is horizontal, but a minimum of 5° should be used to facilitate PV modules self-cleaning by rain. In general, when the mean monthly load is variable, the optimization should be focused in the month most affected. As an example, many crops require more water in summer, near to solstice. The optimal tilt is that to minimize the incidence angle of the sun at noon in this period; it can be easily estimated as the local latitude, in absolute value, minus 23.5°.[17]

Going back to the issue of surface availability, a common question amongst architects and engineers willing to incorporate PV in buildings is about the appropriateness of the buildings' envelopes in terms of the solar resource. In this respect, it is useful to evaluate the theoretical solar potential of a surface with azimuth α and tilt β relative to the maximum potential at the same location, by means of the "Irradiation factor", FI:

$$FI = \frac{G_a(\alpha, \beta)}{G_a(\alpha_{opt}, \beta_{opt})}, \tag{1}$$

where $G_a(\alpha, \beta)$ is the annual global solar radiation incident on the surface and $G_a(\alpha_{opt}, \beta_{opt})$ is the maximum annual global solar radiation incident on the optimal surface (maximum solar gains for fixed surfaces). Obviously, $FI \leq 1$.

Particularly interesting is the development of the Irradiation Factor for different tilt and azimuth angles and different locations, as shown in Fig. 3. In the figure, the results for two locations representative of the European

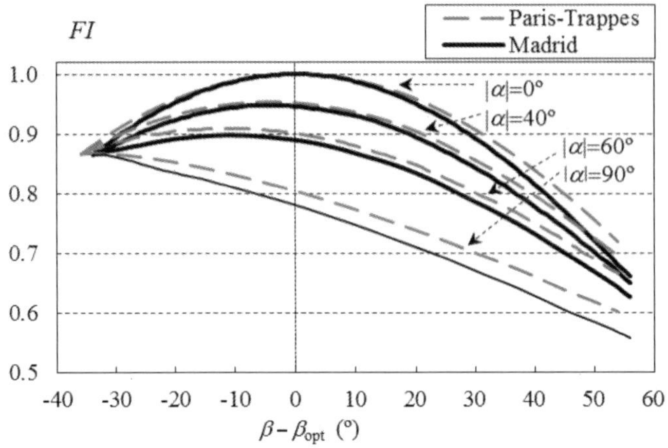

Fig. 3. Irradiation factor for two locations representative of European solar resource.

solar resource are shown viz.:

- Madrid (latitude 40°27′ North), representative of Southern Europe, with a temperate warm climate characterized by minimum summer rains and hot summers.
- Paris–Trappes (latitude 48°46′ North), representative of Central Europe with a temperate humid climate characterized by uniform rains and cool summers.

In both cases solar radiation data considered are from the corresponding Typical Meteorological Years, together with the best performing models mentioned in Sec. 2.1.

It is worth mentioning the similarity between the curves corresponding to the same azimuth angle, with differences below 0.05 between both locations. It can be demonstrated that similar results are obtained when different locations of mid latitudes (ranging from 30 to 55°) are considered. Consequently, the results corresponding to Madrid can be used as representative of the relative solar potential of fixed surfaces. From these results, it is implicit that a horizontal surface (left end of the curves) receives more than 85% of the maximum solar potential ($FI > 0.85$), whereas vertical surfaces (right end of the curves) receive more than 65% of the maximum solar potential if south oriented ($\alpha = 0$, $FI > 0.65$) and about 55% if east or west oriented ($\alpha = \pm 90°$, $FI \sim 0.55$). Even more interesting is the fact that for optimally oriented surfaces the solar potential is very high for a wide range of tilt angles ($\alpha = 0$, $|\beta - \beta_{\mathrm{opt}}| \leq 20° \Rightarrow FI > 0.95 \leftrightarrow$ losses smaller than 5%

compared to the local maximum solar gains). For south-east and south-west surfaces, although the potential is slightly lower, it can still be considered quite high ($\alpha = 40$, $|\beta - \beta_{\text{opt}}| \leq 20° \Rightarrow FI > 0.9 \leftrightarrow$ annual losses lower than 10%). Obviously, the optimal position for a fixed surface varies with the location, as well as the corresponding annual solar potential, $G_a(\alpha_{opt}, \beta_{opt})$. For instance, for the two representative locations considered, using solar radiation data and appropriate models[8,12] the results are:

$$G_a(\alpha_{opt}, \beta_{opt})|_{\text{Madrid}} = G_a(0°, 34°) = 1940\,\text{kWh/m}^2,$$
$$G_a(\alpha_{opt}, \beta_{opt})|_{\text{Paris}} = G_a(0°, 36°) = 1230\,\text{kWh/m}^2.$$

It should be noted that the results could differ with different initial solar radiation data and models used. However, the fact that many surfaces commonly used in building envelopes are suitable for solar PV use has been clearly demonstrated.

2.3 *Shading Losses*

The calculation procedures described in Sec. 2.1 that allow estimating the theoretical solar resource available on a given surface, assume that no obstacles exist that cast shadows over the PV modules. However, with an increasing integration of PV technologies in built areas, shadows have become a frequent phenomenon that have to be considered from the early stages of the PV generator design, in order to minimize their effect on the PV system's electricity production.

To achieve this aim, several methods exist to quantify the effect of the surroundings on the solar gains of a static surface.[18] They usually consist of representing the obstacles profile affecting the potential PV generator surface together with the band of the sun's trajectory and calculating how much the resulting solar resource is affected. It should be noted that not all components of solar radiation are equally affected by shadows: in fact, only those with directional characteristics (the beam component and the circumsolar part of the diffuse component) will be intercepted by the obstacles profile. In addition, the effect of a given obstacles profile will also depend on the receiving surface characteristics. In order to illustrate this aspect, Fig. 4 shows the sun trajectory diagrams of three locations with latitudes ranging from 19 to 51°, where the daily trajectories of the December solstice, equinoxes and June solstice are shown (in continuous, long-striped and short-striped lines respectively), together with two additional ones that allow defining "bands" denoted as A, B, C, and D in the figure. Table 1

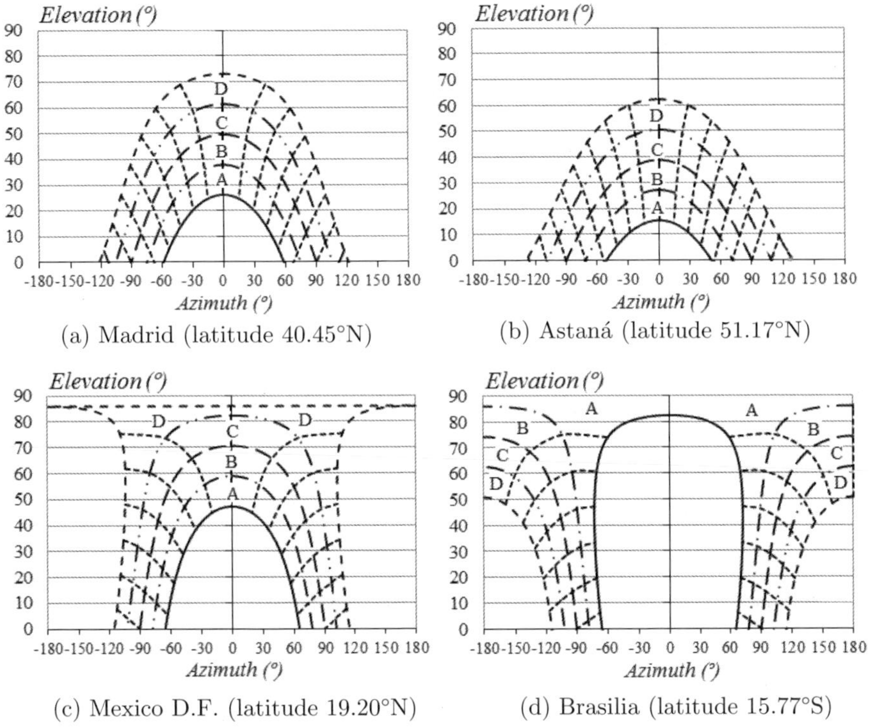

(a) Madrid (latitude 40.45°N)

(b) Astaná (latitude 51.17°N)

(c) Mexico D.F. (latitude 19.20°N)

(d) Brasilia (latitude 15.77°S)

Fig. 4. Sun's trajectory diagrams for locations with different latitudes.

shows, for different surfaces — (α, β) combinations, the importance of each band in terms of interceptable solar radiation, expressed as a percentage of the annual global solar resource.

As an example, let us consider a typical distribution of PV modules in parallel rows, with a distance between rows that guarantees no shadows at noon. Under the assumption of rows constituted of 11 PV modules of dimensions 1.5 m × 1 m each (the shorter dimension being parallel to the horizontal plane) optimally oriented and tilted, Fig. 5 shows the corresponding obstacles profile "seen" by the most affected points of each row (central PV module, central point of lower profile). Table 2 summarizes the corresponding annual global irradiation losses.

Note that quantification of these losses requires identifying, from the obstacles profile, the adjacent celestial region affected by shadows, in this case the one with elevation angles below the obstacles profile (area with vertical stripes).

Table 1: Sun's trajectory bands relevance for the interception of solar radiation by different surfaces in the locations of Fig. 4.

Location (latitude)	Receiving surface position		Energy relevance (% of $G_a(\alpha, \beta)$)			
	α (°)	β (°)	A	B	C	D
Madrid (40.45°N)	0	5	14	12	16	41
	0	38 (β_{opt})	21	14	16	32
	0	90	30	15	12	14
Astaná (51.17°N)	0	5	10	11	16	44
	0	47 (β_{opt})	19	14	16	33
	0	90	26	15	14	20
Mexico D.F. (19.2°N)	0	5	27	15	15	28
	0	25 (β_{opt})	32	15	14	23
	0	90	42	13	6	1
Brasilia (15.77°N)	180	5	23	12	16	30
	180	24 (β_{opt})	19	11	15	34
	180	90	<1	3	11	43

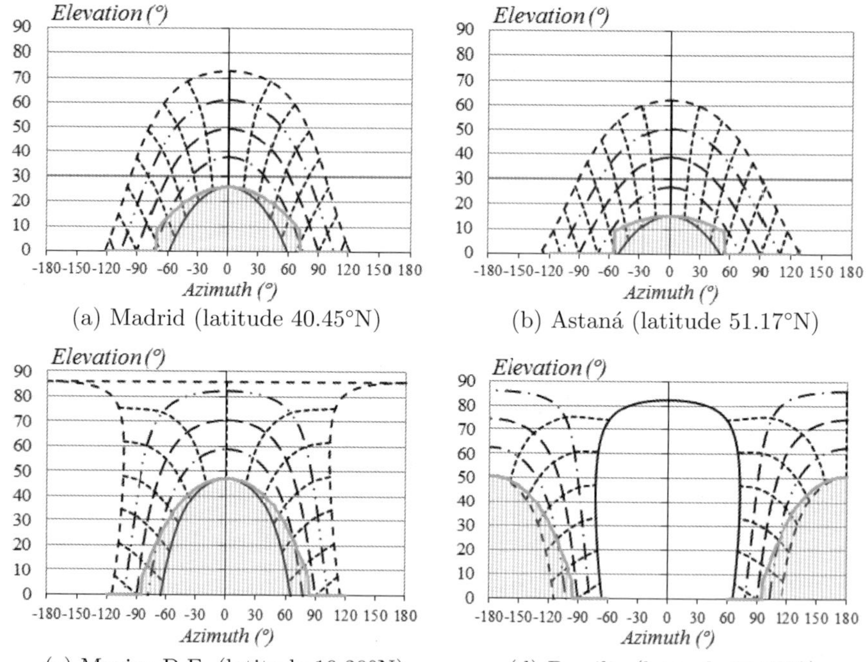

(a) Madrid (latitude 40.45°N)

(b) Astaná (latitude 51.17°N)

(c) Mexico D.F. (latitude 19.20°N)

(d) Brasilia (latitude 15.77°S)

Fig. 5. Example of obstacle profile for rows of PV modules with optimal position separated so that no shadows occur at noon.

Table 2: Annual solar resource losses due to the obstacle profiles of Fig. 5.

Location (latitude)	PV modules position α (°)	$\beta = \beta_{opt}$ (°)	G_a ($\alpha_{opt}, \beta_{opt}$) (kWh/m^2)	Annual shadow losses (% of G_a (α,β))	(kWh/m^2)
Madrid (40.45°N)	0	38	1950	6.5	127
Astaná (51.17°N)	0	47	1758	4.5	79
Mexico D.F. (19.2°N)	0	25	2140	10.5	225
Brasilia (15.77°S)	180	24	1921	11.9	229

Obviously the previous analysis should be performed in one or several points of the projected PV generator surface, depending on its size and proximity to the obstacles. For relatively small PV generators (up to a few kilowatts) the most affected point could be selected as representative of the whole surface. For larger generators several points are recommended, which can provide valuable information for the electrical configuration of the PV generator (series-parallel association of PV modules).

Even if the losses of available solar resource due to shading have been properly identified, determination of their effect on the PV generator production is not an easy task. Several aspects influence the electrical behavior of a PV installation under shadowing conditions, such as:

- Electrical configuration of the solar cells and by-pass diodes within the PV modules.
- Electrical configuration of PV modules within the PV generator.
- The inverter capability to extract the maximum power available from the PV generator (maximum power point tracking strategy and voltage range).

Due to such complexity, currently most design and simulation software tools estimate electrical losses proportionally to the shaded area of PV strings or generators.[18] It is expected that appropriate models resulting from ongoing research and development efforts[19] will be incorporated in the near future.

3 Performance Assessment of Small-Scale PV Systems

Due to the intrinsic decentralized feature of PV, the generation potential of PV systems differs enormously. Particularly, the specific location is a determining factor of PV systems operational conditions and, therefore, of their electricity production. Due to this fact, the assessment of small scale

PV systems performance should not be done exclusively on the basis of the number of kWh produced but also in terms of the efficiencies of the main components. In this sense, the following parameters defined in the early 1990's by the international PV community are useful for this purpose:[20]

- Array yield, or electricity produced by the PV generator per unit of installed or nominal power ($P_{nomG} = 1$ kWp, or 1 kW at Standard Test Conditions[d]):

$$Y_A = \frac{E_{PVG}}{P_{nomG}} \quad [\mathrm{h}]. \tag{2}$$

- Final yield, or useful electricity produced by the PV system per unit of installed power:

$$Y_F = \frac{E_{PV}}{P_{nomA}} \quad [\mathrm{h}]. \tag{3}$$

- Reference yield, which characterizes the PV generator available solar resource in terms of "equivalent peak hours" (equivalent to a constant incident irradiance of 1 kW·m^{-2}):

$$Y_R = \frac{G_S(\mathrm{kWh{\cdot}m^{-2}})}{1\,\mathrm{kW{\cdot}m^{-2}}} \quad [\mathrm{h}], \tag{4}$$

  where G_S is the solar irradiation measured by a specific sensor, that should be expressed in kWh·m^{-2}.
- Performance ratio, or ratio between the normalized electricity production (Y_F) and the available solar resource (Y_R):

$$PR = \frac{Y_F}{Y_R} = \frac{E_{PV}}{P_{nomA} \cdot Y_R}. \tag{5}$$

Note that the position of Y_R in the denominator of Eq. (5) means that the actual electricity production (E_{PV}) is "normalized" by the available solar resource which could be interpreted as an elimination of the local operational conditions effects. This assumption explains the fact that PR is the most widely used parameter for assessing PV systems performance.[21] However, by doing so the second order effect of solar cells temperatures on electricity production (a reduction between 2% and 5% per 10°C increase above 25°C) is not eliminated, which "penalizes" PV systems operating in

[d]Defined by normal in-plane irradiance of 1 kW·m^{-2}, solar spectrum 1.5G, solar cells temperature of 25°C and wind speed 1 m·s^{-1}.

warm/hot ambient conditions. Two additional parameters allow the effective elimination of the temperature effect:

- A temperature corrected reference yield, in which temperature losses are considered equivalent to a reduction of the available solar resource for electricity conversion:

$$Y_{RT} = Y_R \cdot [1 + \gamma \cdot (T_C - 25°C)] \quad [\text{h}], \tag{6}$$

  where γ is the thermal coefficient of the PV modules maximum power (typically between -0.4 and $-0.5\%°\text{C}^{-1}$ for crystalline Silicon technologies, around half of this for thin-film technologies) and T_C is the solar cells equivalent temperature (see Eq. (7)).
- A temperature corrected performance ratio:

$$PR_T = \frac{Y_F}{Y_{RT}}. \tag{7}$$

The use of the previous parameters can refer to different operating periods (one day, one month, one year), the monthly average daily values and annual values being the most commonly used. It is convenient at this point to make some comments on the practical measurements of the associated variables leading to a quality monitoring of PV systems:

- Recommended variables that should be measured in real time for grid-connected PV systems include operating conditions (irradiance, temperature, wind speed), PV generator and inverter output (voltage, current, power), and availability (PV system down-times).
- For SAPVS the battery (voltage, current), DC output and diesel generator output (if existing) should also be measured.
- The in-plane irradiance should be measured with a crystalline silicon reference device which complies with international standards.[22] The use of more precise, thermopile-based instruments such as pyranometers is not recommended given the increased cost and complexity of use.
- For the ambient temperature measurements, devices protected from solar radiation should be used. The most precise instruments are platinum thermoresistances (typical accuracy between ± 0.03 and ± 0.3°C). Less precise and less costly sensors with accuracy of ± 1°C are also acceptable.
- The measurement of solar cells equivalent temperature is highly recommended, especially when PV generators are building integrated, since operation temperatures and consequently thermal losses can be relevant.

Indirect methods based on the simple measurement of an electrical variable, namely, the open-circuit voltage of a PV reference device is recommended.[23] For free-standing or well ventilated PV generators the following estimate can be used from measurements of irradiance and ambient temperature (typical accuracy of ±5°C):

$$T_C(°\mathrm{C}) \cong T_a + G(\mathrm{W{\cdot}m^{-2}}) \cdot \frac{NOCT(°\mathrm{C}) - 20}{800\ \mathrm{W{\cdot}m^{-2}}}, \tag{8}$$

where T_a is the ambient temperature, G is the in-plane irradiance and the *NOCT* parameter is provided by the PV modules technical data sheets.

In order to show the operational improvements achieved in PV systems over the first two decades of grid-connected applications, the following figure summarizes the performance of PV systems installed worldwide, compiled by the International Energy Agency — Photovoltaic Power Systems Programme, Task 2: Performance, Reliability and Analysis of Photovoltaic Systems.[24] The dataset consists of 339 PV systems installed between 1986 and 2002, with nominal PV generator power between 1 kW and a few hundred MW (87% below 25 kW). Figure 6 shows the distributions of the annual Final, Reference yield and Performance ratio parameters. In statistical terms, average Final and Reference Yield values were 864 h and 1235 h respectively. Concerning the Performance Ratio, a change in the distribution shape can be seen comparing Figs. 6(c) and 6(d): whereas in the first decade the average annual Performance Ratio was 0.66, for systems installed after 1996 the average performed 6% better (average *PR* of 0.70).

Figure 7 shows the range of PV potential demonstrated by PV systems installed in different countries: Germany, Austria, The Netherlands, Italy, Japan, Switzerland and others (grouping Belgium, Spain, Israel, Mexico, Poland, Portugal, United Kingdom, and Sweden). As can be seen, the PV systems installed in the second period (see Fig. 7(d)) show a smaller variation range in terms of Performance Ratio and higher average values, thus demonstrating a better use of the existing solar potential. This tendency has been observed also in more recently installed PV systems.[21]

The use of these parameters for SAPVS is not very useful. Consider a SAPVS designed to power an isolated telecommunication service, with a constant daily load all year. The PV generator and battery should be sized to guarantee the service in the worst month, which is the month

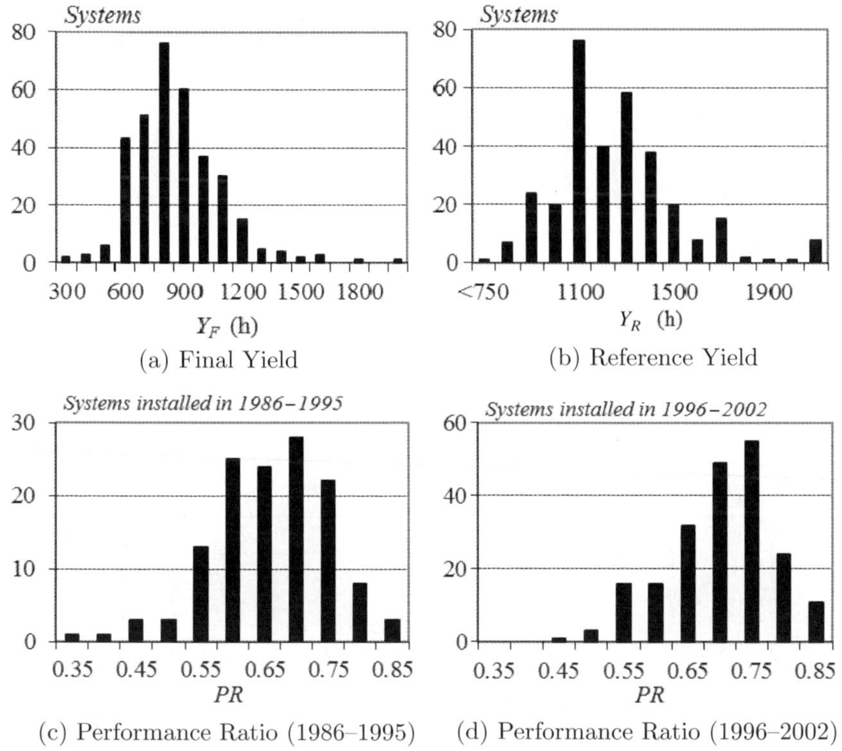

(a) Final Yield (b) Reference Yield

(c) Performance Ratio (1986–1995) (d) Performance Ratio (1996–2002)

Fig. 6. Distribution of annual characteristic parameters of 339 PV grid connected systems installed in IEA countries between 1986 and 2002.

Source: Ref. 24.

with minimum mean daily solar irradiation value. Probably, all the energy generated in this month will be consumed in the load or stored in the battery. As the energy generated in this month is the lowest of the year, the parameters Y_A and Y_F for this month will be also the lowest, but PR will be the highest. However, for the month with highest solar irradiation, the PR will be the lowest because part of the energy generated could not be stored when the battery is full and there is not consumption. This fact produces lower values than expected for Y_A and Y_F, while Y_R is the highest of the year. To evaluate the performance of SAPVS, the best parameter is the loss of load probability (LLP), unfortunately there is not any physical or mathematical model capable of predicting the LLP value with accuracy. This parameter only can be evaluated by software simulation if long-term solar radiation and load consumption data series are available.

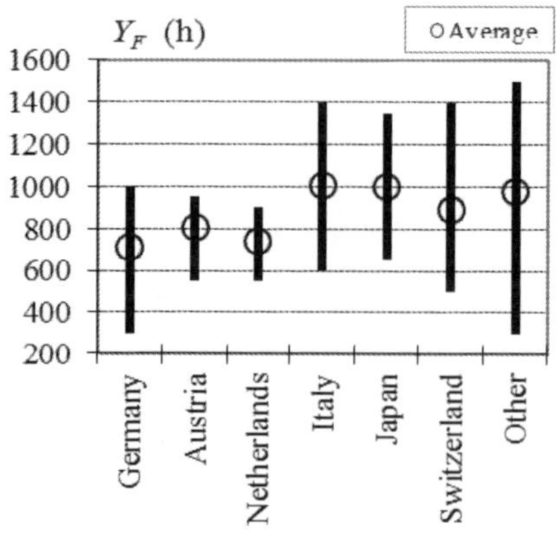

(a) Final Yield

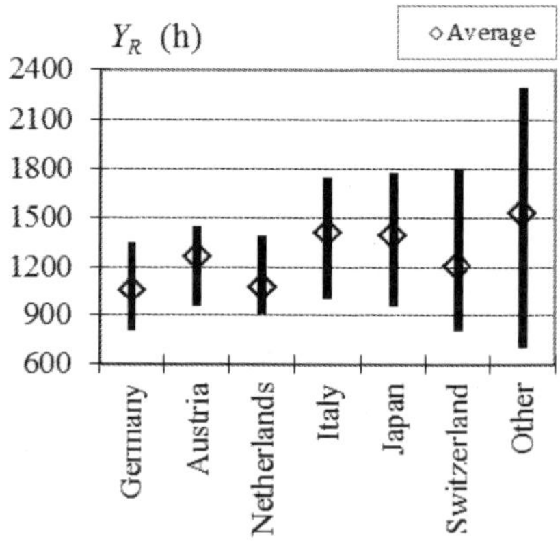

(b) Reference Yield

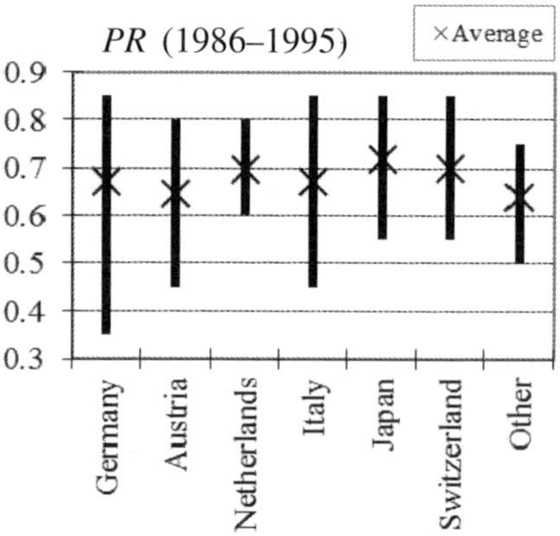

(c) Performance Ratio (1986–1995)

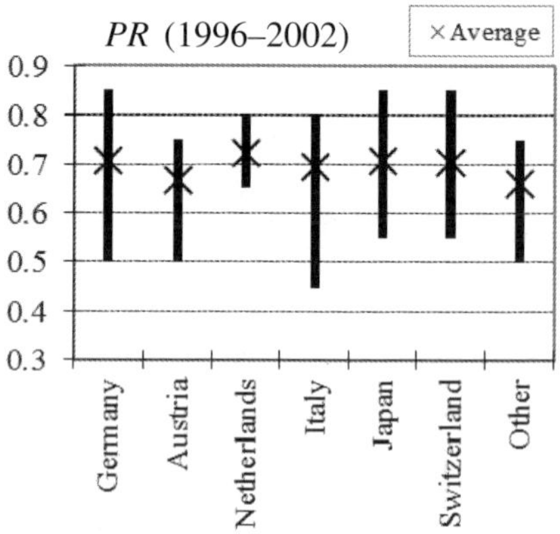

(d) Performance Ratio (1996–2002)

Fig. 7. Annual characteristic parameters range of 339 PV grid connected systems installed in IEA countries between 1986 and 2002.

Source: Ref. 24.

3.1 *PV Systems Losses*

The combination of the previous parameters allows identifying the main PV system losses:

(1) Capture losses associated with the PV array:

$$L_C = Y_R - Y_A = L_{CT} + L_{OE}, \tag{9}$$

which can be divided into Thermal and Optical-Electrical Capture losses, corresponding respectively to solar cells temperature (differences with reference 25°C), irradiance sensor (shading, spectral and angular response) and electrical parameters (power mismatch between PV modules, operation point differences from maximum power point, DC wiring voltage losses):

$$L_{CT} = Y_R - Y_{RT}, \tag{10}$$

$$L_{COE} = Y_{RT} - Y_A. \tag{11}$$

(2) Conversion losses associated with the inverter:

$$L_I = Y_F - Y_A. \tag{12}$$

The combination of these losses allows the calculation of the PV system's final yield and, consequently, the Performance ratio parameters:

$$PR = \frac{Y_F}{Y_R} = \frac{Y_R - (L_{CT} + L_{COE} + L_I)}{Y_R} = 1 - \frac{L_{CT} + L_{COE} + L_I}{Y_R}, \tag{13}$$

$$PR_T = \frac{Y_F}{Y_{RT}} = \frac{Y_{RT} - (L_{COE} + L_I)}{Y_{RT}} = 1 - \frac{L_{COE} + L_I}{Y_{RT}}. \tag{14}$$

Figure 8 shows examples of annual losses of 15 PV grid-connected systems installed in buildings worldwide,[24] with different typologies of architectural integration including free-standing elements, integration in roofs

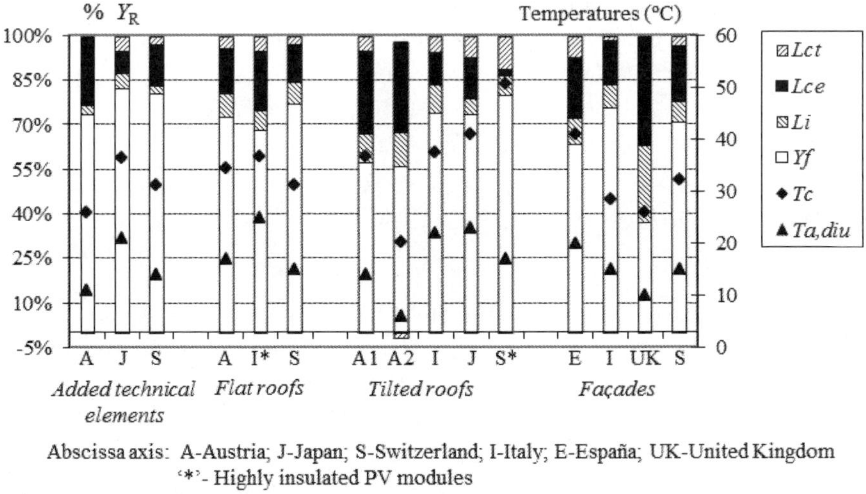

Fig. 8. Annual losses, final yield and temperatures of 15 PV grid-connected systems worldwide with different typologies of architectural integration.

Source: Ref. 24.

and façades. Note that the different variables are represented as percentages of the solar resource (Y_R). Consequently the height of the Final Yields (Y_F, white columns) shows the Performance Ratio parameters. Also shown in the figure are the average solar cells temperatures and diurnal ambient temperatures (T_C and $T_{a,diu}$ respectively). Note also that the heights resulting from adding the columns Final Yield and Thermal Capture losses equal the temperature corrected Performance Ratios.

As can be observed, annual Thermal Capture losses in PV generators (L_{CT}) vary between 11% (tilted roof, S*) and 2% gains (tilted roof, A2, representing a PV generator operating at an average temperature below 25°C, typical of high mountain areas) of the available solar resource (Y_R). This is consistent with the fact that highly insulated PV modules operate at higher temperatures, and consequently with higher thermal losses, than free-standing ones. Optical-electrical capture losses (L_{COE}) are higher than thermal capture ones, ranging from 8% to 37% (the last representing a PV generator highly affected by shadows connected to an inverter with a poor tracking function of the PV generator maximum power). Concerning inverters conversion, annual losses vary between 3% and 25%. As a result, annual Final Yield parameters range from 37% to 81% of the available solar resource (equivalent to PR values between 0.37 and 0.81). It is worth mentioning the fact that the performance of inverters has improved significantly, as shown in Ref. 24, where more than 70% of the 339 PV systems analyzed operated with inverter annual conversion efficiencies higher than 90%.

This tendency has been continued in PV systems installed in the last decade, demonstrated by reported representative PR values of PV systems operating in Ref. 25: Germany ($PR = 0.85$), France ($PR = 0.76$), Belgium ($PR = 0.78$), and Taiwan ($PR = 0.74$).

4 PVs in Buildings

4.1 *Introduction*

Although PVs were already used in the 1950s and 1960s as a power supply for telecommunication applications,[26] it was not until the mid-1970s that the first technical recommendations for the use of PV modules in buildings were issued in the United States.[27] Soon after came the first real applications in residential and commercial buildings which, together with firm governmental support led to a rapid growth of demonstration projects.[28] In Europe the first examples of PV systems in buildings took place in the late-1980s in Switzerland,[29] Germany,[30] and Austria.[31] Initially characterized by technological demonstration projects, new examples appeared in other countries (Japan, The Netherlands, Spain, Italy, United Kingdom, France, and Sweden) through the 1990s. Nowadays PV is increasingly present, not only in buildings, but also in the public spaces of many towns and cities thus making PV the renewable electricity technology with the most distributable potential. In this section basics on PV systems types, design and sizing rules, operation and maintenance provisions are described. Also, a description of the most successful business models and international experience is provided.

4.2 *Types of PV Systems*

Several types of PV systems in buildings can be defined, according to the following approaches: electrical configuration and building integration. From an electrical perspective, two main configurations are used:

- Distributed topologies for Maximum Power Point Tracking (MPPT), where PV modules operate typically connected to small power-conditioning devices, namely microinverters and power optimizers, also known as module-level power electronics (MLPE) (see Fig. 9). These solutions, within the 200–600 W range, can provide certain advantages over

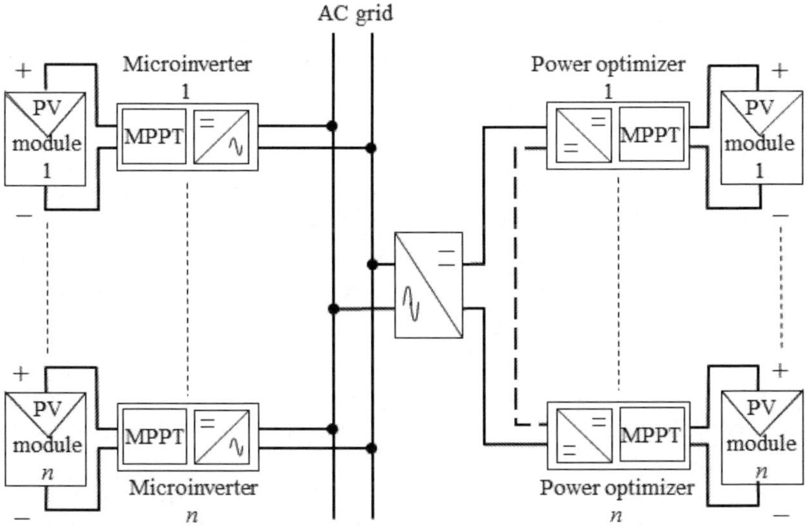

Fig. 9. Distributed topologies for MPPT: microinverters and power optimizers.

centralized MPPT present in conventional inverters like: higher electricity production in mismatched situations (due to shading, for example), advanced monitoring allowing for relatively complex fault detection routines (monitoring has to be done at the individual PV module level), protection against hot-spots, system design flexibility (combination of PV modules with very different location are possible), better safety against fire, etc.

In shaded PV generators, recent research has confirmed[32] that, although some power gains can be obtained compared to centralized MPPT (typically 10–15%, with maximum reported values up to 34%), they will not translate to the same amount of annual energy gains due to several factors like: the stationary nature of the obstacles, the moving position of the sun throughout the year and, in some cases, the amount of the diffuse fraction of the solar resource.

The main drawback, namely the reduced reliability, was experienced in the first commercial microinverters in Europe (1990s), mostly related to high working temperatures (the MLPE devices are usually placed behind the PV modules) and, consequently, low operational lifetimes. Also, more complex replacement operations are to be expected in architecturally integrated PV generators if the MLPE devices lifetime is shorter than that of the PV modules. Since 2010, after continuous R&D efforts, new microinverter designs and power optimizers have increased their market share steadily (especially in the U.S.) with better guarantees. Although they are not limited to urban environments, these are their natural market since in large ground-mounted PV systems the extra cost has been difficult to justify up till now.

- Centralized topologies for MPPT, based on string- and central inverters are shown in Fig. 10. The selection of one or the other is far from an easy task, since both cost and technical issues must be taken into account. In energy efficiency terms, the efficiency of both solutions is practically the same (string inverters often being slightly more efficient). Other factors may also influence the final selection, such as: PV modules/arrays in multiple locations, space or weight constraints (string inverters are more compact and lighter), reliability (specially demonstrated in central inverters), etc.

From the perspective of how the PV modules are incorporated into buildings, the novelty of the use of PVs in the building industry has not

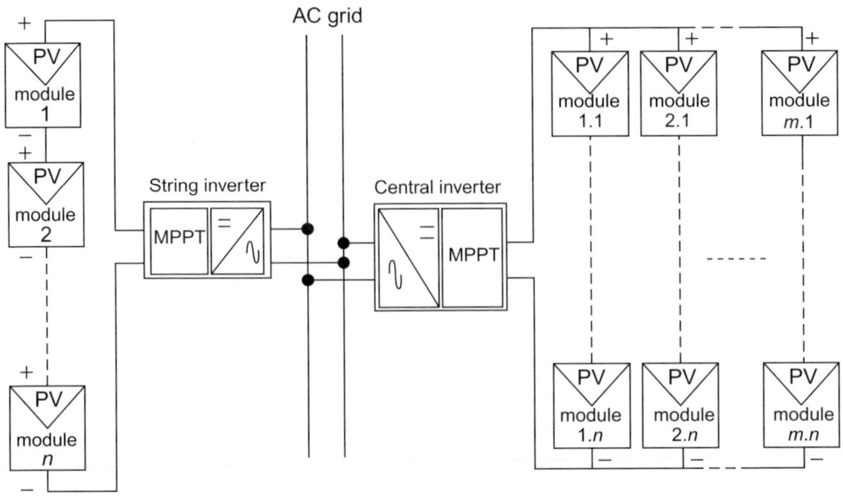

Fig. 10. Centralized topologies for MPPT: string inverters and central inverters.

allowed it to reach a consensus yet in defining what is "building integration". As has happened with other technological innovations, definitions continuously evolve as knowledge and experience increase. It is worth mentioning at this point the joint efforts done under the International Energy Agency framework, particularly through its Solar Heating and Cooling and Photovoltaic Systems Programs (respectively, Task 16: Photovoltaics in Buildings,[33] and Task 7: Photovoltaic power systems in the built environment[34]). In parallel, the following distinction was increasingly used:

- "Building Added/Attached Photovoltaics" (BAPV), when the PV modules are incorporated over the building envelope, on top of the building skin.
- "Building Integrated PV" (BIPV), when the PV modules play dual (or multiple) functions by replacing conventional construction elements in roofs, façades, glass construction systems, shading and canopy systems, etc.[35]

Continuous international collaborative work done under the International Energy Agency umbrella has allowed developing publicly accessible case studies databases, so that cumulated experience can be shared, such as: the "PV Database" of building integrated and urban PV solar energy projects (jointly developed with European Union resources)[36]; and the "Case study collection",[36] where projects using passive solar and active

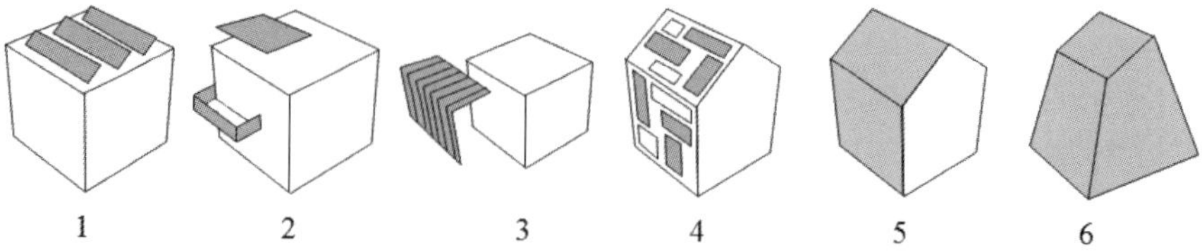

[1- added technical elements with single function; 2- added technical elements with double function; 3–free standing structures; 4- part of surface composition; 5- complete façade/roof surfaces; 6- form optimized envelopes for solar energy use]

Fig. 11. Solar energy integration typologies, according to the IEA-SHC, Task 41: Solar Energy in Architecture (Subtask C: Concepts, Case Studies, and Guidelines).

solar thermal technologies can also be found. This approach has clearly demonstrated the variety of PV integration possibilities to be used in the envelope composition (see Fig. 11 for schematic illustration, the grey elements being PV generators).

The previous classification is obviously not final and appears to be a candidate for continuous evolution over the next few years, particularly if urban and buildings design are successful in providing solutions to growing concerns internationally about the impact of climate change, sustainability, rising energy prices and interdependence, etc. These concerns are, in fact, behind European Union legislation[37,38] concerning buildings' energy efficiency and many member States national regulations (see for example the cases of Germany, France,[39] and Spain[40]), as well as getting behind international cooperation initiatives under the auspices of the International Energy Agency, Solar Heating and Cooling Program (Task 40: Net Zero Energy Solar Buildings[41]).

4.3 *Design and Sizing*

As mentioned at the beginning of Sec. 2.1, small to medium PV grid-connected systems can be designed according to different objectives ranging from maximum annual production to optimization of the generation profile at certain time periods, either in daily or seasonal terms (e.g. at times of particularly high building electrical loads, in order to partially or totally meet the demand).

In any case, design and sizing of a grid-connected PV system entails:

- Concerning the PV generator; determination of the nominal power, technology and electrical configuration.

- Regarding the inverter; determination of its nominal and rated output power and typology.
- Selection of the suitable protection measures to guarantee electrical safety for people, other electrical equipment and the electricity grid.

In the next paragraphs, basic recommendations are provided for the previous aspects.

4.3.1 *PV generator*

The decision about the PV generator nominal power depends, in the absence of legal (maximum acceptable value) or economical limitations, on the surface available which is reasonably free of shadows and on the physical dimensions of the PV modules to be used.

Regarding the electrical configuration of the PV generator, the following alternatives are used[42]:

- Functionally earthed PV generator, where one of the DC conductors is intentionally connected to earth for performance reasons (for example, to avoid "Potential-induced degradation", designation for power losses due to leakage currents induced by the system voltage bias together with certain environmental conditions such as humidity and temperature[e]).
- Floating PV generator, where both active DC conductors are isolated from ground. This alternative is intrinsically safer for people (see comments about safety below), although the absence of a reference earth voltage may expose the PV modules to high voltages above ground potential — to which the metallic frames are usually connected).

Both alternatives entail different protection measures, as described below.

Concerning the maximum voltage of the PV generator, although internationally (International Electrotechnical Vocabulary) "low voltage" is limited to 1500 V, values below 1000 V are typically used, particularly when

[e]Initially detected in n-type monocrystalline solar cells, evidence has been also found in multi- and monocrystalline p-type cells, as well as in some thin-film technologies. Research and Development efforts done by research institutions and manufacturers have led to significant improvements and an international technical standard for experimental detection (IEC/TS 62804: "Test methods for detection of potential-induced degradation of crystalline silicon photovoltaic modules").

the PV systems are installed in buildings. This maximum voltage establishes a limit for the open circuit voltage at Standard Test Conditions:

$$V_{OC,STC} = \frac{V_{MAX}}{k}, \quad (15)$$

where V_{MAX} is the maximum design voltage and k is a constant depending on the minimum historical local ambient temperature (for example[42]: 1.12 for temperatures between −1 and −5°C; 1.16 for temperatures between −11 and −15°C; 1.25 for temperatures between −36 and −40°C).

In practice, several other factors influence also the selection of the PV generator open circuit voltage and consequently the number of series connected PV modules, such as:

- Inverter input voltage characteristics (maximum acceptable voltage and MPPT voltage range) and conversion efficiency curve (input voltage for which the inverter performs best according to its technical sheet).
- PV modules Physical distribution (reasonable layout of DC cabling) and open circuit voltage.
- Safety considerations (for example, design according to Extra Low Voltage provisions).

Regarding the maximum current of the PV generator, the inverter maximum acceptable input current is the main limiting factor. This value, together with the PV modules short circuit current determines the maximum number of PV strings to be connected in parallel.

4.3.2 *Inverter*

Accumulated experience in grid connected PV systems, together with continuous technological improvements in power electronics has led to several inverter design concepts:

- Inverters with transformer. Their main advantage is the possibility of operating with functionally earthed PV generators. They also generally offer a wide range of input voltages, and due to the transformer, the transmission of electromagnetic interference to the electricity grid is reduced. The main disadvantage is the losses introduced by the transformer negatively influencing the conversion efficiency. Also size and weight are higher than in transformerless devices (especially in the case of low frequency −50/60 Hz — transformers, less in high frequency transformers). Typical maximum conversion efficiencies achieved by these inverters nowadays are between 91% and 96%.

- Transformerless inverters. Their main advantage is the increased efficiency, reduced size and weight. As disadvantages, higher electromagnetical interferences are induced in the electrical grid unless specific filtering is used. Also, in some topologies a minimum PV generator voltage is required (of around 400 V), which reduces PV generator configuration possibilities. Also, only floating PV generators can be used. Currently, typical maximum conversion efficiencies achieved by transformerless inverters are between 94% and 98%.

At this point, it is important to note that when speaking about efficiency, two aspects should be considered:

- Instantaneous conversion efficiency or ratio between the output power and the input power (P_{AC} and P_{DC} respectively):

$$\eta = \frac{P_{AC}}{P_{DC}}. \tag{16}$$

  Given the dependence of this parameter on the input voltage and power, information about conversion efficiencies measured at different input voltages covering the MPPT range[43,44] should be provided in the devices' technical data sheet[f] and be considered together with the PV generator expected voltage range.
- Energy conversion efficiency or ratio between the output energy and input energy during a specific time period, typically one year. One of the most widely used parameters are the "Euro" (typical of Central European conditions) and "CEC" weighted (issued by the California Energy Commission and representative of sunnier conditions) annual energy efficiencies. These are calculated considering optimal PV generator position, optimal inverter size and nominal input voltage and are included in most inverter technical data sheets.

Concerning other important technical requirements, compliance with safety standards,[45,46] national/local grid connection standards and codes[47,48] is essential.

Currently, the main technological trends of grid connected inverters include: multiple MPPT ("multistring" concept), where each PV string

[f] Although required by international standards since the early–mid 2000s, unfortunately it is not usually included in the inverters technical data sheets.

has its own input and independent MPPT function; active and reactive power control; "master-slave" concept; active filtering; and "hybrid" power conditioning where, besides grid-connection, a dedicated input for electrical storage is provided allowing a more efficient and flexible energy management of PV electricity.

4.3.3 *Operation and maintenance*

Operation of small-scale PV systems requires a minimum maintenance, as any electrical installation does; PV generators usually need more maintenance as they are located outdoors. The question of cleaning the PV modules has no simple answer, since it depends on the environment (frequency of soiling phenomena such as dust, pollution, etc.) and other determining factors (e.g. aesthetics in building integrated PV generators). In this respect, mechanisms should be planned to facilitate cleaning, access and substitution of potentially faulty PV modules. Several options can be used, ranging from sliding cranes mounted in buildings to external platforms, automatic mechanisms and self-cleaning glass.

In addition, maintenance routines should include:

- Monitoring provisions to guarantee the correct PV system operation (ideally, a daily check of inverter faults and analytical monitoring[25]).
- Preventive plans to maintain within acceptable limits the operation, safety and durability of the installation: visual inspections, verification of protections performance, cleaning and revision of the PV modules, supporting structure, inverter, cables, enclosures, etc.

4.4 *Business Models and International Experiences*

Market development of small-scale PV grid connected systems has been associated with different business models, the evolution of which has been dependent on policy issues, market regulations and, especially, PV technology costs:

- Financing incentives consisting of subsidies or tax rebates applicable to the PV systems investment costs. Initially adopted in the U.S. and some European countries in the 1990's, their main purpose was to reduce the principle non-technical barrier identified at that time. One of the best known examples was the German "1000 Roofs Program (1990–1995)", promoted by the German Federal and States governments, in which subsidies amounted to 70% of the investment costs. Besides the realization of more than 2200 PV systems, the program produced other relevant results such as fostering the development of a national inverters industry and the formulation of the first regulations concerning installation and grid connection of decentralized PV systems. A detailed monitoring campaign of the installed PV systems allowed the identification of one of the main drawbacks of this model for opening new markets and attracting pioneer customers, namely the fact that users are usually more interested in obtaining the subsidies than in the continuous performance of the systems. In addition, success of this model strongly depends on the availability of funding supported by policies and/or regulations.

- "Rate-based" incentives (also known as "feed-in-tariffs", FIT), where the PV owners receive economic incentives for every kWh fed into the electricity grid, the origins of which come from the grid customers' bills. Pioneered by the German city of Aachen and later adopted in many countries in Europe (Austria, Belgium, Czech Republic, France, Greece, Italy, Portugal, Spain, Sweden, Switzerland, United Kingdom), it is considered the most effective policy at stimulating the rapid development of renewables, and more specifically of PVs. The success of this mechanism to steadily increase the PV market has shown a clear dependence on the difference between the incentives with commercial electricity prices and electricity production costs. Also important for the design of these incentives is the capacity to dynamically adapt the FIT tariffs to market development. Among the most successful experience is the German Renewable Energy Sources Act. Issued in 2000, it has been revised and amended several times in order to adapt the feed-in-tariffs and other legal provisions to the market evolution and European Union legislation concerning state aid rules. Similar but less successful incentives are "feed-in-premiums", where the remuneration for PV electricity consists of a general market price (e.g. wholesale electricity price) and a premium, the sum of which is less than the retail electricity price.

Over the last two decades, PV technology cost competitiveness has improved significantly due to remarkable cost and price reductions in PV modules (a price decrease of 20% every time the cumulative sold volume of PV modules has doubled, and a price reduction of around 85% between 2007 and 2014) and in the so-called Balance-of-system (BOS: inverter, cables, protections, etc.). This fact, together with decreasing rate-based incentives and rising prices of retail electricity has generated a growing interest in the concepts of "self-consumption" and "self-sufficiency"[g], physically possible when using the electrical block diagram shown in Fig. 2(a). Besides the availability of the solar resource and PV system costs (including financing), other relevant aspects for the profitability of self-consumption schemes are:

- The cost of grid electricity. Under most common electricity billing practices used in (theoretically) liberalized electricity markets, commercial electricity tariffs for residential and small-commercial users are the highest, the main reasons being related to the full use of electrical infrastructure required (generation, transport and distribution services). For this reason, this sector is generally considered to be able to benefit first from the so-called "grid parity" condition, where the cost of generating 1 unit of the user's own electricity equals the unitary market value (commercial electricity). As an indicator, the parameter "Levelized cost of electricity" (LCOE) is most often used. Assuming a constant value per year it can be calculated, in currency units per kWh, as:

$$LCOE = \frac{\sum_{t=0}^{T} C_t/(1+r)^t}{\sum_{t=0}^{T} E_{PV,t}/(1+r)^t} \quad [\text{c·u/kWh}], \tag{17}$$

  where C_t is the net cost of PV system investment for a time t; $E_{PV,t}$ is the net PV electricity generated by t; r is the discount rate for t; and T is the lifetime of the PV system.

Although criticized for the difficulty and frequent lack of clarity of the many underlying assumptions (discount rate, PV system costs, financing and incentives) of the reported values and for ignoring social and economic externalities associated by the different electricity supply mix options, it has been used to demonstrate academically the "turning" point for the cost effectiveness of solar PV, both in America and Europe (starting in California, Mexico, Spain, Italy). Unfortunately, still existing technical and non-technical barriers, also not accounted for in LCOE calculations, still hinder an important share of potential investments.

- The potential self-consumption in relation to the user's electricity demand. Here the time coincidence between generation and load and the existence (or not) of electrical storage are of paramount importance.

[g]Calculated as follows: the "self-consumption" rate is the ratio between PV electricity locally consumed and total PV generation; the "self-sufficiency" rate is the relation between PV electricity locally consumed and the local demand. Both rates can be calculated in different time-frames, the annual period being the most widely used.

Whereas the first can be improved with strategies of demand-side management (DSM[h]), the second allows daily shifting of PV electricity to supply evening-night loads by using only relatively small storage capacities, thus increasing both self-consumption and autarky rates. In a recent study done with data from 10 countries and the most important PV markets, for typical residential load and PV generation profiles (respectively, annual electricity demand of 3500 kWh and average PV system size of 3.5 kWp with no storage), the expected share of self-consumption lies in the range between 29% and 43%. Small scale storage (providing less than one day of "autonomy") combined with DSM can provide relatively high self-sufficiency levels (around 80%) at moderate LCOE increase. For the commercial and industrial consumers sectors of the previously mentioned European countries, assuming standardized load profiles and PV system (annual demand of 150 MWh and PV system size around 150 kWp), the share of annual self-consumption lies between 30% and 52%.

- The value of the PV electricity excess or surplus after meeting the local demand, which is fed into the grid. In this respect, a particularly interesting model for residential and small commercial customers is "net-metering", where the PV excess value equals the price of commercial electricity; in other words: when it is fed into the grid the electricity meter runs backwards. A similar model can be used when following the electrical block diagram of Fig. 1, in the so-called "net billing" schemes, where the PV electricity excess generates the right to consume an equal amount of electricity within a pre-defined period (within the next one or two billing periods, in the following year, etc.). Examples of these schemes can be found currently in America and Europe (several U.S. States, Brazil, Belgium, Italy, The Netherlands).

Another relatively common business model for medium-large PV grid-connected systems in built areas are Power Purchase Agreements, by which the PV system owner sells the electricity within the same building or via a direct line to a nearby consumer, so that the consumer reduces his grid electricity demand and consumes PV electricity at a (usually) lower price. In this model the profitability is mainly influenced by the grid electricity costs, in terms of both power and energy demand. Large commercial and

[h] Although there is no commonly accepted definition, Demand Side Management is considered here as the set of actions that influence the way consumers use electricity, leading to savings and a more efficient use.

industrial buildings can profit easily from this model, given the also usually available large surfaces for solar use; an environment-friendly image can also provide marketing benefits.

As a final conclusion, it can be stated that the accumulated experience with PV grid-connected systems over the last 2 decades has demonstrated that PV technologies have matured significantly. With reliable components being nowadays commercially available, technical and non-technical challenges still lie ahead including an expected sustained effort in research and development (as has been the case in the past). PVs could be the paradigm of DG for the much needed sustainable electricity systems worldwide.

5 SAPVS

5.1 *Introduction*

In this section, we describe briefly the design and the sizing of SAPVS in order to supply the energy requested by a specific autonomous application. As was commented previously, PV technology is often the best solution to electricity supply for a wide range of autonomous applications since it can provide energy from Wh to MWh cost-effectively. Nevertheless, it should be mentioned that PV electricity generation depends on an intermittent source and requires particular procedures to guarantee the reliability in the electrical service.

At the beginning of PV terrestrial use, the message was that this technology does not require maintenance, but the reality is quite different. Like other isolated electrical generators that have high maintenance requirements, SAPVS also needs to implement O&M procedures.

The market for SAPVS could be divided into professional applications (located throughout the world), and domestic applications which are mainly installed in regions without electricity access. Many countries of the developing world in Africa, Asia and Latin-American are implementing SAPVS to facilitate electricity access to rural populations. Some discussion about financial and business models applied in these countries will be included.

5.2 *Types of SAPVS*

Before describing the different types and topologies it is convenient to describe the particularities of PV modules and BOS components of PVSAS.

PV Modules: The PV technology most commonly used is crystalline silicon. Also a-Si is present in low power PV systems, but the share is less than 10%; higher degradation rates and lower reliability are the reason. The CdTe technology is not used at present. Because of the contamination risk due to Cadmium, the manufacturers of these thin film modules are required to recycle it at the end of life. But the operational cost to implement this with low power generation units and isolated systems is not economically viable at present.

As the first PV terrestrial applications were isolated, and the common batteries used were car or truck batteries, which have a 12 Vdc nominal voltage, the first PV modules had between 34 to 36 solar cells connected in series (or multiples of this value). The voltage for the maximum power point is between 16.5 V and 18 V with this number of crystalline solar cells. Therefore, the working point of the PV module (imposed by the battery) is located in the high current region of their curve for all ambient temperature ranges. This is why, in some PV module specification sheets, there is a parameter named "Nominal Voltage". In fact this value indicates the voltage of the battery to be connected to this module.

The difference with grid connected systems is that also very small modules are employed with peak power lower than 100 Wp, i.e. solar lanterns use PV modules of a few watts. These small modules are made with pieces of bigger solar cells.

Support structures: Practically all PV array supports have fixed orientation throughout the year. Sun tracking structures are not used because of their lower reliability in these environments. They require higher maintenance, which is very expensive in isolated locations.

Storage: With the exception of solar pumping, in which the storage is in potential energy with water reservoirs, SAPVS accumulate the not consumed energy generated in electrochemical batteries: the most common is the lead-acid battery, but the Li-Ion battery is being introduced more and more into the PV market. A battery bank is composed of a set of battery elements connected in series; in Pb-acid batteries each element has a nominal voltage of 2 V, which, depending on the state of charge (SOC), fluctuates between 1.9 and 2.3 V. Commercial batteries have several tens to thousands of ampere-hour capacity.

Li-Ion technology has important advantages: a high energy density, an efficiency of 95% and these batteries can be discharged by 95% without damage. The main drawbacks are the cost, although this is decreasing quickly, and its time degradation rate.

The battery life time depends on the number of charge–discharge cycles and their operation regime; frequent overcharge and/or over-discharge dramatically reduces the useful life of the battery. This is why the use of control devices is compulsory.

Charge controller or Battery Management System: Its main function is to protect the battery against overcharge and over-discharge. The identification of the SOC is made with the battery voltage. While quiescent value voltage is a good indicator of the SOC, the dynamic voltage value is not very precise because it depends on the equivalent internal resistance, which changes with the SOC and the number of charge–discharge cycles. To overcome this, some charge regulators integrate control algorithms based on the history of the battery to modulate the voltage set points.

The charge controller can be classified as parallel or series depending on the strategy to protect from overcharge. Figure 12 shows the schematic of these configurations.

Parallel charge controllers have a better efficiency because in normal operation (both switches on) the losses in control devices are lower than in series charge controllers. Nevertheless, a major inconvenience is that a PV module could be short-circuited for long periods when there is not consumption. If the short-circuited PV module has some solar cell with different I_{SC} parameters or is partially shadowed, a "hot spot" failure could appear.[i] The terms of reference of some tenders explicitly forbids the use

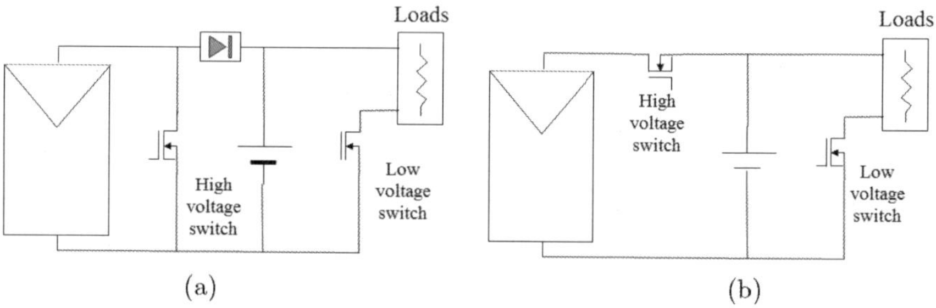

Fig. 12. Charge controller with (a) parallel and (b) series control for overcharge.

[i]When a solar cell works in different condition such that it has to dissipate part of the power generated by the remaining solar cells. If the temperature is higher than the encapsulate fusion temperature, the PV module will be damaged quickly. This failure is denominated "Hot Spot".

of parallel charge controllers. Besides the "hot spot" failure is linked to the module quality.

In addition to its battery energy control function, the charge controller integrates status indicators to improve the user management. Many charge controllers integrate also a maximum power point tracking MPPT to extract the maximum power to the PV generator. With this control device the PV power could increase by a mean value of 8%, and in cold months could achieve 14%.[49] Another advantage of MPPT charge controllers is that they increase the range of PV modules to be used in SAPVS, because they increase the voltage range of PV generators suitable to connect with batteries.

Inverter: This equipment is essential to supply AC loads in SAPVS, but this design is different from grid connected inverters because these have to synchronize with the grid (and switch off if there is a blackout). The stand-alone inverters have to produce AC electricity without external references and have to supply electricity to not purely resistive loads, i.e. they have to support transient power requested by loads such as motors just to arrive at a permanent regime. This is why it is important to verify the overload capacity in the specifications sheet. While the size of a grid connected inverter depends on the PV array size, the size of a stand-alone inverter only depends on the maximum current supply to the AC loads. The efficiency of stand-alone inverters is typically below grid-connected ones, mainly because the input DC voltage is lower: the usual values are between 0.92 and 0.95.

There is a particular inverter, named bidirectional or charger-inverter that permits power flux from AC side to battery, besides battery to AC loads; it is employed to charge the battery from an AC supply. This equipment is also used in a particular topology, an AC bus, which will be described later.

In Chapter 2, Sec. 5.3 were represented all the main SAPVS topologies. In this section, we analyze in more detail the topologies used for the different autonomous applications. The simplest SAPVS is composed of a PV module, a battery and a charge controller to supply energy to a DC load, as shown in Fig. 13.

Most SAPVS have this topology. The typical loads are: lights (fluorescent, LED lamps), electronics, small kitchen appliances, etc; in general, low consumption equipment. In the case of domestic rural electrification, the terms used are: pico-solar (picoPV)[50] for systems with a PV module lower than 5 Wp; solar home systems (SHS) composed by a PV module between

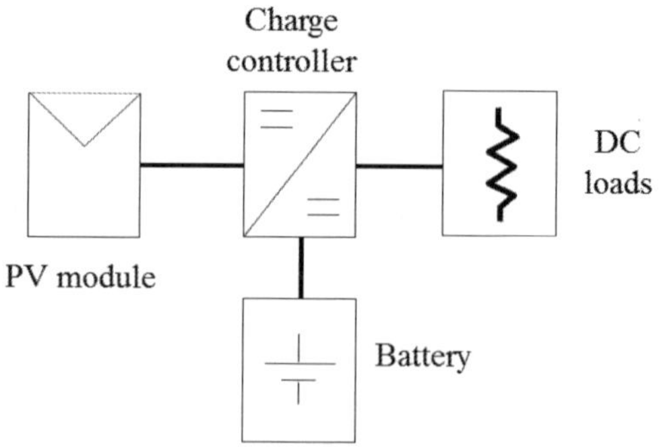

Fig. 13. PV system diagram for DC loads.

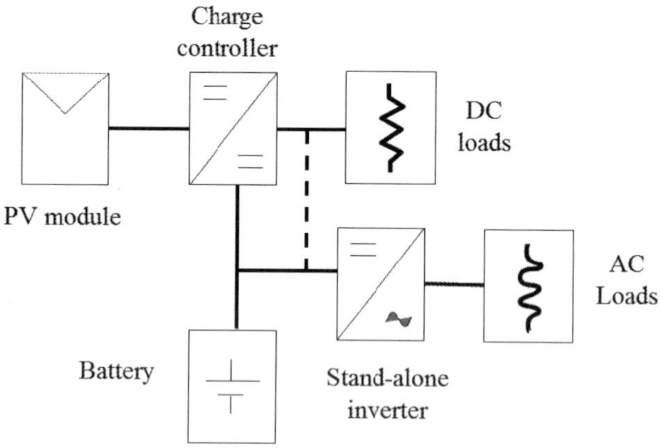

Fig. 14. PV system for DC and AC loads.

50 to 100 Wp, a Pb lead battery with capacity about 100 Ah to 150 Ah to supply fluorescent lamps and small appliances; and, recently, microSHS, that are composed of a PV module between 10 to 20 Wp, a Li-Ion battery between 40 to 90 Ah to supply LED lamps and phone chargers.

When the PV system has to supply AC loads the previous scheme is completed with a stand-alone inverter, as it is shown in Fig. 14.

The inverter should be plugged into the PV system through the charge controller (dotted line in Fig. 14), but many charge controllers do not work properly in this case. Power transients, when an AC load is switch on, produce a temporary large reduction in the battery voltage, lower than the voltage set point for the discharged battery. Consequently, the charge controller disconnect the loads. Some manufacturers integrate a delay of some minutes before disconnecting or adding a data line between the two pieces of equipment (valid only if both have the same manufacturer). To avoid this inconvenience it is usual to connect the inverter directly to the battery, but this requires that the inverter contains a protection for battery over-discharge.

A particular topology for SAPVS is the case of water PV pumping where the electrochemical storage is replaced by potential energy storage with reservoirs. The motor is pumping water while there is generation and the tank is not full. Depending of the water quantity per day and the height to be pumped, DC or AC pumps are used. The capacity is higher for AC pumps, but if a special inverter is used that can change the voltage and the frequency, then the global efficiency is improved. PV pumped water was one of the first competitive applications in comparison with more conventional off-grid pumping procedures such as diesel engines.[51]

In general, when the requirements for energy delivered are higher, the SAPVS integrate additional power sources such as a diesel generator, wind machine, hydropower, etc.; they are called Hybrid Generation Systems. Figure 15 shows a PV-diesel hybrid system, the only difference with the PVSAS represented in Fig. 14 is the added engine. It is a DC bus topology because the main power line is direct current.

Another option, when the systems increase in size, is AC coupled hybrid systems, as shown in Fig. 16. In these systems DC coupling is completely avoided and the battery is equipped with a bidirectional inverter. This provides great modularity, simplifying the addition of any new equipment,

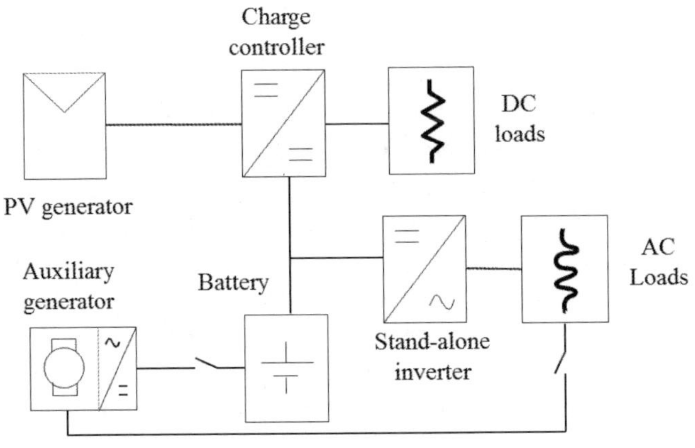

Fig. 15. Hybrid PVSAS.

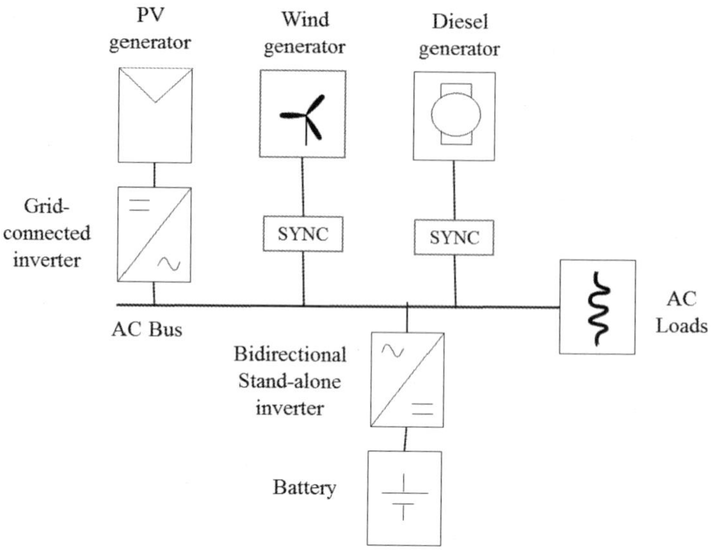

Fig. 16. Hybrid system coupled in an AC bus.

with the cost of having additional power conditioning equipment and, thus, additional power conversion losses.[52]

In AC coupled hybrid systems an island grid is built. Normally, one of the components acts as a voltage source and constitutes the grid, setting voltage and frequency; the other AC generators feed in their power, acting as current sources. This results in a relatively simple control structure. However, it also has some drawbacks. Because the grid must always be kept alive, a permanent power supply has to be provided.

If a diesel generator is used for building the grid and is kept running during long periods only for this purpose, the overall efficiency of the system will be affected. In systems without a diesel generator, if the battery inverter is used, the whole system could fail during low battery periods. When the sun comes out again, the PV inverter, which is a grid-connected type, will not work without the grid formed by the battery inverter and will not be able to charge the batteries. At this point the system is blocked and to unblock it the batteries would have to be charged independently of the system, with the obvious drawback and system down time. In hybrid systems with a diesel backup some strategies could be implemented to avoid this blackout.

In addition, the power produced by the PV generator has to be transformed to AC and the excess power with respect to AC load consumption has to be transformed to DC for battery storage. When the power to the AC loads is delivered from the batteries two unnecessary energy transformations are taking place: AC–DC to the battery and DC–AC. Considering a 92% weighted efficiency of the battery inverter,[53] this results in a maximum efficiency loss of 15%.

From the combination of the two previous systems and extracting the advantages of each system, a combined AC–DC coupled system is possible. Basically, this topology consists in coupling all DC sources and loads in DC and all AC sources and loads in AC. Figure 17 shows a schematic of this topology.

In the combined AC–DC topology the bi-directional inverter is the brain of the system. It is in charge of building the AC grid and regulating the charge and discharge of the battery. It is a proven robust technology,[54]

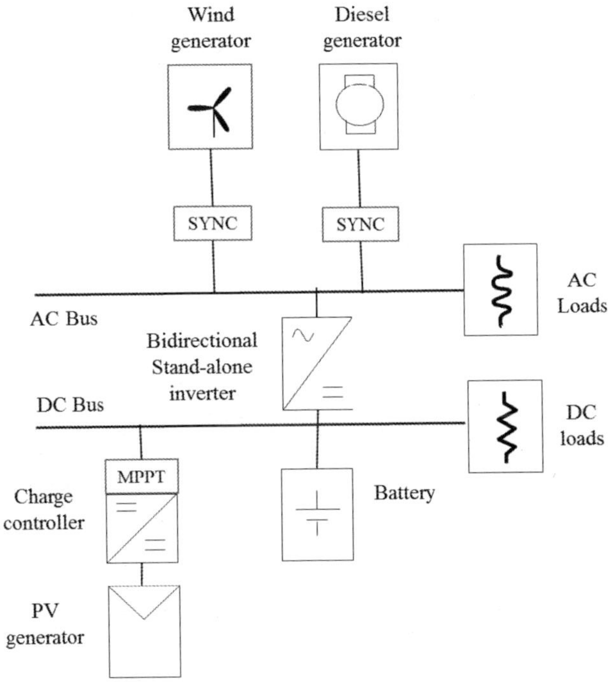

Fig. 17. Combined AC–DC coupled hybrid mini-grid.

plus it solves the problem of blocking the system as in AC coupling because the batteries can be charged without the grid, and the modularity advantage remains. It is suitable for large systems and it gets rid of unnecessary efficiency losses in the battery inverter of AC coupled systems.

5.3 *Design and Sizing*

Planning the design of a stand-alone power generator requires us to specify the load demand curve and their evolution in the future. With this information, we can proceed to estimate the nominal power of the generator. In the case of PV generators there is a supplementary effort: to estimate the energy resource, the solar irradiation, which is a non-deterministic variable.

The merit of a SAPVS should be judged in terms of the reliability of the electricity supply to the load. This is usually quantified by the concept of loss of load probability, *LLP*, defined as the ratio between the energy deficit and the energy demands on the load, over a long period of time.[55] Because of the random nature of the solar radiation, the *LLP* of even a trouble-free PV system is always greater than zero.

The size of a PV system is a general concept including the dimensions of the PV array and the accumulator. A useful definition of such dimensions relates to the load: On a daily basis, the PV array capacity, C_A, is defined as the ratio between the mean PV array energy production and the mean load energy demand. The storage capacity, C_s, is defined as the ratio between the available energy from the accumulator fully charged and the mean load energy demand. As will be justified later, it is convenient to consider values for both parameters averaged over a month. Thus:

$$C_{Am} = \frac{E_{FVdm}}{L_{dm}} = \frac{P_{NOM,G}G_{dm}(\alpha,\beta)/G^*}{L_{dm}}PR,$$

$$C_{Sm} = \frac{C_U}{L_{dm}},$$

where E_{FVm} is the useful energy generated by the PV array, which has a nominal power of $P_{\mathrm{NOM,G}}$, when the mean monthly daily irradiation value on the array plane is $G_{\mathrm{dm}}(\alpha,\beta)$ and the efficiency is PR, the performance ratio of the PV system. The mean daily energy consumption is represented by L_{dm} and the useful storage capacity is C_U, which could be estimated as the product of the nominal capacity and the maximum depth of discharge allowed by the charge controller. The minimum monthly value of

C_{Am} defines the reliability of the SAPVS and this will be the "worst" month. When the load is constant for all the year, the worst month is that with the lowest mean daily irradiation value.

Note that C_{Am} depends on the meteorological conditions of the location. That means that the same PV array for the same load can be "large" in one site and "small" in another site with lower solar radiation.

Given a location and a load, two general ideas are intuitive: First, it is possible to find many different combinations of C_{Am} and C_{Sm} leading to the same *LLP* value. Second, the larger the PV system size, the greater is the cost and the lower the *LLP*.

The task of sizing a SAPVS consists of finding the better trade-off between cost and reliability. Very often, the reliability is an *a priori* requirement from the user, and the PV engineering problem is formulated as follows: Which pair of C_{Am} and C_{Sm} values leads to a given *LLP* value at the minimum cost? To find the solution depends on the available solar daily irradiation data for the location and the load evolution. A minimum time for daily data needed for accuracy estimation is 7 years. In this case it is easy to simulate the daily energy flows between generation, consumption and storage for a pair of C_{Am} and C_{Sm} values; the ratio between energy not delivered by the generator and energy delivered to the load is the loss of load probability.

Unfortunately, the number of locations with such information is scarce, particularly for isolated regions. Nor has an algorithm based on solar radiation statistical parameters been developed to estimate the pairs of C_{Am} and C_{Sm} with which a specific *LLP* value is achieved. When the information about solar radiation is limited to daily mean monthly values, one value for each month, the relation between C_{Am}, and *LLP* depends upon the intuition of the PV engineer. The PV array dimension is calculated to ensure that the generation during the worst month exceeds the consumption by a security factor directly established by the PV engineer according to the type of application and his own experience. A similar procedure is used to estimate the battery size. To summarize:

$$\min(C_{\text{Am},m=1\,\text{to}\,12}) = F_{S1} \quad \text{and} \quad C_{\text{Sm}} = F_{S2},$$

where F_{S1} and F_{S2} are arbitrary factors. For instance, a typical F_{S1} value for domestic SAPVS is 1.2, which means that the mean daily PV energy production, in the worst month, will be the 120% of the energy consumed. Normal values for F_{S2}, when there is not another power supply, are between 2 and 5. This method is very simple and useful to calculate the PV system dimensions, but they do not allow us to quantify the reliability. Nevertheless this is the method most generally used.

The steps to design a SAPVS with the "worst" month procedure are as follow:

(a) Estimate the monthly mean daily load consumption.
(b) Calculate the mean values of a month's solar irradiation.
(c) Select the month with the lower ratio between solar irradiation and load consumption, once the tilt plane is optimized (see Sec. 2.2).
(d) Decide the values for F_{S1} and F_{S2}. Then the nominal power of the PV generator, $P_{NOM,G}$, and the useful battery capacity, C_U, could be calculated considering a performance ratio as 0.65 as a typical value for SAPVS.
(e) Select the specific PV modules and battery elements in combination with the charge controller; the number of series batteries depends of the nominal DC voltage, V_{NOM}. The parallel-series connection of PV modules depends upon the charge regulator selected, but if this does not have maximum power point tracking depends of V_{NOM}.[j] The nominal capacity of the battery will be the useful capacity divided by the maximum depth of discharge allowed by the charge controller. In this phase it is fundamental to verify the electrical compatibility between the PV generator and the charge controller in terms of voltage and current.
(f) Select the stand-alone inverter, if the PVSAS has to supply AC loads. If there is no information about the hourly load distribution, the designer should postulate it in order to select the nominal power of the inverter.
(g) Estimate more exactly the performance ratio, if possible, to correct the $P_{NOM,G}$ obtained in step d. Also it is interesting to reevaluate F_{S1} and F_{S2} with the definitive values of $P_{NOM,G}$ and C_U.

In the case of hybrid systems the sizing task is more complex and it is indispensable to appeal to multivariable algorithms.[56] As an example, consider the design of a PV-diesel hybrid system in which the diesel engine has a backup role, i.e. the yearly energy delivered by the engine is lower than 10% of the total energy consumed. With this hybrid system, the selection of

[j]The electrical engineering rule to select the nominal DC voltage is to minimize the thermal losses in wiring, but in SAPVS possible conflicts could be presented when there are DC loads that have to be supplied with specific voltage values (12, 24, 48, 120 V).

battery storage capacity is fully dependent on investment costs, the replacement costs of the battery and the fuel cost, but not dependent on the loss of load probability required: the deficit of energy production, a consequence of the solar radiation variability, is provided by the diesel backup.

To optimize the design of a hybrid system it is very suitable to use software tools such as HOMER (Hybrid Optimisation Model for Electric Renewables), which is freely available as HOMER Legacy Version 2.68.[57]

5.4 *Operation and Management*

Although the system is installed according to the guidelines and is thoroughly inspected prior to handing it over to the user, its short-term as well as long-term performance is considerably influenced by the user. There are some simple but basic guidelines, all of them compulsory, to be followed in system operation:

- The PV module should be without shadows and properly cleaned.
- Limit consumption according to the technician recommendations.
- Connect only allowed line extensions and new load appliances (attention to power and voltage).
- Do not by-pass the charge controller.
- Pay attention regularly to charge controller warnings.
- Do not short-circuit the fuse location with a wire.
- If load replacements or any other task is done on the installation, pay attention to polarity.

Users' training is important for sustainable SPV stand-alone electrification effort.

- Maintenance is a key factor for PV systems sustainability. Maintenance strategy, resources and responsibilities should be clarified at the project design phase, before the installation. Some points to be defined are: Substitution of old equipment;
- Preventive actions: frequency and tasks;
- Time of response after users warning;
- Maintenance tasks distribution; expert technicians, local technicians, users;
- Sanctions in case of maintenance delay;
- Costs;
- Training.

Table 3: PV generator maintenance guidelines.

General item	Specific item	Action (if failure is detected)
Location	Position changes, shadowing	Relocation, avoiding shadowing as much as possible.
Cleanliness	Appreciable soiling (visual inspection)	Cleaning, with a frequency depending on outdoors conditions
Connections	Tightness, box enclosure, signs of corrosion	Reconnect wiring Change terminals by new ones

Technical guidelines for stand-alone system maintenance are given below. Local factors and project design considerably affects system operation, performance, failures and maintenance needs. These guidelines can be suitably modified for any particular project conditions. The specific points to be checked by the technician and the recommended corrective action are given in Table 3. All maintenance calls should be recorded (the type of incidence and date) by the technician. This is very useful for the long-term performance of the system and also as reference document for the installations to follow.

(A) PV generator:

The PV generator does not usually need frequent actions, although they are basic to achieve an optimum energy generation. Particular conditions determine this PV generator maintenance.

(B) Battery:

Battery ageing is one of the main concerns in PV system performance and energy supply. Battery checking, maintenance and replacement are essential tasks in long-term operation. However, it is not easy to state specific criteria for decisions, since a bad battery state could have different, but sometimes dependent, causes. The experience of the technician is crucial for good maintenance. In the Table 4 there is a proposed set of battery maintenance guidelines.

(C) Charge controller:

Charge controllers should be periodically checked, because of their influence on the system operation. Some specific points should be always checked (see Table 5).

(D) Load appliances:

Loads are obviously the components that change most frequently between technicians' visits. Special attention need to be paid not only to particular

Table 4: Battery maintenance guidelines.

General item	Specific item	Action (if failure is detected)
Labeling	Battery type and capacity	Ask for battery characteristics and labels
Location	Safety, ventilation, temperature	Relocate the battery or rebuild its protection enclosure
General conditions	Liquid spillage, box breakage, left cup	Ask for causes of bad conditions (internal or external) → change battery if energy supply is not correct
Connections	Tightness, signs of corrosion	Reconnection, cleaning Ask for energy supply conditions
Electrolyte level	At minimum level signal (above plates)	Refilling with distilled water → too frequent water refilling can be caused by a poor battery or a high voltage regulation
	Below the plates	Battery can be seriously damaged → verify energy supply and check its evolution
Electrolyte density	Mean value → it depends on the SoC (never below 1.10 g/cc)	Recharge battery
	Differences between cells (never above 0.03 g/cc)	Equalization charge → Density correction (to be made only by technicians) → Battery change, due to cell damage
Voltage	Open-circuit voltage (never below 11.4 V)	Check the actual irradiation conditions and the energy supply evolution → battery ageing, charge controller failure.
Capacity	High number of load disconnections (users complaints if no monitoring equipment is installed)	Check system conditions (load consumption, component conditions) → battery change → verify remaining capacity of replaced batteries (or samples)

lamps conditions but on the number and type of the appliances, due to their influence on the energy balance of the system (see Table 6).

(E) Wiring and accessories:

Load line extensions made by users to supply electricity to new rooms are common practices in the field. Special attention need to be paid to assure some basic requirements (see Table 7).

(F) Inverter:

Inverter maintenance tasks are similar to the ones for charge controllers (see Table 8).

Table 5: Charge controller maintenance guidelines.

General item	Specific item	Action (if failure is detected)
General conditions	Breakage, burnt signs	Substitution
Warning signals	Active in maintenance time	Substitution
Power transmission: voltage measurement	Values out of specifications	Substitution due to internal component failure
Regulation thresholds[k] (field measurement with a potentiometer)	Out of specifications	Readjustment if possible Substitution → cause of battery damage
Fuse conditions	Rating, Blown, short-circuit fuse position	Connect new fuses (including spare parts)
Connections	Tightness, terminals	Reconnect

Table 6: Loads maintenance guidelines.

General item	Specific item	Action (if failure is detected)
Type and power	Out of project specifications	Retirement of incorrect appliances → Advise for low consumption if no low power appliances are available (provisional measure)
Light performance	No light	Replacement → Verify replacement frequency for further causes
	Excessive blackening	Spare parts needed → check time in operation → quality
Connections	Tightness	Reconnect

(G) PV system:

After all components have been checked, the system operation needs to be verified (see Table 9). If there are no failures or if the failures can be repaired at the moment, the system is ready to operate. If not, provisions have to be made to assure the fastest repair possible.

To conclude this section, the paramount importance of O&M in SAPVS has to be stressed; the long term reliability is linked to the implementation of the best O&M practices, but this is not an easy task in rural electrification

[k]This test is the substantial importance for the long-term operation of the PV system. The procedure is easy, but for a technician. Only a voltmeter, a potentiometer and the battery itself are enough to measure the voltage set points of the charge regulator.

Table 7: Wiring maintenance guidelines.

General item	Specific item	Action (if failure is detected)
Type	Incorrect type, no polarity identification	Substitution
Electrical behavior	High voltage losses, insufficient capacity	Substitution, (optionally: double wiring, as interim solution)
Line extensions	Permission, conditions	Retirement if not allowed Substitution if out of specifications
Plugs and sockets	External conditions, polarity, connections	Reconnect Substitution
Switches	Bad connection, type, rating	Reconnect Substitution

Table 8: Inverter maintenance guidelines.

General item	Specific item	Action (if failure is detected)
General conditions	Breakage, burnt signs	Substitution
Warning signals	Active in maintenance time	Substitution
Self-consumption	Values out of specifications	Substitution
Voltage regulation (AC voltage at real DC input)	Out of specifications	Substitution → loads damage
Fuse conditions	Rating, Blown, short-circuit fuse position	Connect new fuses (including spare parts)
Connections	Tightness, terminals	Reconnect

Table 9: PV system maintenance guidelines.

Item	Criteria	Action (if failure is detected)
System performance	Switch all loads	Return to component verification

programs with a high degree of dispersion. In fact, many rural electrification programs fail because the operation costs were not sustainable.

5.5 *Business Models and International Experiences*

To analyze the business models for SAPVS it is convenient to separate the PV systems depending on the particular economic regions where the PV systems are located. For SAPVS installed in the developed world the sales model, direct or credit, is practically the only one present. However, for rural electrification in developing countries the most extended business model are microcredits and fees for service, also denominated "Energy Service Company".[58]

Rural electrification programs with Solar Home Systems based on microcredits are the most extended, probably because it is the preferred model in Asia (in Bangladesh more than 3.5 SHS millions have been installed at the end of 2014[59]), but there are many electrification programs based on fee for service: 58.000 SHSs in Laos,[60] 25.000 in Morocco[61] 24.000 in the program funded by the Foundation Rural Energy Services — Mali, South Africa, Burkina Faso, Guinea-Bissau and Uganda — and at a lower-scale Botswana, Ethiopia, Kiribati (the first example of Energy Service Company at the end of 80s), Namibia, Zambia, Fiji, Ghana, India, Cape Vert, Tuvalu, Benin, etc.

Also in Latin America there are fee for service rural electrification programs. Some examples are: Argentina, Bolivia, Brazil, Chili, Honduras, Peru, and Dominican Republic. It should be remarked that the Argentinian program PERMER,[62] was started in 1996 and is still running. One of the companies working in this program is EJESA. This company has 7.000 clients disseminated in the Jujuy Department, who receive electricity produced by thermal generation, PV-diesel hybrid systems, hydroelectricity and solar homes systems (4,000 are installed). Other interesting case study is the program *Luz para todos*, in Brazil. In the Bahia Department, Coelba, alone, the concessionary company installed 20.000 SAPVS for AC loads at the end of 2012. After reviewing the extended bibliography about business models for SAPVS the conclusion is that the business model selected does not guarantee the success of the program.[63]

In an interesting reference, the success and failure factors are analyzed for ten energy access programs supplied by renewable energy in Asian countries (eight with PV technology).[64] The methodology used for the study is based on a multifactorial qualitative analysis from a selected bibliography and many interviews with the stakeholders. The results show ten key elements for success: appropriate technology, income generation, financing, political leadership, capacity generation, programs flexibility, marketing and awareness, stakeholders implication, real appropriateness of the project as seen by the users community and technical quality. Once again the study shows that the business model is not critical for the success of the rural electrification program.

6 Safety and Protections

Before describing the main recommendations on safety, it is worth mentioning the particular characteristics of PV generators, that are different from other DC electricity generators. First, PV generators are intrinsically current limited sources (with the rated current I_{SC} being approximately just 20% higher than the optimal operating point I_M for crystalline silicon technologies). Secondly, they are spatially distributed and permanently operate outdoors, which increases the exposure to ageing agents. And third, the previous fact, together with their particular energy source (the sun) makes PV generators especially difficult to be "turned-off" in case of faults. Faults that, in cases such as electric arcs, may persist for long periods before being noticed when the damage is already done. All these aspects influence the required safety measures to be implemented in the design and sizing of PV systems.

Potential faults in PV systems include faulty electrical connections, insulation losses, short circuits, overcurrents and overvoltages, the causes being common to conventional electrical installations: ageing, degradation, mechanical damages, ill-functioning connections, adverse meteorological conditions, etc. In any case, a PV system should be designed to guarantee safety for people, electrical installations and the electricity grid, if it is grid connected. In the next paragraphs basic design criteria and the most frequently used methods are described.

The main principles underlying electrical protection for human beings are to prevent direct access to active (live) parts and to make accessible conductive parts safe in normal operation conditions or in the presence of a single insulation defect. In this sense, consideration of the effects of DC and AC current on human beings has led to the definition of limit contact voltages applicable to PV generators and inverters under which no dangerous currents occur,[65] respectively: 120 V (maximum open circuit voltage[66]) and 50 V (dry/humid AC electricity supply). These limits would involve large currents of PV generators in systems of just a few kW and, additionally, would be not compatible with grid connection requirements. Consequently, protection measures must be taken in order to guarantee:

- Protection against direct contact, or physical contact with an electrical material (e.g. a conductor, or equipment terminal) that is live in normal circumstances.

- Protection against indirect contact, or contact with an exposed-conductive-part that is not normally alive, but has become so accidentally (due to an insulation failure or some other cause).

Particularly, effective in this sense is the use of double or reinforced insulation, especially in the DC circuit. This requires using PV modules and other electrical materials (cables, connectors, connection boxes, etc.) equivalent to class II insulation (class A according to a specific PV modules standard[67]) and results in floating PV generators, which are intrinsically safer than functionally earthed ones. However, since insulation faults cannot be fully avoided in the DC circuit, automatic disconnection of supply can be implemented[1,42]:

- For inverters with galvanic isolation by means of continuous measurement of the PV generator insulation resistance. Disconnection should be triggered if this resistance falls below a specific limit; re-connection with the grid should occur when insulation resistance returns to acceptable values.
- For inverters without galvanic isolation (e.g. transformerless) upon continuous measurement of insulation resistance and residual currents (leakages to ground). In addition to an insulation resistance limit, a residual current limit should also trigger the disconnection.

For functionally earthed PV generators that should be only connected to inverters with galvanic isolation, implementation of automatic disconnection of supply is also based on insulation resistance monitoring prior to connection and continuous residual currents monitoring during normal operation.

Automatic disconnection of supply on the AC side can be implemented by using circuit breakers and residual current devices, depending on the AC earthing/grounding configuration (TT, TN, IT).[68–70]

Concerning protection measures of electrical installations, they should include:

- Short circuits and overcurrents. In addition to selecting the DC and AC cables' adequate sections (given the expected maximum currents), cable layout should prevent contact between opposite polarity conductors and

[1]Combining inverter shut down and physical disconnection of the PV generator–inverter or inverter–grid interfaces.

terminals. Protection devices to interrupt faulty currents should be also used, complying with existing standards (e.g. fuses,[71] circuit breakers, switches,[72] etc.).

- Overvoltages. A PV generator does not necessarily increase the risk of the direct impact of lightning strikes. However, if PV is installed in buildings, in case of lightning strikes, the electrical facilities may experience electromagnetic interference induced in the PV conductors, cable raceways, etc. In any case, a lightning strike risk assessment should be done according to existing standards[73] and results considered for the electrical design. In addition, surges in PV systems may be induced by coupled voltages due to lightning discharges and switching operations in the electrical grid. Consequently, overvoltage protection must be implemented at the PV generator–inverter or charge controller interface (usually before DC cables enter the building and at the inverter's input) and at the inverter–grid connection interface (at the inverter's output and at the building's low-voltage power supply).[74,75]
- Equipotential bonding and grounding. These are paramount for the correct operation of protection measures against electrical shock, lightning and overvoltages; consequently, they should be carefully planned taking into account their particular operation conditions (especially in the PV generator). Grounding should be also compatible with existing AC earthing/grounding configuration. Functional earthing of PV generators should be implemented according to recommended practices (standards).[42]
- When batteries are present in the installation, additional safety provisions have to be made. Batteries are potentially dangerous, because of their high energy density: short circuit and overcharge could produce an explosion; additionally, in the case of vented lead-acid the overcharge could cause the dissociation of water into oxygen and hydrogen and both gasses are highly flammable. To avoid critical situations it is compulsory to add protective measures for lead-acid[76] and Li-Ion[77] batteries that are preferred for PV systems.

Last, but not least, small grid connected PV systems must maintain the electricity grid quality parameters, described in the corresponding national/local grid connection standards,[47,48] an explicit responsibility allocated to the inverters. In addition, PV systems should not transfer to the grid any potential defects and disturbances.

References

1. R. Barlow, B. McNelis and A. Derrick, Status and experience of solar PV pumping in developing countries, in *Proc. 10th European Photovoltaic Solar Energy Conf.*, Lisbon, Portugal, pp. 1143–1146 (1991).
2. T. Ackermann, G. Anderson and L. Soder, Distributed generation: A definition, *Electric Power System Research* **57**(3) (2001), pp. 195–204.
3. S. Wilcox and W. Marion, Users' manual for TMY3 data sets, Technical Report NREL/TP-581-43156, National Renewable Energy Laboratory (2008).
4. World Meteorological Organization, World radiation data centre. Available at: http://wrdc.mgo.rssi.ru/wrdc_en_new.htm 1 September 2014.
5. European Commission, Joint Research Centre, Institute for Energy and Transport, Photovoltaic Geographical Information System (PVGIS), Available at: http://re.jrc.ec.europa.eu/pvgis/index.htm. Accessed 1 September 2014.
6. Solar and Wind Energy Resource Assessment (SWERA). Available at: http://en.openei.org/apps/SWERA/. Accessed 1 September 2014.
7. R. J. Aguiar, M. Collares-Pereira and J. P. Conde, Simple procedure for generating sequences of daily radiation using a library of Markov transition matrices, *Solar Energy* **40**(3) (1988), pp. 269–279.
8. M. Collares-Pereira and A. Rabl, The average distribution of solar radiation correlations between diffuse and hemispherical and between daily and hourly insolation values, *Solar Energy* **22**(1) (1979), pp. 155–164.
9. D. G. Erbs, S. A. Klein and J. A. Duffie, Estimation of the diffuse radiation fraction for hourly, daily and monthly — average global radiation, *Solar Energy* **28**(4) (1982), pp. 293–302.
10. R. Perez, P. Ineichen, E. Maxwell, R. Seals and A. Zelenka, Dynamic global to direct irradiance conversion models, *ASHRAE Transactions-Research Series*, pp. 354–369 (1992).
11. J. E. Hay, Calculation of monthly mean solar radiation for horizontal and inclined surfaces, *Solar Energy* **23** (1979), pp. 301–330.
12. R. Perez, R. Seals, P. Ineichen, R. Stewart and D. Menicucci, A new simplified version of the Perez diffuse irradiance model for tilted surfaces, *Solar Energy* **39**(3) (1987), pp. 221–231.
13. J. Cronemberger, E. Caamaño-Martín and S. Sánchez, Assessing the solar irradiation potential for solar photovoltaic applications in buildings at low latitudes — making the case for Brazil, *Energy and Buildings* **55** (2012), pp. 264–272.
14. T. Nordmann and L. Clavadetscher, Understanding temperature effects on PV system performance, *Proc. 3rd World Conf. Photovoltaic Energy Conversion*, **3** (2003), pp. 2243–2246.
15. J. A. Duffie and W. A. Beckman, *Solar Engineering of Thermal Processes* 3rd edn. (Wiley, New York, 2006).
16. E. Lorenzo, Energy collected and delivered by PV modules, in eds. A. Luque and S. Hegedus, *Handbook of Photovoltaic Science and Engineering* (Wiley, Great Britain, 2003), pp. 905–970.

17. M. Iqbal, *Introduction to Solar Radiation* (Academic Press, Canada, 1983).
18. Deutsche Gesellschaft für Sonnenergie, *Planning and Installing Photovoltaic Systems*, 3rd edn. (Routledge, London, 2013).
19. J. Solórzano-Moral, D. Masa-Bote, M. A. Egido-Aguilera and E. Caamaño-Martin, A model for the simulation of energy gains when using distributed maximum power point tracking (DMPPT) in photovoltaic arrays, *Progress in Photovoltaics: Research and Applications*, **23**(2) (2015), pp. 182–193.
20. G. Blaesser and D. Munro, Guidelines for the assessment of photovoltaic plants document. A photovoltaic system monitoring, Commission of the European Communities, Joint Research Centre, Ispra, Italy, EUR 16338 EN, Issue 4.2 (June 1993).
21. International Energy Agency — Photovoltaic Power Systems — Task 13: Performance and Reliability of Photovoltaic Systems, Analysis of long-term performance of PV systems, IEA-PVPS T13-05:2014 (2014).
22. IEC 60904-2, Photovoltaic devices — Part 2: Requirements for photovoltaic reference solar devices.
23. IEC 60904-5, Photovoltaic devices — Part 5: Determination of the equivalent cell temperature (ECT) of photovoltaic (PV) devices by the open-circuit voltage method.
24. International Energy Agency — Photovoltaic Power Systems Programme — Task 2: Country reports on PV system performance, Report IEA-PVPS T2-05:2004 (2004).
25. International Energy Agency — Photovoltaic Power Systems — Task 13: Performance and Reliability of Photovoltaic Systems, Analytical monitoring of PV Systems, IEA-PVPS T13-03:2014 (2014).
26. K. Butti and J. Perlin, *A Golden Thread: 2500 Years of Architecture and Technology* (Van Nostrand Reinhold Company, New York, 1980).
27. S. J. Strong and R. H. Wills, Building integration of photovoltaics in the United States, in *Proc. 11th European Photovoltaic Solar Energy Conf.*, Montreaux, pp. 1672–1676 (1992).
28. E. Kern, Residential utility connected systems, in *Proc. 19th IEEE PV Specialists Conf.*, New Orleans, pp. 1007–1011 (1987).
29. C. Meier, C. Schaffner and W. Blum, 1996 Photovoltaic energy statistics of Switzerland. Overall yield, quality assurance and Results of monitoring, in *Proc. 14th European Photovoltaic Solar Energy Conf.*, Barcelona, pp. 901–906 (1997).
30. H. Gabler, K. Heidler and V. U. Hoffmann, Market introduction of grid-connected photovoltaic installations in Germany, in *Proc. 14th European Photovoltaic Solar Energy Conf.*, Barcelona, pp. 27–31 (1997).
31. H. Wilk, Grid interconnection of photovoltaic systems in Austria, in *Proc. Seminar*, PV grid-interconnection, Task V of International Energy Agency — Photovoltaic Systems Programme, Zurich (1997).
32. J. Solorzano, Distributed control strategies for Photovoltaic systems in urban environments, PhD, Escuela Técnica Superior de Ingenieros de Telecomunicación, Universidad Politécnica de Madrid (2014).
33. F. Sick and T. Erge (eds.), *Photovoltaic in Buildings. A Design Handbook for Architects and Engineers* (James & James (Science Publishers), 1996).

34. D. Prasad and M. Snow (eds.), *Designing with Solar Power. A Source Book for Building Integrated Photovoltaics (BIPV)* (Earthscan, London, 2005).
35. T. H. Reijenga, PV in Architecture, in eds. A. Luque and S. Hegedus, *Handbook of Photovoltaic Science and Engineering* (Wiley, Great Britain, 2003), pp. 1006–1042.
36. International Energy Agency — Solar Heating and Cooling Programme — Task 41, Solar energy and architecture, subtask C: Concepts, case studies and guidelines. Available at: http://task41.iea-shc.org/casestudies/. Accessed on 22 February 2015.
37. Directive 2002/91/EC of the European Parliament and of the Council of 16 December 2002 on the Energy Performance of Buildings, Official *Journal of the European Union*, 04.016.2002, L.1/65–L.1/71.
38. Directive 2010/31/EU of the European Parliament and of the Council of 19 May 2010 on the Energy Performance of Buildings, Official *Journal of the European Union*, 18.06.2010, L.153/13–L.153/35.
39. F. Crassard and J. Rode, The evolution of building integrated photovoltaics (BIPV) in the German and French technological innovation systems, Chalmers University of Technology, Report 2007, 16 (2007).
40. Royal Decree 314/2006, of 17 March 2006 approving the Spanish Technical Building Code. [In Spanish]. Available at: http://www.boe.es/diario_boe/txt.php?id=BOE-A-2006-5515. Accessed on 15 February 2015.
    See also a revised version of parts related to energy use, including solar active technologies (solar thermal and PV): Order FOM/1635/2013, of 10 September updating Basic Document DB-HE "Energy Savings", of the Spanish Technical Building Code. Available at: http://www.boe.es/diario_boe/txt.php?id=BOE-A-2013-9511. Accessed on 2 February 2015.
41. A. Athienitis and W. O'Brien (eds.), *Modelling, Design, and Optimization of Net-Zero Energy Buildings*, 5th edn. (Ernst & Sohn, Berlin, 2015).
42. IEC/TS 62548, Photovoltaic (PV) arrays — Design requirements.
43. IEC 61683, Photovoltaic systems — power conditioners — procedure for measuring efficiency.
44. W. Bower, C. M. Whitaker, W. Erdman, M. B. Behnke and M. Fitzgerald, Performance test protocol for evaluating inverters used in grid-connected photovoltaic systems, Sandia National Laboratories (2004).
45. IEC 62109-1, Safety of power converters for use in photovoltaic power systems — Part 1: General requirements.
    See also: IEC 62109-2, Safety of power converters for use in photovoltaic power systems — Part 2: Particular requirements for inverters.
46. UL 1741, Inverters, converters, controllers an interconnection system equipment for use with distributed energy resources.
47. EN 50438, Requirements for micro-generating plants to be connected in parallel with public low-voltage distribution networks. *Note*: This European standard deals with installations rated up to 3×16 A, or a maximum output of approximately 11 kW. A new standard for larger generators is under development.

48. IEEE Standards Association, IEEE 1547 Standard for Interconnecting Distributed Resources with Electric Power Systems.
49. M. Camino-Villacorta, M. A. Egido-Aguilera and P. Díaz, Test procedures for maximum power point tracking charge controllers characterization, *Progress in Photovoltaics*, **20**(3) (2012), pp. 50–62.
50. J. Keane, *Pico-solar Electric Systems: The Earthscan Expert Guide to the Technology and Emerging Market*, Earthscan expert series (Routledge, New York, 2014).
51. B. Van Campen, D. Guidi and G. Best, Solar photovoltaics for sustainable agriculture and rural development, (Rural Development, FAO Publication, Rome, 2000).
52. J. Solorzano and M. A. Egido, *Simulation of AC, DC and AC–DC Coupled Hybrid Mini-grids — In Search of the Most Efficient System* in *6th European Conference PV Hybrids and Mini-grids* (Chambery, France, 2012).
53. A. Chouder and S. Silvestre, Automatic supervision and fault detection of PV systems based on power losses analysis, *Energy Conversion and Management* **51**(10) (2010), pp. 1929–1937.
54. P.-O. Moix and C. Ruchet, Partial AC-coupling in minigrids, in *5th European Conf. PV Hybrids and Mini-grids* (Tarragona, Spain, 2010).
55. M. Egido and E. Lorenzo, The sizing of stand-alone PV-system: A review and a proposed new method, *Solar Energy Materials and Solar Cells* **26**(1) (1992), pp. 51–69.
56. D. Neves, C. A. Silva and C. Stephen, Design and implementation of hybrid renewable energy systems on micro-communities: A review on case studies, *Renewable and Sustainable Energy Reviews* **31** (2014), pp. 935–946.
57. S. Sinha and S. S. Chandel, Review of software tools for hybrid renewable energy systems, *Renewable and Sustainable Energy Reviews* **32** (2014), pp. 192–205.
58. *Meeting the Challenge of Rural Electrification in Developing Nations — The Experience of Successful Programs*, ed. D. Barnes, ESMAP (2005).
59. www.idcol.org. Accessed in March 2015.
60. M. J. Bambawale, A. L. D'Agostino and B. K. Sovacool, Realizing rural electrification in Southeast Asia: Lessons from Laos, *Energy for Sustainable Development* **15**(1) (2011), pp. 41–48.
61. B. Allali, *TEMASOL: Providing Energy Access to Remote Rural Households in Morocco*, GIM Case Study No. B093 (United Nations Development Programme, New York, 2011).
62. Chapter 10: A case study on exclusive concessions for rural off-grid service in Argentina, Energy Services for the World's Poor, ESMAP (2000).
63. M. Bazilian, P. Nussbaumer, C. Eibs-Singer, A. Brew-Hammond, V. Modi, B. Sovacool, V. Ramana and P. Aqrawi, Improving access to modern energy services: Insights from case studies, *The Electricity Journal* **25**(1) (2012), pp. 93–114.
64. B. K. Sovacool, A qualitative factor analysis of renewable energy and Sustainable Energy for All (SE4ALL) in the Asia-Pacific, *Energy Policy* **59** (2013), pp. 393–403.

65. IEC/TR 60479-5, Effects of current on human beings and livestock — Part 5: Touch voltage threshold values for physiological effects.
66. IEC 60364-7-712, Low-voltage electrical installations — Part 7-712: Requirements for special installations or locations — Solar photovoltaic (PV) power supply systems.
67. IEC 61730-1, Photovoltaic (PV) module safety qualification — Part 1: Requirements for construction.
68. IEC 61008, Residual current operated circuit-breakers without integral overcurrent protection for household and similar uses (RCCBs).
    See also: IEC 61009, Residual current operated circuit-breakers with integral overcurrent protection for household and similar uses (RCBOs).
69. IEC 60898-1, Electrical accessories — Circuit breakers for overcurrent protection for household and similar installations — Part 1: Circuit-breakers for a.c. operation.
    See also: IEC 60898-2, Circuit-breakers for overcurrent protection for household and similar installations — Part 2: Circuit-breakers for a.c. and d.c. operation.
70. IEC 60947-2, Low-voltage switchgear and controlgear — Part 2: Circuit-breaker.
71. IEC 60269-6, Low-voltage fuses — Part 6: Supplementary requirements for fuse-links for the protection of solar photovoltaic energy systems.
72. IEC 60947-3, Low-voltage switchgear and controlgear — Part 3: Switches, disconnectors, switch-disconnectors and fuse-combination units.
73. IEC 62305-2, Protection against lightning — Part 2: Risk management.
74. IEC 62305-3, Protection against lightning — Part 3: Physical damage to structures and life hazard.
75. DEHN+SÖHNE, *Lightning Protection Guide*, 3rd edn. (DEHN + SÖHNE GmbH + Co.KG., Neumarkt (Germany), Germany, 2014).
76. IEEE Std 937-2000, IEEE recommended practice for installation and maintenance of lead-acid batteries for photovoltaic (PV) systems.
77. IEC/TS 62257-9-5, Recommendations for small renewable energy and hybrid systems for rural electrification — Part 9.5: Integrated system — Selection of stand-alone lighting kits for rural electrification.

Chapter 8

Low Temperature Solar Thermal Applications

Brian Norton
Dublin Institute of Technology
Grangegorman, Dublin 7
president@dit.ie

Hans Martin Henning
Division Thermal Systems and Buildings
Fraunhofer-Institut für Solare Energiesysteme ISE
Heidenhofstrasse 2, 79110 Freiburg, Germany
hans-martin.henning@ise.fraunhofer.de

Daniel Mugnier
TECSOL SA, 105 avenue Alfred Kastler — BP90434
66004 Perpignan — France
daniel.mugnier@tecsol.fr

The applications of harnessing solar energy as heat are presented. The emphasis is on the practical uses of these technologies and their potential for more extensive adoption. Solar water heating, the space heating of buildings, industrial and agricultural process heating together with the many uses of solar cooling are discussed.

1 Outline of the Chapter

A range of solar energy applications are considered that produce heat or cooling from about −20°C to +200°C. They include domestic hot water production, the provision of heating and cooling for buildings and industrial processes and agricultural uses such as drying. The emphasis in this chapter is on applications on the left-hand-side of Fig. 1.

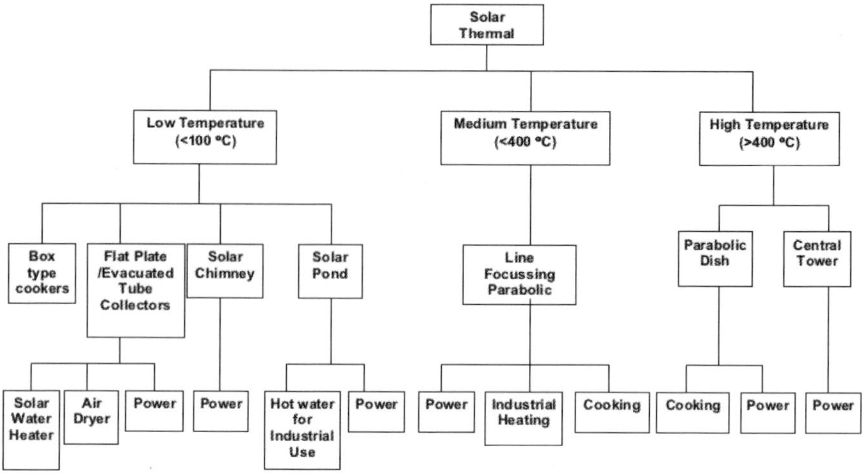

Fig. 1. Taxonomy of solar thermal applications.

2 Brief History of the Low Temperature Uses of Solar Thermal Energy

Mankind's conscious intentional harnessing of solar energy probably commenced with solar drying of harvested crops, animal skins and of clay to make bricks. Appropriate orientation of buildings for solar heat gain featured in China from 6000 BC, in Greece from 500 BC[1] and 2,000 years ago Roman baths included glazings to retain solar heat.[2]

In 1974, Organization of Petroleum Exporting Countries action led to an increase in oil prices. This led to increased use of solar water heaters with novel designs arising from a general upsurge in solar energy research and development.[3] For example, the heat pipe evacuated tube collector and improved flat plate collectors using solar selective surfaces were developed in this period.[4,5] Such urban planning for solar access has long antecedents, for example La Chaux de Fonds in Switzerland; shown in Fig. 2 was rebuilt (after a fire) so that minimal mutual overshading of buildings ensued.

The first solar water heater, the "The Climax Solar Water Heater", shown in Fig. 3 was patented in 1891.[6] It comprised four 2 liter galvanized iron dull-black tanks in a single-glazed felt-insulated wooden box.[4]

The first solar water heater which separated the collector and hot water tanks was patented in 1910. This thermosyphon unit[7] consisted of a single serpentine absorber tube flat-plate collector. Header-and-riser collector absorber tube arrangements, that gave both less pressure resistance to fluid

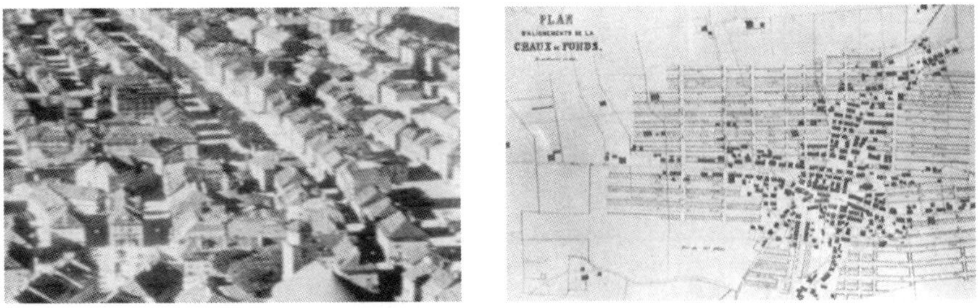

Fig. 2. Urban planning to avoid overshading in La Chaux de Fonds.

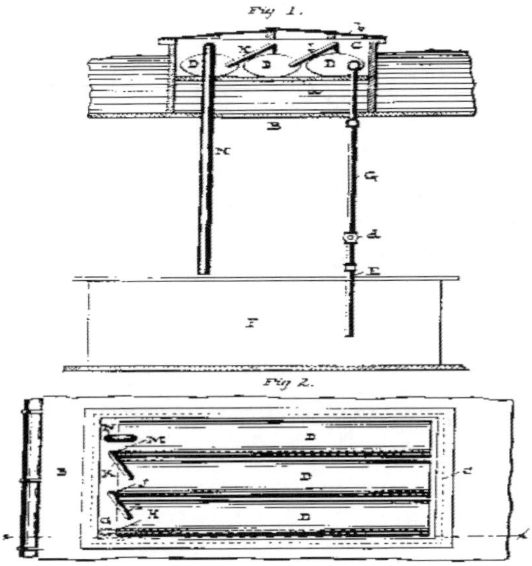

Fig. 3. Kemp's "Climax" solar water heater.[6]

flow and more effective heat transfer, emerged later but spread rapidly as can be seen from an example of such a latter system installed in London U.K. in 1947 shown in Fig. 4.

Such thermosyphon solar water heaters which stored heated water for use at night, represented a significant advance over integral collector store solar water heaters that had poor nocturnal heat retention. In the 1950s and 1960s, solar water heaters of varying design had niche markets in, for example, Japan, the Caribbean, and Cyprus. Significant markets also emerged in Israel and Australia.

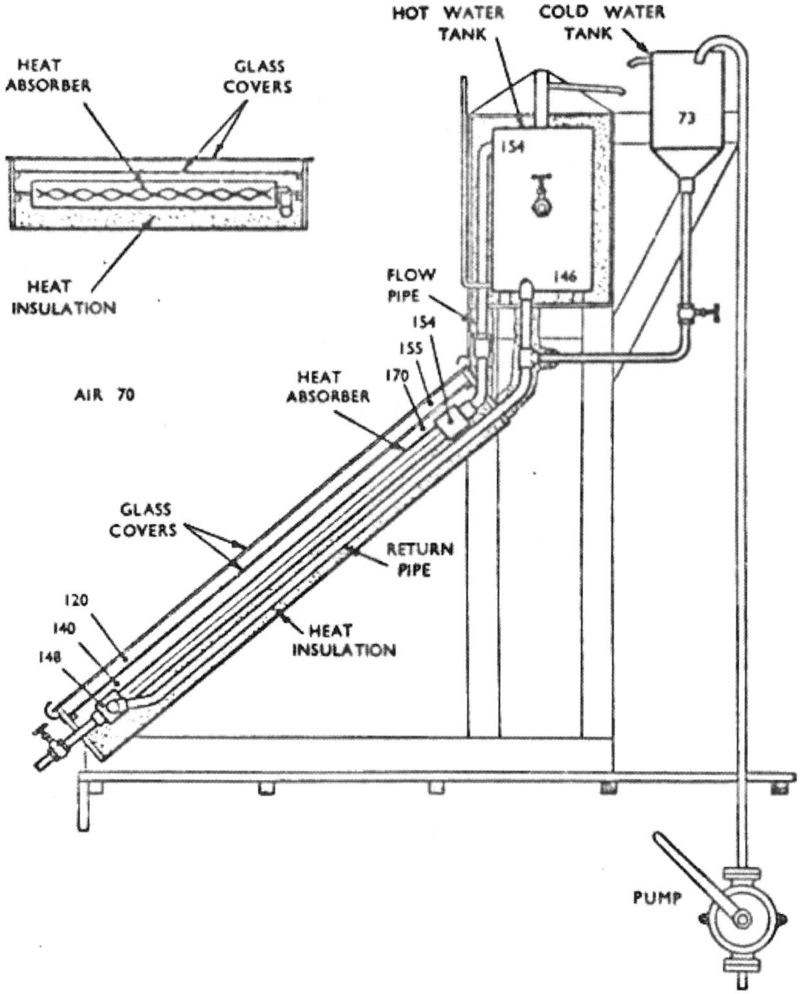

Fig. 4. Diagram of a typical early header-and-riser absorber thermosyphon solar water heater with system temperatures in Fahrenheit.

3 Solar Water Heating

Solar water heaters shown in Fig. 5, are termed either direct or indirect depending on whether the liquid in the collector constitutes the hot water withdrawn or is a separate liquid that passes through a heat exchanger. The heat exchanger can (i) be part of another hydronic circuit or outside the hot water tank,[8] (ii) surround the store[9,10] or (iii) most commonly, is

Fig. 5. Water heating solar collectors. Clockwise from top left, a flat-plate thermosyphon system, an unglazed collector for swimming pool heating, a photovoltaic-thermal collector and a evaluated tube thermosyphon.

immersed in the tank.[11] The terms passive or active apply when the heat transfer liquid circulates buoyantly or by the action of a pump respectively. In a thermosyphon solar-energy water heater, the difference in the densities of the hot water in the solar-heated collector and the cooler water in the store induces gravity-driven buoyancy forces in a closed pipework circuit comprising either (i) in a direct system comprising a solar collector, and hot-water store or (ii) in an indirect system, in an aqueous anti-freeze solution passing through a solar collector and heat-exchanger located in the tank. In regions with "hard" water, the use of soft water as the anti-freeze solvent in an indirect system avoids calcium carborate scale deposits on the inner walls of collector tubes that restrict flow and inhibit heat transfer. Cold water can be supplied at mains water pressure or at reduced pressure via either a pressure reduction valve or a header tank.

Though pumped circulation solar water heaters neither provide more hot water nor heat water more efficiently than a comparable thermosyphon system,[11] they allow for greater layout flexibility as; (i) the hot water tank can be located beneath the collectors, (ii) only the collectors need be located external to the building so the roof does not need to support the weight of a hot water tank and (iii) particularly for larger systems, installation is easier.

In pumped circulation systems, temperature sensors at the collector's inlet and outlet feed signals a controller so that the pump ceases operation at night and during periods when the solar heat gained by the collector is less than the heat lost from it. To achieve this, various control algorithms have been developed[12] though in practical systems pumps are activated usually by an unmodified temperature difference across the collector. Systems using a photovoltaic-powered circulation pump operate only when the solar energy intensity is sufficient to meet the power requirement of an optimally-selected[13] direct current pump. Most pumped-circulation systems have roof-mounted collectors, through collectors have been wall-mounted[14,15] and formed free-standing ground-level collector arrays[16,17] for very-large-scale solar district heating applications.

Satisfying parts of both a building's space heating demand and domestic hot water, a solar combi-system can provide high energy savings when a well-insulated thermally-stratified small-capacity storage tank is operated at a low set-point temperature.[18] Combi-system configurations including (i) tank-in-tank solar combi-systems that have a domestic hot water tank integrated into the space heating hot water store: solar heat is transferred by an internal spiral heat exchanger located in the lower part of the tank with domestic hot water withdrawn directly from domestic hot water tank in the store; (ii) "bikini" tank combi-systems comprise two separate mantles encircling the hot water tank circumferentially at upper and lower levels, these circulate domestic hot water and solar heated water respectively, both transferring heat to the space heating water with domestic hot water being withdrawn directly from the tank. Bikini tank combi-systems are particularly suited to low-energy buildings because they operate effectively at set-point temperatures.[19]

In integral collection storage solar water heaters, the hot water tank absorbs solar energy directly beneath a glazed aperture. Nocturnal radiative heat losses to clear night skies can lead to only luke-warm water being available early the next day. Techniques to retain heat better overnight include; (i) perforated inner sleeves,[20–27] (ii) transparent foils separated by a honeycomb structure[28,29] (iii) "Breadbox" integral collector storage systems,[30] in which an insulated manually or automatically moveable aperture cover is deployed to reduce night-time heat loss and is opened during the day when it can double as a reflector[30] and (iv) evacuation of the space between the store and the glazed aperture.[31]

Passive indirect closed two-phase transfer of heat from the solar collector to the store avoids scaling, fouling and the need for freeze protection, With suitable fluids, corrosion is far more limited than with aqueous systems. Working fluid options include acetone, petroleum ether,[32] trichlorofluoromethane,[33] R134a,[34] R407C,[35] and R410A.[36] In a novel integral collector storage solar water heater with an outer absorbing vessel and an inner storage vessel, a small pool of water is placed in the space between the two vessels? When heated this water evaporates to condense on the colder surface of the inner vessel, thus transferring thermal energy to the store.[37]

In a Combi^{+} system, heat provided by solar collectors may be used to evaporate the working fluid in a heat pump in a combi-system that transfers heat from a colder reservoir to a warmer reservoir. During compression, the temperature of the heat pump as[38] working fluid increases to well above the temperature provided by the solar collector. During condensation, heat is rejected at a higher temperature to the hot water store.[39,40]

As swimming pools' water temperatures are modest, unglazed flat plate collectors are used with simple arrangements of black plastic pipes often being employed. Swimming pool heating is currently the main solar water heater type in the USA, Canada, and Australia.[41] Systems almost always use pumped circulation and together with the pool itself, collectors are drained in winter in locations at higher latitudes to avoid freezing. Parametric analyses have been conducted to optimize solar heated swimming pool configurations.[42]

PV solar water heating ensues when electricity from a PV array is dissipated as heat from a resistive heater immersed in the hot water store.[43,44] The maximum power point of the photovoltaic array's current–voltage characteristic is maintained by varying optimally the electrical load imposed by the heating element. Solar energy conversion efficiencies of PV systems are lower than solar thermal collectors so for a PV solar water heater the collection area is larger. However as only a cable connects the heating element in the store to the PV array, the installation process is simple and the use of electricity to transfer energy obviates any need for mains water pressure reduction and/or freeze protection. Many building integrated PV systems act as embedded generators that supply a grid. Where photovoltaic-to-grid feed-in tariffs for electricity sold to the grid are not in place or do not provide a sufficient return on investment, the best use of excess PV energy may be to heat water.

The use of solar heated air passing through an air-to-water heat exchanger obviates the need for freeze protection. Air heating collectors are lighter than those containing water and air heating systems require less specialist installation expertise. However the lower efficiency of air heating solar collectors and significant heat exchanger losses, mean that collector areas must be very large.

A taxonomy of solar water heaters based on key attributes is provided in Table 1.[11] Both solar heated air-to-water systems and the use of PVs specifically for water heating are uncommon. All the other types are available in a very wide range of commercial products.

Water freezing in the collector can lead to the need to replace the collector and where joints located internally burst can cause significant water damage to a building. Both of these incur costs that adversely affect economic viability and consumer confidence. Use of evacuated collector tubes,[45] flat plate collectors with multiple glazing layers[15] or transparent insulation materials[28] can provide collector freeze protection. Similar measures can be applied to integral passive solar water heaters, though collector performance is reduced by optical losses through too many multiple glazed layers.[46] Thermal mass provides significant freeze protection for integral pressure solar water heaters.[22] Systems of all types may also be drained in winter.[5,11]

The most common protection against potential damage arising from water freezing within the collector pipe work is to have an aqueous glycol solution flowing in a closed pipe work loop comprised of the collector and a heat exchanger, located in the associated hot water tank. Requiring neither automated nor manually-operated drain valves, in winter a glycol-filled indirect system can produce warm water unlike a seasonally drained-down system. This advantage is counteracted in summer by the slightly lower thermal efficiency[47] due to (i) the higher viscosity of aqueous glycol solutions (compared with water) thus reducing the natural circulation flow rate or incurring higher pumping power in a forced circulation system, (ii) the additional (compared with a direct system) flow resistance introduced by the heat exchanger, (iii) additional heat transfer resistance in the heat exchanger, and (iv) the lower specific heat capacity of the heat transfer fluid compared with water. For the freezing points of different concentrators of aqueous propylene-glycol solutions and ambient temperature distributions for each of the months when frost is recorded, critical concentrations for system survival can be determined. For a particular water heater located in London UK,[47] the critical aqueous propylene-glycol concentration for year-long survival is 25%. An indirect system with critical antifreeze concentration operates throughout the year. It's output (in the London, UK, climate and for typical systems) is approximately 12% greater than that from a direct system as the latter is used from March to October only and drained-down in winter.

4 Building Heating and Cooling with Solar Energy

A building mediates between the seasonal and diurnal variation of temperature, illuminance and wind speed to create the more stable environmental conditions that are normally required for human comfort. By means of the design and form of the building itself, together with additional specialist mechanical components, solar energy can reduce heating ventilation and/or lighting loads. The best use of solar energy in a building is specific to a given climate, site, building function, occupancy and use as well as the prevailing cultural and architectural preferences. Climate specific handbooks on solar energy use in buildings are available.[48–50] As any one building's shading and wind-flow is affected by other buildings nearby, in large urban settings, master planning is particularly important.

Direct solar gain refers simply to a room with a southerly-facing window.[47] To have a low heat loss, the window comprises a multiple pane glazing or a glazing in which low heat for convective and radiative heat transfer is reduced by intervention between the glass panes.

For windows in which the space between panes is evacuated or encloses an aerogel, the frame also must exhibit low heat loss if overall window performance is to be commensurate with that of the glazed element. In both cases low long-wave emittance coatings are applied to the glass together with an appropriate framing. When multiple panes are framed, an insulating spacer may be placed between the panes to reduce conductive heat transfer between them.[51,52] In very low heat loss glazing systems, such as evacuated glazing[53–59] and aerogels,[60] the edges of the glass panes are sealed contiguously. As an insulated spacer thus cannot be employed to inhibit thermal conduction between the panes, new forms of insulating frame are therefore required.[61]

The fabrication of a vacuum glazing (VG) was first reported[62] using a low melt solder glass gasket that formed a contiguous edge seal at temperatures of above 450°C. At such temperatures, tempered glass and many types of soft low-emittance coatings degrade. A method for producing an edge seal at less than 200°C has overcome this problem.[54] For low-temperature sealed vacuum glazing, measured[58] heat transmittance of less than 1 W m^{-2} K^{-1} in the center-of-glass area has been achieved. Depending on its emittance value, employing a low-e coating on one or two glass

Table 1: Predominant combinations of attributes distinguish different generic solar water heaters.

				Hot-water store	Prevention of nocturnal reverse thermosyphon
Hydronic	Integral Collector Storage			Single or multiple cylindrical usually horizontal. Plastic pillow or bags	Hot water withdrawal. Double glazing. Movable aperture installation.
	Thermo-syphon	Close-coupled	Single phase	Horizontal cylindrical	Height of collector above store. Pipework arrangement.
			Two phase		Cessation of fluid evaporation
		Distributed		Vertical cylindrical	Height difference between collector and store
	Pumped	Solely Hot water		Vertical cylindrical	Non-return valve Flow resistance of pipework and stationary pump
		Combisystem: space and water heating		Vertical tank-in-tanks. Bikini tanks	
		Solar-assisted heat pump (Combi $^{+}$)		Vertical cylindrical	
		Swimming pool heating		Swimming pool	
Aeolic	Solar Heated Air		An air-to-water heating exchanger		Unnecessary
Electric	Photovoltaic		Vertical cylindrical		

Table 1: (*Continued*)

Collector freeze production	Position of store		Installation	Collector type
	Relative to collector	Relative to building		
Heat retained in thermal mass of the store	Same item	On roof exterior	Installed as a complete factory — built unit	Integral
Indirect. Drained.	Directly above			Evacuated tube. Flat-plate. Photovoltaic/ Thermal.
Indirect.	Above	Inside loftspace	Assembled from components during installation	
Indirect.	Below	Variety of possible locations inside building		
Outdoor pool collectors drained in freezing weather	Pool usually located below	Solar heated pools are indoor and outdoor		Unglazed. Glazed flat plate.
	Various			Air-heating
	Any	Interior	Water heating is an optional use of installed photovoltaic electricity	Photovoltaic

surfaces within the vacuum gap reduces radiative heat transfer across vacuum glazing significantly.[53,61] A low-e coating also reduces solar and visual transmittance. Nevertheless, the use of a low-e coating increases thermal resistance much more than it reduces the solar gain.

An electrochromic film can be used to control visual light transmittance. Simulation studies of window systems that have considered the effect of control strategies on energy saving[63] have demonstrated that in a cooling-dominated climate, the inclusion of an electrochromic double glazed window can enable a building to consume less cooling energy by controlling visual light transmission and solar heat gain coefficient when compared to conventional double glazing. In a heating-dominated climate, as an electrochromic window should remain in its bleached state during the heating season, it does not improve energy performance. An electrochromic window plus vacuum glazing provides both very low heat loss with visible light transmittance whilst allowing control of solar gain and thus thermal comfort for the building's occupants while reducing peak energy demands both for cooling during summer and heating during winter. In a study of the effect of insolation intensity on the thermal behavior of electrochromic vacuum glazing shown in Fig. 6,[61] it was found that to avoid intolerable overheating discomfort, the electrochromic layer between the vacuum glazing and the third glass pane must face the outdoor environment.

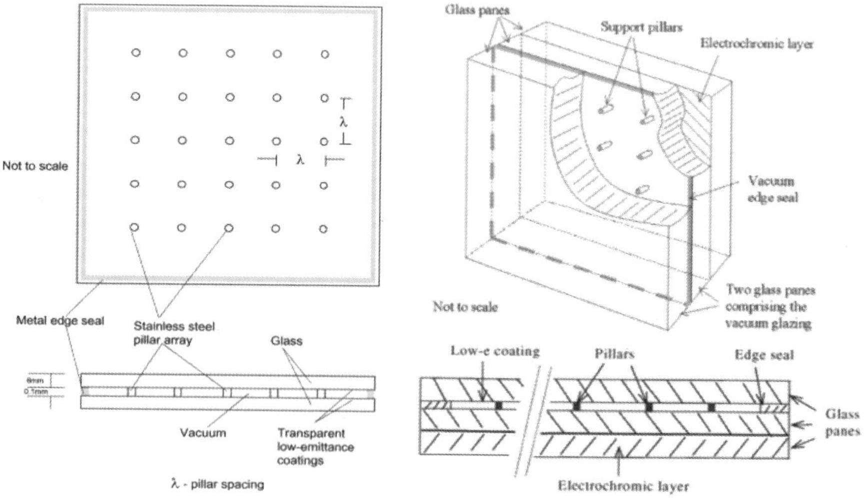

Fig. 6. Electrochromic vacuum glazing.

A conservatory located on the south side of a building provides indirect solar heating by (i) conduction only through the separating wall (ii) natural circulation of air between the conservatory and the heated building and/or (iii) forced circulation of air between the conservatory and the building (iv) a net flow of air from outside into the heated building via the conservatory.[46] Appropriate sensors, controls and actuators can be used to optimize the energy collection. In well insulated buildings, in climates with short heating seasons, as temperatures in an attached conservatory are rarely above the building's internal temperature, the annual heating contribution is likely to be small. A contribution to a reduction of heat load however, does ensue from the reduction in the rate of thermal conduction out through the separating wall. If the air entering the building via the conservatory forms the major constituent of the air required for ventilation, then the net flow of air from outside into the heated building via the conservatory can provide significant solar ventilation pre-heating. This is particularly true for very well insulated buildings as most energy is used to heat ventilation air.

For a Trombe–Michel wall, as shown in Fig. 7, the indirect thermal coupling to the heated space are the same virtually as with conservatories. However, the conductive heat transfer is generally larger than the convective heat transfer. As Trombe–Michel walls delay the delivery of solar heat, they

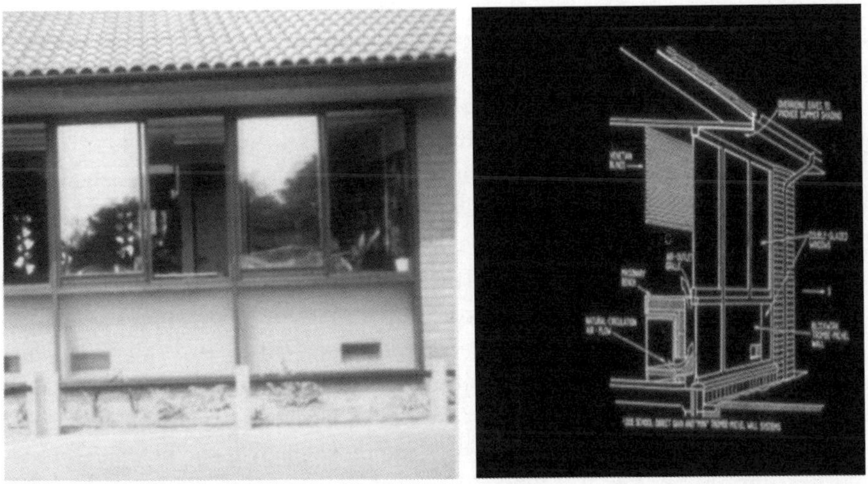

Fig. 7. Trombe–Michel Wall located beneath a south-facing window.

are ideal for providing heating in the early part of a cold night after a sunny day, as encountered frequently in arid and mountainous regions. Alternative design for Trombe–Michel walls have attempted to provide more controlled heat output.[64]

Isolated solar gain, decoupled thermally from the building, may be accomplished via an insulating separating wall, as in thermosyphoning air panels, or by location above the building, as in roof space collectors. These approaches enable controllable solar heat gain combined with no summer overheating. Isolated gain solar collectors such as the thermosyphoning air panel,[46] overcome some of the disadvantages of indirect gain collectors by dispensing with heat storage and relying totally on convective heat gain. Heat input is almost immediate whilst heat losses during non-gain periods when the collector is isolated from the heated space, are low. This design is ideally suited to the task of providing daytime heat in cool or cold climates as it operates in the same manner as the natural convection mode of a Trombe–Michel wall. The absorber is often made of metal, usually aluminium or steel, and the unit is insulated to prevent heat loss to, or from, the building. There are many system variants that are typically some combination of: (i) a low thermal capacity dual-pass absorber, (to maximize the heat transfer area between air and absorber) decoupled from the south wall of the building with a (ii) buoyancy-driven or fan driven distribution of heated air throughout the dwelling via air channels within the ceilings, walls and floors (iii) thermal energy storage within these ceilings and walls and floors.

A roof-space solar-energy collector, is essentially a pitched-roof which is partially or fully-glazed on its southerly aspect.[46] Solar-heated air from the roof space collector is conveyed by an automatically controlled fan via a duct either directly into the living space or as a pre-heated supply to a warm-air space-heating system. The roof-space collector is replenished with air either from within the dwelling or from the outside ambient environment. A roof-space collector has a low initial capital cost as its physical construction does not differ greatly from that of a conventional pitched roof. A low initial cost can arise from the employment of components (i.e. fans and controls) that would be already present in an auxiliary air heating system. In urban locations exhibiting high housing densities, a dwelling may often experience levels of overshading at lower sun angles by neighboring buildings, but roof-space collectors are more likely to be unshaded most of the time. Figures 8 and 9, show domestic and non-domestic roof-space collectors.

Fig. 8. Domestic roof-space collectors.

Fig. 9. Roof-space collectors on a school building.

5 Low Temperature Solar Heating in Industrial Applications

Water extracted from streams, rivers, and lakes in tropical locations frequently contains pathogens and thus requires disinfection before human consumption. A solar ultra-violet disinfection process, has been developed for this purpose.[68] However as some bacteria have enzymes that repair their DNA after ultra-violet radiation damage, ultra-violet treated water can only be consumed safely within one day. Solar thermal pasteurization, in which the water is heated to 348 K for 10 minutes destroys, rather than damages, all pathogens, does not require the water to be filtered and has fewer operational constraints than chlorination. Though solar pasteurization is effective and incurs little requirement for maintenance, it has been shown to have high life-cycle costs when compared with competing options.[69]

Many arid regions have underground brackish water resources or are close to sea water and have high annual levels of insolation. The production of potable water using solar energy has thus been well-researched[70] and, in remote or isolated regions adopted practically. Fundamentally there are three potable water extraction processes that can employ solar energy: (i) distillation in which water is evaporated using solar heat, it condenses separated from its mineral content. This can be at atmospheric pressure in various diverse forms of passive basin stills[71] or in a multiple effect process. In a vapor compression system, water vapor is compressed adiabatically producing a superheated vapor. This is first cooled to saturation temperature and then compressed, using mechanical energy, at constant pressure, (ii) reverse osmosis, in which a pressure gradient across a membrane causes water molecules to pass from one side to the other.

Farm-built metal solar air-heaters range from unglazed low-temperature transpired units for animal husbandry through to double-glazed medium temperature systems for crop drying.[72] The principal attraction of farm-built air-heating collectors is the low initial investment required. The disadvantages are potentially poor fabrication quality control leading to poor performance and the frequent lack of optimized sizing of system components.

The potential use of solar heated water in industrial and agricultural processes remains largely unrealized.[73]

The use of solar energy in a thermal non-domestic application should ideally be designed, installed, and operated to meet the specific energy and temperature requirements of the particular industrial or agricultural

Table 2: Process temperatures in low to medium temperature solar industrial process applications.

		Process	Required temperature range (°C)
Sector	Food and Beverages	drying	30–90
		washing	40–80
		pasteurising	80–110
		boiling	95–105
		sterilizing	140–150
		heat treatment	40–60
		pre-heating of feedwater to boilers	30–80
		space heating of factories	30–100
	Textiles	washing	40–80
		bleaching	60–100
		dyeing	100–160
		pre-heating of feedwater to boilers	30–80
		space heating of factories	30–100
	Chemicals	boiling	95–105
		distilling	110–300
		ancillary processes	120–180
		pre-heating of feedwater to boilers	30–80
		space heating of factories	30–100

context via an optimal combination of efficient performance, high solar fraction, low initial and running costs, robustness and durability, safety and environmental sustainability. As can be seen in Table 2, industrial processes vary greatly as to their required processing temperatures.

6 Use of Solar Energy to Dry Crops

The objective in drying is to reduce the moisture content of an agricultural product to prevent deterioration for a time frame regarded as the safe storage period. Drying is a dual process of (i) heat transfer to the product from the heating source; and (ii) mass transfer of moisture from the interior of the product to its surface and from the surface to the surrounding air.[36] In solar drying, solar energy is used either as the sole source of the required heat or as a supplemental source, and the air flow can be generated by either forced or natural convection. Figure 10 shows a taxonomy of solar dryers.

The heating procedure involves the passage of pre-heated air through the product, by directly exposing the product to solar radiation or often

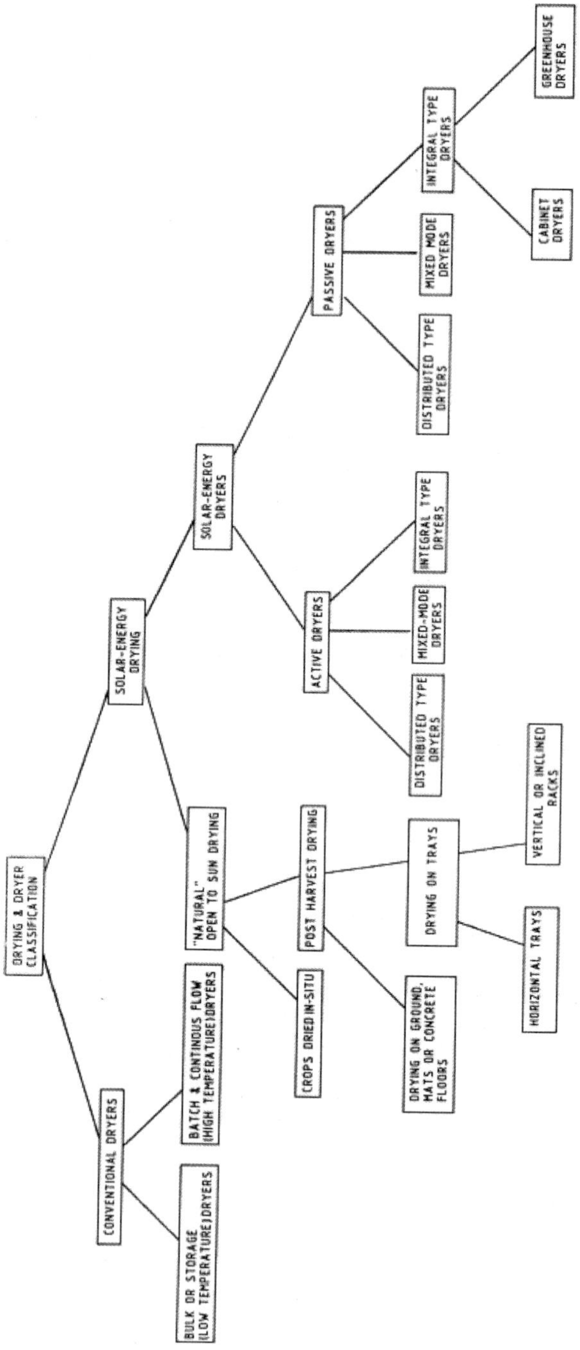

Fig. 10. A taxomony of solar dryers.

a combination of both. The major requirement is the transfer of heat to the moist product by convection and conduction from surrounding air mass at temperatures above that of the product, by radiation mainly from the sun and/or to an extent from surrounding hot surfaces, or conduction from heated surfaces in contact with the product.[74] Water starts to vaporize from the surface of the moist product when the absorbed energy has increased its temperature sufficiently for the water vapor pressure of the crop moisture to exceed the vapor pressure of the surrounding air. The rate of moisture replenishment to the surface by diffusion from the interior depends largely on the nature of the product and its moisture content.[75] If the diffusion rate is slow, it becomes the limiting factor in the rate of the drying process, but if it is sufficiently rapid, the controlling factor becomes the rate of evaporation at the product surface. Both the moisture diffusion and convective mass transfer coefficients increase with temperature though the rate will depend on how the crop is prepared for drying, that is whether it is peeled and/or sliced. For example large differences, particularly for convective mass coefficients, have been found between the drying kinetics of cylindrical and sliced potatoes. The solar absorptance of the product is important in direct solar drying; most agricultural materials have relatively high absorptances of between 0.67 and 0.81. Heat transfer and evaporation rates must be controlled closely for an optimum combination of drying rate and acceptable final product quality.[76]

Direct absorption of solar radiation enhances the proper color 'ripening' of greenish fruits by allowing the decomposition of residual chlorophyll during dehydration.[76,77] For certain varieties of grapes and dates, exposure to sunlight[78] is considered essential for the development of the required color in the dried products. A period of exposure to sunlight of Arabica coffee is thought to give full flavor in the roasted beans.

The advantages of solar dryers over traditional open sun drying include (i) a smaller area of land in order to dry similar amounts of crop, (ii) relatively high quality of dry crop, because insects and rodents are unlikely to infest it during drying, (iii) shortened drying period (iv) protection from sudden rain, and (v) low capital and running costs.[79–82]

To avoid localized over-heating and relatively slow overall drying rates, a solar chimney, as shown in Fig. 11, is often employed to provide enhanced buoyant force on the air stream thereby increasing the rate at which dry air enters or fans are used to convey air through solar collectors[83] as shown in Fig. 12.

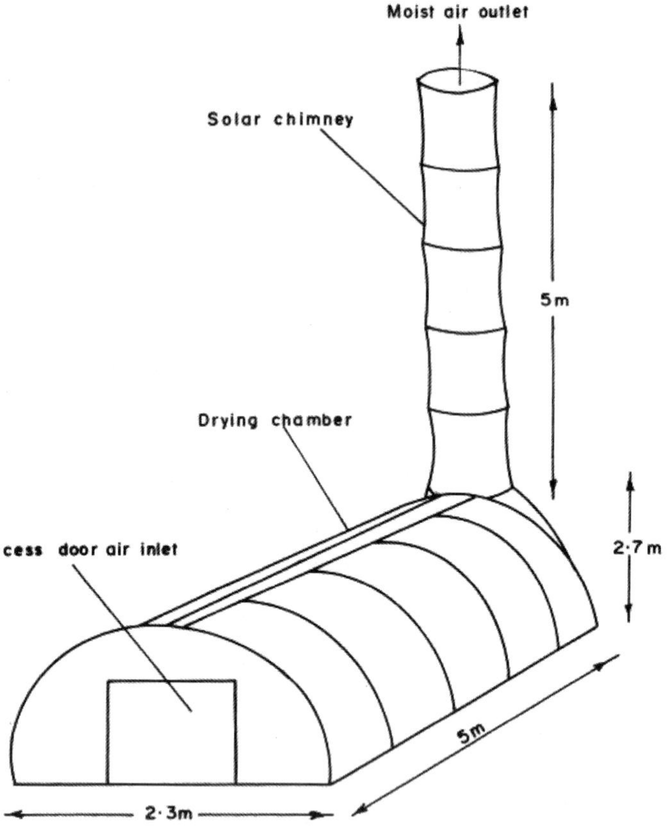

Fig. 11. A solar drying using natural circulation enhanced by a chimney.

Cabinet dryers are usually relatively small units used typically to preserve domestic quantities of fruits, vegetables, fish and meat. Solar radiation is transmitted through the cover and is absorbed on the blackened interior surfaces as well as by the product itself. Holes located at both the upper parts and the base of the cabinet's sides allow warmed moist air to leave and replenishing fresh air to be drawn in respectively, under the action of buoyant forces. Shallow layers of the product are placed on perforated or mesh trays inside the enclosure.

Solar dryers performance can be compromised by very high wet season ambient humidity[84]: as can be seen in Fig. 13, at night the "dry" air temperature can fall below the prevailing ambient temperature.

Fig. 12. A fan-driven solar drying system.

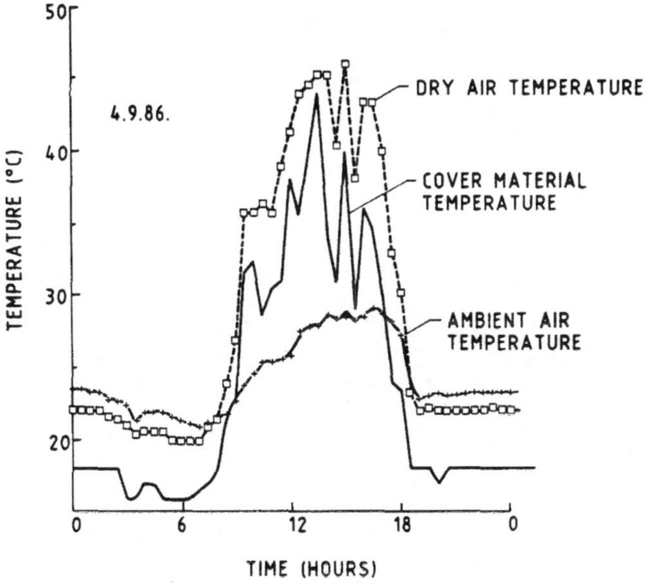

Fig. 13. Typical diurnal variation of temperatures in a natural circulation solar dryer in Nigeria in September.

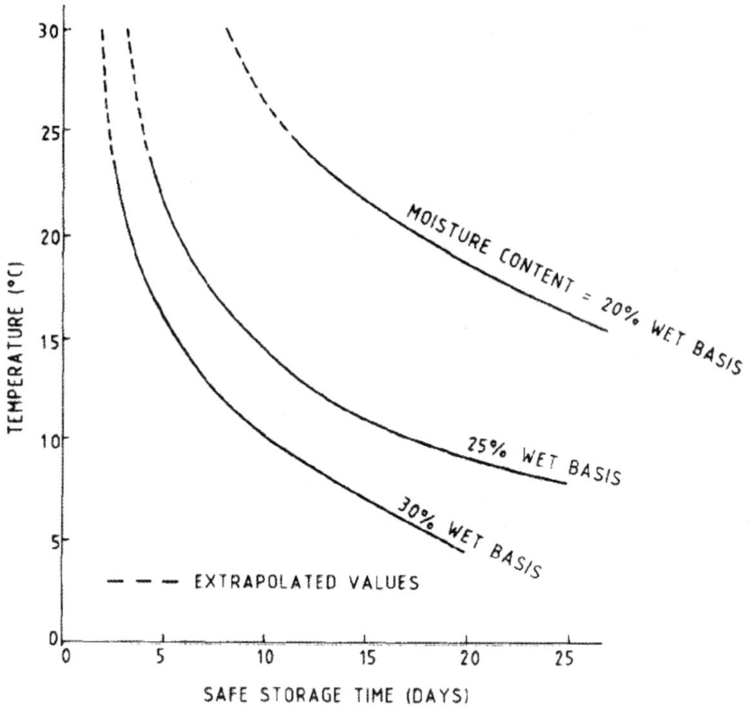

Fig. 14. The variation of the drying duration to achieve safe storage with drying temperatures and product moisture contents.

This leads to relatively moist air being entrained into the dryer and nocturnal reabsorption of moisture by the product.[85]

For crops for which low safe storage moisture content is required, drying times can be many days as shown in Fig. 14 depending on the dryer's operating temperature.[86–93]

7 Solar Air Cooling Conditioning and Refrigeration

7.1 *Passive Cooling of Building*

The discomfort of unwanted solar gains can be avoided by preventing the initial solar gain by means of shading devices,[50] switchable glazings[63] rejecting the solar gains by means of ventilation and/or absorbing the solar gain in thermal mass.[48] To keep the temperature elevation above ambient temperature small by ventilation alone, very large ventilation rates can be required.

Fixed shading devices use the seasonal geometry of solar radiation to control unwanted solar radiation. However for buildings where solar energy through windows provides significant heating, more flexible control of solar gain is needed. This cannot be provided by fixed shading devices. The most efficient shading is provided by external devices (e.g. awnings), as the solar energy is rejected before it enters the building.[48] However, weather proof external shading devices can be expensive.

Natural cooling systems dissipate heat from the building to environmental heat sinks which are at lower temperatures than that of the building interior.[65,66] These heat sinks may include ventilation to achieve the ambient dry bulb temperature, evaporative cooling to the ambient wet bulb temperature, night sky radiative cooling and using the ground as a conduction heat sink.[50] If the night-time ambient dry bulb temperature is low enough it is possible to cool the thermal mass of a building sufficiently for it to absorb the daytime cooling load without the interior temperature raising to an uncomfortable level. During the day it will usually be advantageous to ventilate by wind capture, a particular vernacular example of the latter is shown in Fig. 15. Window placement, landscaping, and the microclimate are important design considerations if natural ventilation is to be used.[67]

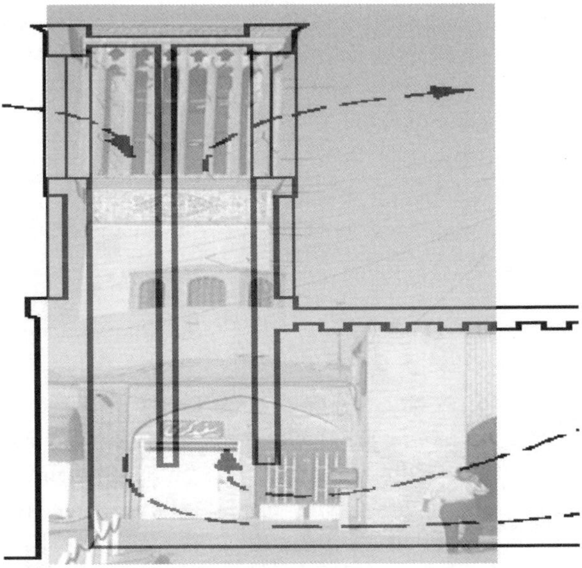

Fig. 15. Air flow in a wind capture tower.

Burying a building in the ground either partially or completely, can reduce the cooling load either by (i) the earth acting as extra insulation, especially since it tends to be dry during the summer, and also tends to reduce air infiltration; and (ii) if the temperature below ground level is significantly lower than the required internal temperature, then heat will flow out of the building into the surrounding earth. If the cooling load is small enough and the ground is sufficiently cool, then no further cooling will be required.

Direct evaporative cooling can be provided when, outside air is blown through a fibrous pad soaked in water. The air is abiabatically cooled and humidified and its dry bulb temperature approaches its wet bulb temperature. The air is then supplied to the building. In an indirect evaporative cooling system, the evaporatively cooled air cools the air in the building via a heat exchanger rather than being supplied directly. This has the advantage that water vapour is not added to the room air, hence extending the applicability of evaporative cooling to semi-humid climates.

In more humid climates, the wet bulb temperature is high enough to limit the amount of evaporative cooling that can be obtained without employing uneconomically large air flow rates and/or heat exchangers. In such climates, the lowest temperature environmental heat sink is the night sky. A cloudless night sky has a radiant temperature up to 10°C below the ambient dry bulb temperature. However, if the sky is covered with low clouds its radiant temperature will be close to ambient. The sky temperature can be predicted from the ambient dry bulb temperature and humidity and the fraction of the sky covered by clouds. A horizontal surface exposed to the night sky will dissipate 5 W per m^2 for every degree of difference between the surface temperature and the sky temperature. If the roof can be coupled to thermal storage, nocturnal cooling can be used to keep the building cool throughout the day.[94,95]

7.2 *Principles of Active Solar Cooling*

Using solar energy to provide active cooling is not an alternative to either energy efficiency measures or passive systems that use environmental heat sinks. The application of active solar heating and cooling technologies should always go hand-in-hand with energy efficiency measures. Holistic design of the building and the solar assisted heating, ventilation and air-conditioning (HVAC) system will lead to an optimal solution in terms of energy balance and cost.

Solar Air-Conditioning and Refrigeration technology (SAC) converts solar energy into cooling via two main routes:

- Electricity generated with PV modules can be converted into cooling using well-known refrigeration and air-conditioning technologies that are mainly based on vapor compression cycles.
- Heat generated with solar thermal collectors can be converted into cooling using thermally driven refrigeration or air-conditioning technologies. Most of these systems employ the physical phenomena of sorption in either an open or closed thermodynamic cycle. Other technologies, such as steam jet cycles or other cycles using a conversion of heat to mechanical energy and of mechanical energy to cooling are less significant. These routes are illustrated in Fig. 16.

Today, the first approach — solar electricity driven cooling — is mainly used for solar driven refrigerators for cooling medicine in remote, sunny regions as well as in small air conditioners, while the second approach — solar thermally driven cooling — is mainly applied for comfort cooling and air-conditioning in buildings. Some first demonstration installations have been realized for large capacity refrigeration applications. The first approach being installed in buildings is emerging through the use of small

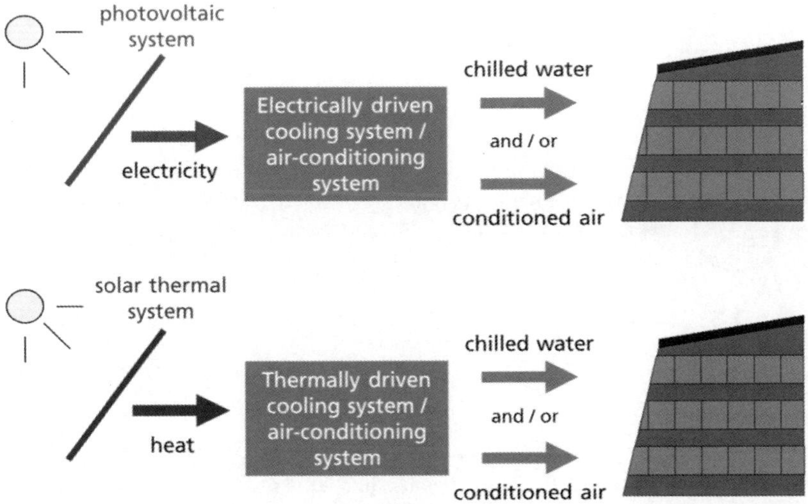

Fig. 16. Principles for solar driven cooling.

air conditioners in Asia. In other cases, it is normally not considered a "solar cooling system" since today most photovoltaic plants are connected to the electric grid and are operated completely independent from the HVAC installations used in buildings or the refrigeration machines used in industrial applications. However, in countries where the electricity price from photovoltaic systems is lower than the price that has to be paid for electricity from the grid, the situation is favorable for a development of large PV cooling systems.

However, this chapter focuses on the widely-used technology for solar cooling using the second approach — heat driven air-conditioning and refrigeration systems using solar thermal energy as the main driving energy. Figure 17 shows the heat fluxes and temperature levels of a thermally driven cooling system and is used for the definition of the thermal Energy Efficiency Ratio $EER_{thermal}$ (often also denoted as thermal Coefficient of Performance, $COP_{thermal}$), which is a key figure to characterize thermally driven cooling systems. Table 3 provides a brief overview the main characteristics of systems using sorption technology.

The main arguments for SAC[96,97] originate from an energy saving perspective:

- Application of SAC saves electricity and thus conventional primary energy sources.

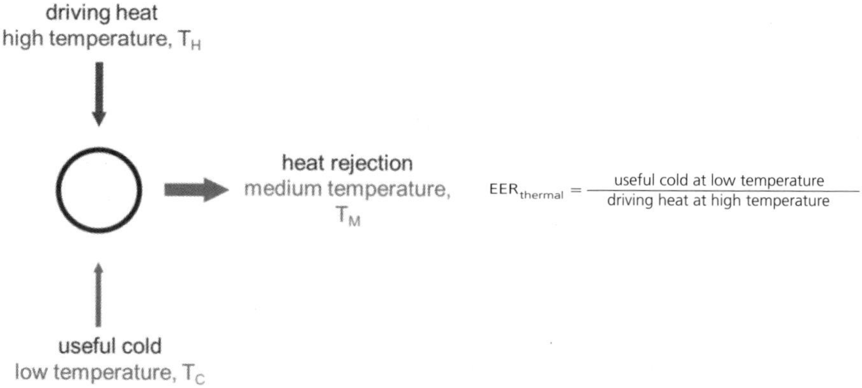

Fig. 17. Heat fluxes of a thermally driven cooling system and definition of the thermal Energy Efficiency Ratio, $EER_{thermal}$. The useful cold is produced at a low temperature level, T_C. The driving heat is supplied at a high temperature level, T_H. Both heat fluxes flow into the machine and have to be rejected at a medium temperature level, T_M.

Table 3: Overview of market available thermally driven cooling systems based on sorption technology.[96,97]

Type of system	Water chillers (closed thermodynamic cycles)						Direct air treatment (open thermodynamic cycles)	
Physical phase of sorption material	Liquid			Solid			Liquid	Solid
Sorption material	Water	Lithium-bromide		Zeolite	Silica gel	Lithium-chloride	Lithium-chloride	Silica gel (or zeolite), cellulose matrix with lithium-chloride
Refrigerant	Ammonia	Water	Water	Water	Water	Water	Water	
Type of cycle[(1)]	1-effect	1-effect	2-effect	1-effect	1-effect	1-effect	Cooled sorption process	Desiccant rotor
$EER_{thermal}$ range	0.5–0.75	0.65–0.8	1.1–1.4	0.5–0.75	0.5–0.75	0.5–0.75	0.7–1.1	0.6–0.8
Driving temperature range, °C	70 ... 100 120 ... 180[(2)]	70 ... 100	140 ... 180	65 ... 90	65 ... 90	65 ... 90	60 ... 85	60 ... 80
Solar collector technology[(3)]	FPC, ETC SAT[(2)]	FPC, ETC	SAT	FPC, ETC	FPC, ETC	FPC, ETC	FPC, ETC, SAHC	FPC, ETC, SAHC

Comments:

(1) 1-effect: single-effect thermodynamic cycle (no internal heat cascade); 2-effect: double-effect thermodynamic cycle (with internal heat cascade)

(2) Valid for production of cold at temperatures significantly below the freezing point of water, i.e. $<0^\circ$C

(3) Abbreviations for solar thermal collector types: FPC = flat plate collector; ETC = evacuated tube collector; SAT = single-axis tracking solar collector (e.g. parabolic trough collectors or Fresnel type collectors); SAHC = solar air heating collector.

- SAC also leads to a reduction of peak electricity demand, which is a benefit for the electricity network and could lead to additional cost savings of the most expensive peak electricity when applied on a broad scale.
- SAC technologies use environmentally sound materials that have no ozone depletion and no (or very small) global warming potential.

Other arguments originate from a more technical perspective:

- Solar energy is available almost at the same time when cooling is needed; this argument holds for both, solar thermal and solar electric based systems.
- Solar thermal systems used for production of sanitary hot water and heating (solar combi-systems) have large collector areas that are not fully used during summer. They can be used for SAC and thereby reduce the risk of stagnation situations of the solar collector system.
- Comparatively low noise and vibration-free operation of thermally driven chillers.

7.3 *Technical Maturity of SAC*

The key components of SAC systems are the solar collector subsystem and the thermally driven cooling subsystem. Additional key components are a heat rejection unit to reject the waste heat of the thermally driven chiller and heat or cold storage.

Solar collectors and solar collection systems have achieved good technical maturity. For SAC systems that operate with temperatures below approximately 110°C there exists a good supply of robust, cost effective solar collectors. Solar collectors have been developed with increased safety and enhanced efficiency of solar collection for solar cooling: examples are solar collectors operated with water and have a drainback system or night recirculation.

Solar collector systems for higher temperatures as needed for SAC are still scarce, but an increasing number of manufacturers are entering the market. Nevertheless, solar collector systems for operation at temperatures above approximately 130°C — typically single-axis tracking with optical concentration — are now commercialized but not so widely used for cooling applications.

Large thermally driven chillers and open sorption cycles have existed for many decades and have been introduced in the market; their main operation today is with waste heat (e.g. from a co-generation system or industrial waste heat) or directly gas-fired. Typically, they are designed for operation to provide base load cooling and not specially adjusted for operation with solar energy.

In the last decade, the main progress was made in the field of small capacity thermally driven chillers and SAC has significantly contributed to stimulate this development. Today, numerous systems from various manufacturers are offered on the market and have reached a considerable technical maturity. However, most of the manufacturers are small start-up companies. Some of those companies set up a manufacturing capacity on an industrial scale and others are moving in the same direction.

Installation of buffer storages is quite common in SAC installations. Sizes range from small buffers to overcome short-term fluctuations up to large buffer stores used to save solar gains for a number of hours (e.g. from noon to afternoon). Storages are applied on the hot and/or cold side and are usually filled with water. In a few applications, ice storage is used on the cold side to increase the storage density (or for applications with cooling demand at temperatures below 0°C). Other phase change materials are still not common in solar cooling.

The main shortcoming of SAC from a technical perspective lies at the system level. Many systems fail to achieve the planned energy savings because of shortcomings in proper design and energy management of systems that result in a high overall electricity consumption of auxiliary components. A particular area where mistakes are made is the heat rejection subsystem, which often has not received sufficient attention in the past. Another mistake made is that many systems were far too complex and as a result created non-optimal control and big maintenance efforts.

The main problems from practical experience in realized installations are:

- Heat rejection: cooling towers often need too much electricity and are not controlled for partial loads. Small capacity wet cooling towers are relatively expensive as they need a high level of maintenance. Dry cooling towers need more electricity and often the re-cooling temperatures are too high for the solar thermally driven chillers. Hybrid systems (dry/wet) seem to be a promising solution, but very few systems are available on the market and they are not adjusted for combination with thermally driven chillers.
- High efficient auxiliary components and a careful hydraulic design are essential. This is particularly important as solar cooling systems need more hydraulic loops than standard systems.

- Another technical barrier is the integration of all components into a complete system. The overall system design requires various professional skills for the different subsystems: solar energy at medium temperature (but higher than standard sanitary hot water applications), hydraulics with pressurized and medium temperature water, and air-conditioning or industrial cooling.

7.4 *Energy Performance of SAC*

Solar cooling systems have proven to be able to save energy in comparison to conventional standard solutions. The amount of achieved energy saving strongly depends on system design and operation. Key figures that determine the energy savings are the solar fraction for the heat needed to drive the thermally driven cooling and the overall electricity demand for auxiliary components, such as the fan (e.g. the fan in the cooling tower) and the pumps in the hydraulic circuits.

Figure 18 shows the dependence of the saved primary energy of a solar cooling system in comparison to a standard system as a function of the solar fraction of the driving heat. Two types of technologies are represented: single-effect machines achieve a lower coefficient of performance. That is they produce less cooling per amount of driving heat. But, they also

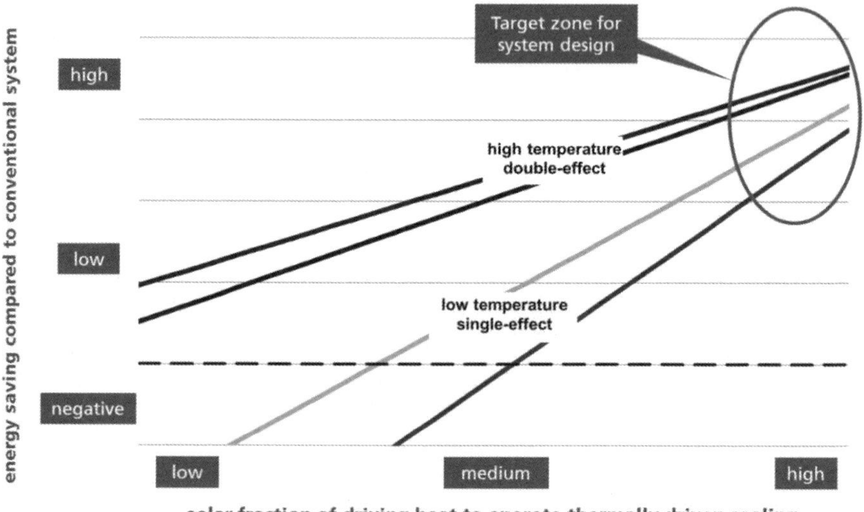

Fig. 18. Energy saving compared to a conventional system versus the solar fraction of the driving heat needed to operate the thermally driven cooling.

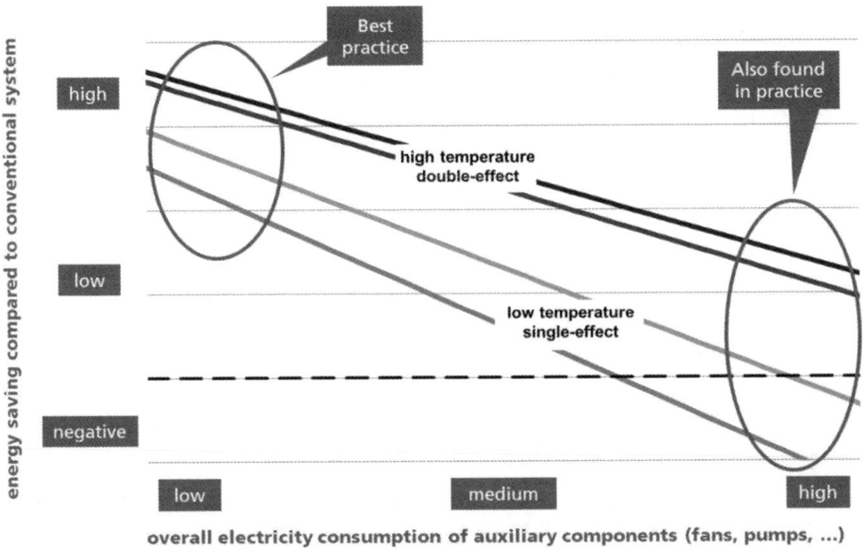

Fig. 19. Energy saving compared to a conventional system versus the overall electricity consumption of all auxiliary components of the solar driven cooling system.

need a low driving temperature, typically 70–110°C, and thus can use flat plate or evacuated tube collectors. Double-effect machines achieve a higher coefficient of performance, but need higher driving temperatures, typically 130–180°C, and thus require more sophisticated solar collector technology, such as single-axis tracking solar systems. System design should aim at achieving a high solar fraction, that is they should use almost no heat from conventional sources, such as natural gas to operate the thermally driven cooling.

In Fig. 19, the influence of the electricity demand of auxiliary components on the achieved primary energy saving is demonstrated. Today, systems are found on the market that work with a relatively high electricity demand and thus do not save much energy or — even worse — lead to a negative energy balance because they consume more primary energy compared to a conventional system.

All the relations shown here are valid for cooling only. Typically, the overall energy balance becomes more favorable for solar thermal systems if other heat needs are covered, such for heating or for sanitary hot water.

In summary, it turns out that SAC has the ability to realize an energy efficient solution and lead to significant energy savings the main requirements to achieve this are:

- Before the choice is made for SHC/SAC it has to be compared to other solutions in terms of energy saving and cost.
- An important design item is to keep systems as simple as possible in order to reduce the risk of errors in implementation, operation, and maintenance.
- A careful design and planning are needed in order to define an optimal sizing of key components and an appropriate design that matches the actual load structure.
- All auxiliary components have to be highly energy-efficient.
- An operation strategy has to be developed that leads to the most energy-efficient operation under full and part load conditions and long-term reliability.
- A careful commissioning phase of the system is necessary to ensure that the system operates as planned.
- An ongoing monitoring ("continuous commissioning") is helpful in order to enable long-term operation at the highest possible performance.

7.5 *Economic Viability and Environmental Benefits of SAC*

First cost (investment cost including planning, assembly, construction and commissioning) for SAC systems is significantly higher than the corresponding cost of best practice standard solutions — this is a very well-known fact for almost all solar energy systems and many other systems using renewable energy. The first cost of actual SAC installations is between 2 and 5 times higher than to a conventional state-of-the-art system depending on local conditions, building requirements, system size, and of course on the selected technical solution. In recent studies, first cost for total systems ranged from 2000 € per kW_{cold} to 5000 € per kW_{cold} and even higher in some particular cases. This large range is due to different sizes of systems, different technologies, different application sectors, and other boundary conditions.

A recent trend is the development of (solar) cooling kits, that is pre-engineered package solutions that consist of all-important components of a system and where the components are well matched to each other. These kits are mainly developed for small capacities up to about 35 kW cooling capacity. Prices excluding installation cost and distribution system to the building for the package solutions dropped from about 6000 € per kW in 2007 to about 4500 € per kW in 2013.

The cost saving during operation very much depends on the boundary conditions. Boundary conditions in favor of a low payback time are:

- High solar radiation leading to high gains of the solar system.
- A long cooling season leading to a large number of hours where the system is used.
- Other heat loads such as for sanitary hot water and/or building heating increase the usefulness of the solar system.
- High prices of conventional energy make a solar alternative more competitive.

When looking at the overall life-cycle cost of an SAC system (excluding any incentives or funds) in comparison to a conventional standard solution the situation looks much better than in the case of first cost. Depending on the particular conditions SAC systems will in many cases amortize within their lifetime. Under promising conditions even payback times of ten years and less can be obtained. However, commercial companies often expect a payback time of five years and less in order to justify an investment. Such low values of payback time will only be achieved under very special conditions.

SAC application has some other advantages that are often difficult to translate to an economic advantage, but may be important for policy makers:

- The use of SAC systems may lead to (primary) energy savings and thus help to reduce the dependence on finite energy fuels, which have to be imported in many countries.
- Correspondingly, the use of SAC systems will lead to reduced CO_2 emissions and thereby contribute to a reduction of climate change and related effects.
- SAC systems using thermally driven cooling cycles show additional environmental benefits since they typically employ refrigerants with no ozone depletion potential and no or a very small global warming potential.
- SAC systems can be used also for all heating applications in a building or industry. The large solar collector field also provides heat for other purposes than cooling and thus helps to avoid consumption of fuel (or electricity) for heating applications.
- SAC systems can contribute to grid stability in regions where a considerable share of daily electricity consumption comes from the grid for air-conditioning with conventional techniques.

7.6 *Market Status of SAC*

Close to a thousand SAC plants were installed worldwide in 2013 covering all types of technologies and sizes. This is of course a very early market in comparison to the tens of millions of compression air-conditioning systems sold worldwide every year. Recently, a number of large and very large installations have been built or are under development. Importantantly a number of projects are completely driven by private investors without major incentive or funding programmes.

The numerous SAC RD&D activities initiated many of those installations and also led to the establishment of a SAC community. An international bi-annual conference focusing on this R&D of SAC has been established and held three times already with an increasing number of participants and papers.[97] Nevertheless there are still only very few companies that offer complete SAC solutions. Most systems are still composed of components or subsystems that are put together by a planning office for the particular project. In addition, there is still a great need for increased expertise and experience on the side of planners, installers, and other involved professionals.

A significant number of companies have been created in the last 10 years that offer thermally driven chillers in the small capacity range up to approx. 35 kW_{cold}. Most of these companies were start-ups that entered into an almost nonexistent market. Therefore, a significant increase in production rates, in particular for these small capacity systems, can be expected and correspondingly significant reductions in cost can be expected here. These systems, which can be operated with relatively low driving temperatures, are not only operated in combination with solar thermal energy, but also as co-generation units, district heat networks and industrial waste heat.

Thus, the overall market for these thermally driven chillers could be much larger than for solar cooling only. Today, there are mature products available and first companies started to establish a manufacturing capacity on an industrial scale. More companies will follow to enlarge their production capacity and thereby develop more automated production processes.

The next important step is to develop today's low volume market channels into a mass market and, at the same time, drive costs down further through standardization of concepts and design methods and development of reliable package solutions.

7.7 *Technical Potentials of SAC*

Potentials on the technical level range from new and advanced materials through improved components to more efficient and more reliable systems.

Main ongoing R&D on the material level is aiming to improve materials for adsorption and absorption (tailor-made adsorbents or Ionic liquids) or absorption cycles and composites mainly using adsorption materials. These materials and compounds have the potential to allow for more compact systems with advanced heat and mass transfer and thus will lead to lowering the cost for installation and broadening the application range. Completely different solutions, such as new thermo-mechanical cycles, promise a significant increase in efficiency are under development but have not yet left laboratory scale.

Work on components concerns advanced cooling cycles on the one hand and advanced solar collectors, which are well adjusted to the needs of thermally driven cooling on the other hand. Important R&D activities addressing advanced components are:

- Integration of the generator of a thermally driven cooling machine in the solar collector will lead to reduced heat transfer losses and to more compact systems. Also, space in a technical room will be saved. Overall, such concepts aim at high efficiency at reduced system cost.
- Double-effect cycle absorption technology, which achieves high efficiency at high operation temperatures, will be extended also for the small capacity range and thus offer solutions with high overall efficiency for applications in the range of small capacity.
- Single-axis tracking solar thermal collectors to produce heat at temperatures in the range of 150°C to 250°C are still a rather new technology and important cost savings may be achieved by development of advanced materials (e.g. for reflectors) and advanced production technologies.
- Non-tracking collectors have achieved a high level of technical maturity. However, improvements towards higher operation efficiency at temperatures of 80°C to 110°C are still possible and advanced production technologies will allow for reduction of cost.

Important work on the system level is required in order to achieve more reliable systems and a high quality in the whole chain of projects ranging from design over planning, construction, and assembly to operation. This work includes measures such as training and education to increase knowledge and experience among professionals active in design and installation of HVAC systems.

7.8 SAC Costs and Economics

Cost reductions are required in order to increase the cost competitiveness of SAC in comparison to conventional solutions and thus increase the market deployment.

On the component level, main cost reductions are particularly expected for small capacity thermally driven cooling machines. Both technology improvements and mass production on an industrial scale will certainly lead to a lower cost of this key component. In principle, a noticeable potential also exists for cost reduction of solar thermal collectors. However, prices of solar thermal collectors remained almost stable over the last decade, which is mainly due to unstable market development and market policies. Another major component is the heat rejection unit. Today, mainly units in the large capacity range exist. The development of heat rejection systems specially designed for their use in SAC will lead to reduced component cost, in particular for small capacity units (e.g. below 100 kW of cooling power).

On the system level two main market areas have to be distinguished:

- Pre-engineered systems: mainly for applications in the small capacity range, that is air-conditioning in residential buildings and systems for small commercial applications, pre-fabricated standard packages will be developed. This pre-fabrication minimizes the planning effort and reduces the effort on the construction side, which also leads to minimized risk of errors in assembly and installation. As was pointed out above a significant cost reduction of pre-fabricated systems, also called kits, has been observed and will certainly continue to take place with growing markets.
- Custom-made systems: for large buildings and industrial applications custom-made systems will remain the appropriate approach. However, also for custom-made systems cost reductions on the system level are possible.
- Increased experience among involved professionals is an important need and guidelines and software tools for the design and layout of systems will help to reduce the effort during design and planning. In particular, standardized measures to assure a high quality of installations is required in order to increase the certainty of energy savings and related operation cost savings.

In the following section, some comments are made on the comparison of solar thermally driven SAC with solar electrically driven SAC using photovoltaic modules and electrically driven vapor compression cycles.

As was pointed out in the introduction, systems using electrically driven vapor compression cycles combined with photovoltaic systems may be an alternative of using solar energy for air-conditioning or refrigeration. Photovoltaic modules experienced a continuous cost reduction with a growing market during the last two decades. Under today's price conditions both solar solutions, that is a solar thermally driven heating and cooling compared to conventional heating and cooling system plus a photovoltaic system, lead to similar economic performance, if no funds or incentive are taken into consideration. Therefore, a decision on the one or the other alternative has to be made based on particular boundary conditions. In particular, when an attractive feed-in tariff for electricity produced by PV systems exists, most probably the economic outcome will be in favor of a photovoltaic system. In these cases, the photovoltaic system will operate completely separated from the heating and cooling system and deliver all electricity into the grid. However, this may change significantly with an increased share of electricity production from fluctuating renewable sources. Onsite solutions that lead to a minimum impact on the grid, such as photovoltaic driven heating and cooling systems using electrically driven vapor compression cycles, may become an alternative. It is difficult to assess which of the two main solar driven heating and cooling solutions — thermally driven or electricity driven — will be more competitive under such conditions. Obviously the thermally driven system always provides a local system solution and thus will always lead to a reduced dependence of conventional energy and has never a negative impact on the grid. Also the easier integration of thermal energy storages compared to electrical storages is in favor of thermally driven cooling systems from a grid point of view. Solar thermally driven systems allow the integration of buffer storage on the "hot" side of the thermally driven chiller and thus offer a degree of freedom in system design and operation, which is not given in the case of photovoltaic driven systems.

7.9 *Market Opportunities for SAC*

Generally speaking most favorable conditions for a successful market implementation for SAC systems are:

- Applications with a high need for heating and cooling (and sanitary hot water).

- Places with a high solar energy potential, that is high solar radiation.
- Conditions characterized by a high coincidence of loads and solar gains since this reduced the need for storage.
- Places with high cost of conventional energy.

A major obstacle is that in many places that fulfill the above requirements no or very little experience with solar energy use exists and often HVAC and refrigeration installations have a comparatively low standard. Nevertheless, under such conditions best economic performance can be expected. Companies that offer overall solutions and have the capability to provide maintenance services (e.g. using remote control approaches) will be able to exploit this market opportunity.

A major argument for the installation of SAC systems is their environmental soundness due to reduced consumption of conventional energy and employment of environmentally friendly refrigerants. This can be an important argument for an investor even if the system is not yet economically competitive, that is the return on investment is above the normal expectations for commercial investments.

Concretely speaking, market opportunities are seen in the following:

- Tertiary buildings, such as office buildings and hotels, in regions with sufficient solar gains: here in particular technologies employing non-tracking solar collector technologies will be employed and the solar system will be used for heating, cooling and sanitary hot water (if needed).
- Production buildings in sunny regions that need cooling for industrial processes (e.g. in the food industry): depending on the required temperature level of the cooling process either non-tracking solar collectors or single-axis tracking solar collector technologies will be employed. Large factory roofs can serve to place the collectors but also systems installed on the ground will be usual, in particular in areas where ground is not a limiting resource.
- In sunny regions in particular a potential for application of SAC in residential houses exists. Heating and cooling solutions making use of solar thermal energy for single family houses as well as for multi-family houses are particularly interesting in new buildings and in combination with highly energy efficient buildings which allow for radiative solutions (e.g. using the floor and/or the ceiling for heating/cooling).

Overall, renewable energies will play an increasingly important role in future energy systems due to the strong need to limit CO_2 emissions originating from conventional energy sources. SAC technology is one of the important solutions applicable on the demand side. Thus, this technology provides a market opportunity for many involved stakeholders including building owners, planners, manufacturers, and installation companies.

8 Economic and Policy Issues Related to Solar Heating and Cooling Thermal Applications

Many solar thermal applications are viable economically now in favorable climate and use contexts. This would be true of even more applications if for those applications nearing economic viability, the economic externalities associated with the potential for CO_2 abatement were given tangible value by appropriate fiscal intervention. Solar materials processing technologies require further research and development. Solar thermal process heat delivery, though a mature technology, still requires intensive programmes of component improvement.

The unreliability and irregularity of supply together with variable and often high cost of fossil fuels and electricity means that in many hot climates, particularly in remote and/or island locations, many thermal applications of solar energy in agriculture and industry are not only viable economically but are the obvious and preferred approach. That they are not ubiquitous is due to two interlinked factors; (i) this lack of widespread system component suppliers and associated design and installation expertise and experience and (ii) the often large initial capital cost. The latter is a particular obstacle where the potential user does not have sufficient available capital and/or is unable or unwilling to borrow funds at favorable interest rates.

However often as a consequence of a diverse range of governmental market stimulation interventions internationally, the influences of such limiting behavioral, trading structure and capital market factors are, in specific favorable contexts, now being superseded by recognition of the tangible commercial advantages of solar energy use. These include for example the often minimal or non-existent recurrent outlays for fuel leading to predictable running costs that are a hedge against inflationary energy costs adversely affecting business competitiveness.

SAC technology is at a critical stage. Mature components are available and many installations have been realized. The technology has shown that significant energy savings are possible, and it has reached a level of early market deployment. However, the financial risk for parties involved in SAC business is still too high:

- Due to no complete assurance on the quality of the system the user cannot be sure about the level of energy savings and related cost savings.
- Most planners and installers have little experience with SAC technology and thus the effort — and related cost — to install those systems is higher than for standard systems.
- Due to unstable and small markets, manufacturers cannot be sure about sale rates and thus cannot bear the risk to significantly increase their production scale.

Therefore, measures to support a sustainable market development are most important. This includes in particular establishment of large-scale solar cooling demonstration programs and implementation of adjusted incentive schemes to lower the risk for investments. In parallel establishment of quality procedures that cover all phases of a project are most critical in order to satisfy the expectations of all involved stakeholders.

Acknowledgments

Solar thermal research at Dublin Institute of Technology has been supported by Science Foundation Ireland and the European Union through the 7th Framework Programme and COST actions. Solar Thermal cooling research at Fraunhofer Institute for Solare Energiesysteme and at TECSOL S.A. has been supported by the IEA Solar Heating and Cooling Programme under Task 38 "Solar Air conditioning and Refrigeration".

References

1. J. Perlin, Let it shine: The 6000 year story of solar energy (2013).
2. J. W. Ring, Windows, baths and solar energy in the roman empire, *Am. J. Archeol.* **100** (1996), pp. 717–724.
3. D. Behrman, *Solar Energy: The Awakening Science* (Routledge and Kegan Paul, London, 1979).
4. K. Butti and J. Perlin, *A Golden Thread* (Van Nostrand, Reinhold Co, New York, USA, 1980).
5. J. A. Duffie and W. A. Beckman, *Solar Engineering of Thermal Processes*, 2nd edn. (Wiley lnterscience, New York, 1991).
6. C. M. Kemp, US Patent No. 451, 384, April 28 (1891).
7. W. J. Bailey, Solar heater, US Patent No. 966, 070 (1910).
8. S. D. Dahl and J. H. Davidson, Performance and modeling of thermosyphon and heat exchangers for solar water heaters, *ASMIE J. Sol. Energ. Eng.* **119** (1977), pp. 193–200.

9. E. Yazdanshenas and S. Furbo, Theoretical comparison between solar water/space-heating combisystems and stratification design options, *ASME J. Sol. Energ. Eng.* **129** (2007), pp. 438–448.
10. Y. C. S. Too, G. L. Morrison and M. Behnia, Performance of solar water heater with narrow, *Solar Energy* **83** (2009), pp. 350–362.
11. B. Norton, Solar water heaters; a review of systems research and design innovation, *Green* **1** (2011), pp. 189–207.
12. M. D. Wuestling, S. A. Klein and J. A. Duffie, Promising control alternatives for solar water heating systems, *ASME J. Sol. Energ. Eng.* **107** (1985), pp. 215–221.
13. A. M. Al-Ibrahim, W. A. Beckman, S. A. Klein and J. W. Mitchell, Design procedure for selecting an optimum photovoltaic pumping system in solar domestic hot water system, *Solar Energy* **64** (1998), pp. 227–239.
14. J. Ji, J. Hun, T.-T. Chow, H. Yi, J. Lu, W. He and W. Sun, Effect of fluid flow and packing factor on energy performance of a wall-mounted hybrid photovoltaic/water-heating collector system, *Energ. Buildings* **38** (2006), pp. 1380–1387.
15. T .T. Chow, W. He and J. Ji, Hybrid photovoltaic-thermosyphon water heating system for residential application, *Solar Energy* **80** (2008), pp. 298–306.
16. T. D. Schmidt, Mangold and H. Müller-Steinhagen, Central solar heating plants with seasonal storage in Germany, *Solar Energy* **76** (2004), pp. 165–174.
17. J. D. Dalenbäck, Take off for solar district heating in Europe, *Polska Energetyka Sloneczna* **1** (2010), pp. 9–13.
18. W. Weiss, *Solar Heating for Houses — A Design Handbook for Solar Combisystems* (James and James, London, 2003).
19. E. Yazdanshenas, S. Furbo and C. Bales, Theoretical comparison between solar combisystems based on bikini tanks and tank-in-tank solar combisystems, *Proc. EuroSun 2008*: 1st Int. Congress on Heating, Cooling and Buildings, Lisbon, Portugal, October (2008).
20. M. Smyth, P. C. Eames and B. Norton, A novel high performance integrated collector/storage solar water heater design, *Renew. Energ.* **16** (1998), pp. 2217–2222.
21. M. Smyth, P. C. Eames and B. Norton, A comparative performance rating for an integrated solar collector/storage vessel with inner sleeves to increase heat retention, *Solar Energy* **66**(4) (1999), pp. 291–303.
22. M. Smyth, P. C. Eames and B. Norton, Evaluation of a freeze resistant integrated collector/storage solar water heater (ICSSWH) for Northern Europe, *Appl. Energ.* **68** (2001a), pp. 265–274.
23. M. Smyth, P. C. Eames and B. Norton, Annual performance of heat retaining integrated collector/storage solar water heaters in a Northern maritime climate, *Solar Energy* **70**(5) (2001b), pp. 391–401.
24. M. Smyth, P. C. Eames and B. Norton, Heat retaining integrated collector/storage solar water heaters, *Solar Energy* **75** (2003), pp. 27–34.

25. M. Smyth, P. C. Eames and B. Norton, Techno-economic appraisal of an integrated collector/storage solar water heater, *Renew. Energ.* **29** (2004), pp. 1503–1514.
26. M. Smyth, P. McGarrigle, P. C. Eames and B. Norton, Experimental comparison of alternative convection suppression arrangements for concentrating integral collector storage solar water heaters, *Solar Energy* **78** (2005), pp. 223–233.
27. M. Smyth, P. C. Eames and B. Norton, Integrated collector storage solar water heaters, *Renew. Sust. Energ. Rev.* **10** (2006), pp. 503–538.
28. C. Schmidt, A. Goetzberger and J. Schmid, Test results and evaluation of integrated collector storage systems with transparent insulation, *Solar Energy* **41**(5) (1988), pp. 487–494.
29. C. Schmidt and A. Goetzberger, Single-tube integrated collector storage systems with transparent insulation and involute reflector, *Solar Energy* **45**(2) (1990), pp. 93–100.
30. S. Baer, *Breadbox Water Heater Plans* (Zomeworks Corporation, Albuquerque, New Mexico, USA, 1975).
31. W. S. Duff and A. Menon, Modelling and validation of integral collector storage evacuated tubular collector DHW systems, *J. Chem. Phys.* **100** (1994), p. 337.
32. R. K. Sion, R. Sangameswar, D. Rao and K. Rao, Performance of a flat plate solar collector with fluid undergoing phase change, *Solar Energy* **23** (1979), pp. 69–73.
33. J. M. Shreyer, Residential application of refrigerant charged solar collectors, *Solar Energy* **26** (1981), pp. 307–312.
34. K. S. Ong and M. Haider-E-Alathi, Performance of a 134a-filled thermosyphon, *Appl. Therm. Eng.* **23** (2003), pp. 2373–2381.
35. M. Esen and H. Esen, Experimental Investigation of a two phase closed thermosyphon solar water heater, *Solar Energy* **79** (2005), pp. 459–468.
36. A. Odaz-Flores, O. Garcia-Valladares and V. H. Gomez, Comparison of the thermal performance of different working fluids in a closed two-phase solar water heating thermosyphon, in *Proc. EuroSun 2008*: 1st International Congress on Heating, Cooling and Buildings, Lisbon, Portugal, October (2008).
37. H. A. De Beijer, Product development in solar water heating, in *Proc. 5th World Renewable Energy Congress*, Florence, Italy, pp. 201–204 (1998).
38. A. Troi, Y. Vangiouklakis, E. Korma, D. Jähnig, E. Wiemken, G. Franchini, D. Mugnier, B. Egilegor, P. Melograno and W. Sparber, Solar Combi^{+}; identification of most promising markets and promotion of standardized system configurations for small scale solar heating and cooling applications, in *Proc. EuroSun 2008*, October, Lisbon, Portugal (2008).
39. G. L. Morrison, Simulation of packaged solar heat pump water heaters, *Solar Energy* **53** (1984), pp. 249–257.
40. B. J. Huang and C. P. Lee, Long-term performance of solar-assisted heat pump water heater, *Journal of Renewable Energy* **29** (2005), pp. 633–639.

41. W. Weiss and Mauthner, Solar heat worldwide, Report of the International Energy Agency, Solar Heating and Cooling Programme (2010).
42. E. Ruiz and P. J. Martinez, Analysis of an open air swimming pool solar heating system using an experimentally validated TRNSYS model, *Solar Energy* **84** (2010), pp. 116–123.
43. A. H. Fanney and B. P. Daugherty, A PV solar water heating system, *ASME Journal of Solar Energy Engineering* **119** (1997), pp. 126–133.
44. A. H. Fanney, B. P. Daugherty and K. P. Kramp, Field performance of PV solar water heating system, *ASME Journal of Solar Energy Engineering* **119** (1997), pp. 265–272.
45. A. Rabl, *Active Solar Collectors and Their Applications* (Oxford University Press, Oxford, UK, 1985).
46. B., Norton, *Solar Energy Thermal Technology* (Springer-Verlag, Heidelburg, Germany, 1992).
47. B. Norton and J. E. J. Edmonds, Aqueous propylene glycol concentrations for the freeze protection of thermosyphon solar energy water heaters, *Solar Energy* **47**(5) (1991), pp. 375–382.
48. N. V. Baker, *Passive and Low Energy Building Designs for Tropical Island Climates* (Commonwealth Secretariat Publications, London, 1987).
49. E. Mazria, *Solar Home Book* (Rodale Press, Emmaus, Pennsylvania, USA, 1980).
50. A. Konya, *A Design Primer for Hot Climates* (Architectural Press, London, 1980).
51. O. S. Aschehoug, Baker, Frame and edge-seal technology — an international view, in *Proc. Window Innovations*, '95, Toronto, Canada (1995).
52. S. Svendsen and P. Fritzel, Spacers for highly insulated windows, in *Proc. Window Innovations '95*, Toronto, Canada (1995).
53. R. E. Collins and T. M. Simko, Current status of the science and technology of vacuum glazing, *Solar Energy* **62** (1998), pp. 189–213.
54. P. W. Griffiths, M. Di, P. Leo, P. C. Cartwright, P. Eames, G. Yianoulis and B. Leftheriotis, Norton fabrication of evacuated glazing at low temperature, *Solar Energy* **63** (1998), pp. 244–249.
55. P. W. Griffiths, B. Norton, P. C. Eames and S. N. G. Lo, Detailed simulation of heat transfer across planar evacuated glazing building, *Res. Information* **24** (1996), pp. 42–147.
56. B. Norton, *Harnessing Solar Heat* (Springer, Heidelberg, Germany, 2013).
57. Y. Fang, P. C. Eames and B. Norton, Influence of insolation level and glass thickness on the thermal performance of evacuated glazing, in *Proc. World Renewable Energy Congress VI*, Brighton, UK, pp. 1878–1881 (2000).
58. Y. Fang, P. C. Eames, B. Norton and T. J. Hyde, Experimental validation of a numerical model for heat transfer in evacuated glazing, *Solar Energy* **80** (2006), pp. 564–577.
59. Y. Fang, P. C. Eames, B. Norton, T. Hyde, J. Zhao, J. Wang, Y. Huang, Low emittance coatings and the thermal performance of vacuum glazing, *Solar Energy* **81** (2007), pp. 8–12.

60. K. Duer and S. Svendsen, Monolithic silica aerogel in superinsulating glazings, *Solar Energy* **63** (1998), pp. 259–267.
61. Y. Fang and P. C. Eames, The effect of glass coating emittance and frame rebate on heat transfer through VG and EC vacuum glazed windows, *Sol. Energ. Mat. Sol. Cells* **90** (2006), pp. 2683–2695.
62. S. J. Robinson and R. E. Collins, Evacuated windows-theory and practice, in *ISES Solar World Congress, International Solar Energy Society*, Kobe, Japan (1989).
63. S. Selkowitz, M. Rubin, E. S. Lee and R. Sullivan, A review of EC window performance factors, *SPIE* **2255** (1994), pp. 226–248.
64. R. Zriken and E. Bilgen, Theoretical study of Trombe–Michel wall solar collector system, *Solar Energy* **39** (1987), pp. 409–419.
65. M. N. Wright, *Solar Radiation as Related to Summer Cooling and Winter Radiation in Residences* (J.B. Pierce Foundation, New York, 1936).
66. M. Wegelin, S. Canonica, K. Mechsner, T. Tleischmann, F. Pasaro and A. Metzler, Solar water disinfection: Scope of the process and analysis of radiation experiments, *J. Water S. R. T. — Aqua* **40** (1994), pp. 154–169.
67. J. D. Burch and K. E. Thomas, Water disinfection for developing countries and potential for solar thermal pasteurisation, *Solar Energy* **64** (1998), pp. 87–97.
68. N. K. Bansal, R. L. Sawhney, A. Misra and A. Boettcher, Solar sterilization of water, *Solar Energy* **40** (1988), pp. 35–39.
69. M. A. S. Malik, G. N. A. Tiwari Kumar and M. S. Sodha, *Solar Distillation* (Pergamon Press, Oxford, U.K., 1982).
70. M. F. Kocher, R. Gerald, P. E. Bodman and M. Lay, Solar grain drying and farrowing house heating with a multi-use collector, ASAE Paper No. 4554 (1981).
71. W. Weiss, H. Schweiger and R. Battisti, Market potential and system designs for industrial heat applications, in *Proc. ESTEC* (2005).
72. S. Kalogirou, The potential for solar industrial process heat applications, *Appl. Energ.* **76** (2003), pp. 337–361.
73. G. N. Kulkarni, S. B. Kedere and S. Bandyopadhyay, Design of solar thermal systems utilising pressurised hot water storage for industrial applications, *Solar Energy* **82** (2008), pp. 686–699.
74. O. V. Ekechukwu, P. O. Fleming and B. Norton, The potential of solar energy agricultural dryers in Nigeria, in *Proc. 2nd Annual Conf. Biotechnology Society of Nigeria*, 16–19 May (1986).
75. P. H. Bailey and W. F. Williamson, Some experiments on drying grain by solar radiation, *J. Agr. Eng. Res.* **10** (1965), pp. 191–196.
76. H. R. Bolin, A. E. Stafford and C. C. Huxsoll, Heated fruit dehydrator, *Solar Energy* **20** (1978), pp. 281–291.
77. H. R. Bolin, A. E. Stafford and C. C. Huxso, Solar through dryer, *Solar Energy* **22** (1979), pp. 455–457.
78. E. A. Arinze, G. Schoenau and F. W. Bigsby, Solar energy absorption properties of some agricultural products, ASAE Paper, No 79-3071 (1979).

79. L. S. Cheema, Solar drying of cassava for alcohol production, *ISES Cong.*, New Delhi, India, 1940–1944 (1978).
80. L. S. Cheema and C. M. C. Ribeiro, Solar dryers of cashew, banana and pineapple, *ISES Conf.*, New Delhi, India, 3, 2075–2079 (1978).
81. P. E. Doe, M. Ahmed, M. Muslemuddin and K. Sachithananthan, A polythene tent drier for improved sun drying of fish, *Food Tech. Australia* **29** (1977), pp. 437–441.
82. R. H. B. Exell, A simple solar rice dryer, basic design theory, *Sunworld* **4** (1980), pp. 186–190.
83. W. E. Ferguson and P. H. Bailey, Solar air heater for near ambient crop drying; Description test methods and 1977 results, *The British Soc. for Res. Agric. Eng., Scotish Inst. Ag. Eng. Departmental Note*, SIN/275, (1979).
84. P. D. Fleming, B. Norton, S. O. Onyegegbu, S. D. Probert and O. V. Ekechukwu, A large-scale facility for experimental studies of natural circulation solar-energy tropical crop dryers, in *Proc. of the 1986 Int. Drying Symp.*, pp. 685–693 (M.I.T. Cambridge, USA, 1986).
85. K. A. McLean, *Drying and Storage of Combinable Crops* (Farm Press Ltd., Ipswich, Suffolk, UK, 1980).
86. B. Norton, P. O. Fleming and O. V. Ekechukwu, Passive autarkic solar drying techniques, physics and technology of solar energy, in *Proc. International Workshop on Physics of Solar Energy*, New Delhi (1986).
87. M. N. Ozisik, B. K. Huang and M. Toksoy, Solar grain drying, *Solar Energy* **24** (1980), pp. 397–401.
88. G. C. Shove, G. W. Barton, M. D. Hall and W. H. F. Peterson, Studies of solar grain drying, ASAE Paper, No 81-4037 (1981).
89. O. P. Singhal and G. P. Gupta, Use of solar energy for parboiling and drying of paddy, *ISES Cong.*, New Delhi, India, pp. 1958–1963 (1978).
90. U. J. Taylor and A. D. Weir, Simulation of a solar timber drier, *Solar Energy* **34** (1985), pp. 249–255.
91. J. F. Thompson and H. E. Studer, Sun drying of prunes continuous trays, ASAE., Paper, No 81-3.051 (1981).
92. J. L. Tschernitz and W. T. Simpson, Solar heated, forced-air, lumber dryer for tropical latitudes, *Solar Energy* **22** (1979), pp. 563–566.
93. B. Brenndorfer, L. Kennedy, C. O. O. Bateman, G. C. Mrema and C. Wereko-Brobby, *Solar Dryers — Their Role in Postharvest Processing* (Commonwealth Science Council (Commonwealth Secretariat Publications), London, 1985).
94. K. Crowther and B. Melzer, The thermosphoning cool pool — A natural cooling system, in *Proc. Third National Passive Solar Conf.*, San Jose, California, USA, 3, pp. 448–451, January (1979).
95. K. Crowther, Cooling from an evaporating thermosyphoning roof pond, in *Proc. Passive Solar Conf.*, Kansas City, Missouri, USA, pp. 409–503 October (1979).
96. www.**iea-shc**.org. Accessed 8/31/2015.
97. http://task48.iea-shc.org/. Accessed 8/31/2015.

Chapter 9

Solar Thermochemical Processes

Roman Bader* and Wojciech Lipiński†
Research School of Engineering,
The Australian National University
Canberra ACT 2601, Australia
**roman.bader@anu.edu.au*
†*wojciech.lipinski@anu.edu.au*

Solar radiation is an abundant energy source with high exergy content, suitable for driving thermochemical processes with high efficiency. This chapter provides a comprehensive review of the field of solar thermochemistry, with focus on production of fuels and other chemical commodities. Material, process, and reactor developments for selected processes are discussed along with pertinent advances in thermodynamics, kinetics, and thermal transport.

1 Introduction

Solar energy can be used as a source of renewable process heat to drive energy-intensive chemical processes. If solar energy is the only source of process heat, the difference between the energy content of the products and that of the reactants corresponds to the amount of solar energy stored chemically during the process.

For example: The endothermic dissociation of $H_2O(l)$ into $H_2(g)$ and $1/2O_2(g)$, with all reactants and products at room temperature and atmospheric pressure, requires an energy input of approximately 286 kJ per mole of dissociated H_2O.

Thus, endothermic chemical processes offer the possibility to absorb and store solar energy. If the resulting products are chemically stable and have sufficiently high energy density, these chemical energy carriers can be used to decouple the time and location of solar energy utilization from the time and location of its collection, thus serving as *solar fuels*.

For example: If the energy density of solar-derived chemical energy carriers is sufficiently high, solar energy can be harvested in rural Australia and transported in chemically stored form to densely populated regions in Australia and overseas to generate mechanical or electrical energy using heat engines or fuel cells.

In addition, solar thermal energy can replace fossil fuels in the production of many other energy-intensive materials, of which the primary use is non-energetic. Such products can be referred to as *solar commodity materials*.

For example: A rapid growth in urban populations in Asia leads to a rapid increase in the demand for construction materials such as cement and steel. Currently, unprocessed iron ore is exported from Australia to China where it is processed to iron using vast amounts of fossil fuels. Instead, iron can be produced in Australia using solar energy as the source of process heat, thereby reducing the amount of raw material to be shipped and the fossil energy used in the production process.

Several high-temperature processes have been considered for the utilization of solar energy in the production of fuels and commodities. This chapter provides an overview of the processes that have been most widely examined to date.

1.1 *Overview of Processes*

Figure 1 gives an overview of the most common solar thermochemical processes, including reactants, main products, and process types. These processes can be broadly categorized according to the type of their products into processes that yield solar fuels and processes that yield solar commodities. Solar fuels processes aim at producing hydrogen (H_2) or syngas, a mixture of mainly H_2 and CO. These processes can be further categorized according to the feedstocks involved. Syngas can be obtained either via H_2O- and CO_2-splitting processes or via conversion of carbonaceous feedstocks other than CO_2. H_2O and CO_2 splitting can be accomplished in a single thermal dissociation step (thermolysis), followed by a high-temperature gas separation step. Thermolysis of H_2O and CO_2 at ambient pressure, while conceptually simple, requires very high temperatures exceeding 2500 K to obtain significant H_2/CO concentrations.[1] In thermochemical redox cycles, intermediate oxygen exchange (redox) materials are used to break the process down into multiple process steps and reduce the temperature required to split H_2O and CO_2. Many of the proposed redox

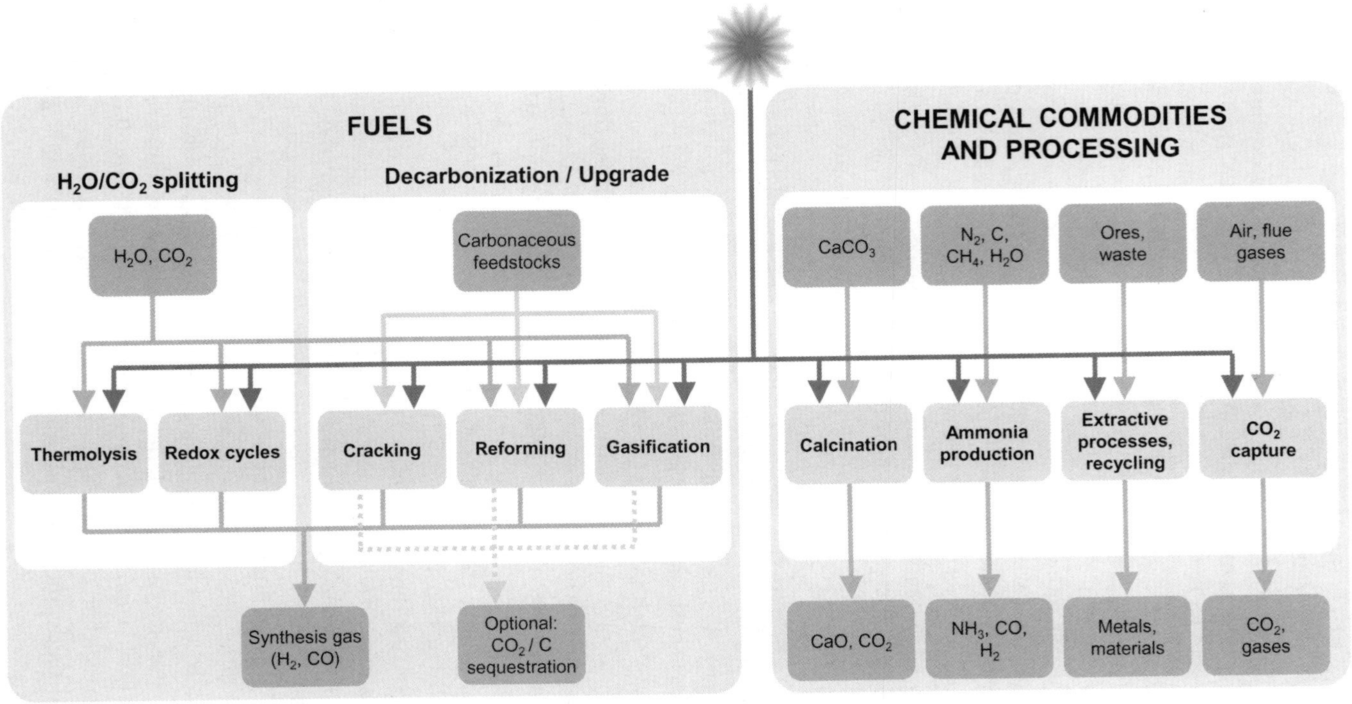

Fig. 1. Overview of process routes for the solar thermochemical production of fuels and commodity materials. This figure is an expanded version of the schematic published in Ref. 2.[3–5]

materials used to split H_2O and CO_2 are metal oxides (e.g. Fe_3O_4, ZnO, CeO_2). These materials are either partially reduced to a lower-valence metal oxide or fully reduced to a pure metal using high-temperature solar heat. Subsequently, the reduced redox material is exposed to H_2O and CO_2 to produce H_2 and CO, respectively. Thereby, the redox material is reverted to its initial oxidation state and is returned to the thermal reduction step. In the carbothermal reduction of metal oxides, reducing agents such as carbon or methane are used to lower the process temperature and co-produce CO or syngas in the reduction step. H_2 or syngas can also be produced from the conversion of carbonaceous feedstocks via methane cracking, methane reforming, or gasification of solid materials (e.g. coal, petcoke, biomass). In conventional autothermal or allothermal processes, part of the feedstock or a separate fuel is combusted to generate the process heat for the endothermic reaction. If solar energy is used as the source of process heat, solar energy in the amount of the enthalpy of reaction of the process is stored in the products.

Syngas produced via solar thermochemical processes can either be used directly as a fuel or further processed to synthetic fuels (e.g. diesel, gasoline, kerosene) via the Fischer–Tropsch process or other catalytic reforming processes. Synthetic liquid hydrocarbon fuels offer high energy density and can be integrated directly into the existing fuel infrastructure.[6] To avoid the CO_2 emissions emerging from the combustion of hydrocarbon fuels, CO_2 can be recovered either at the source, e.g. via a $CaCO_3$-based calcination–carbonation process, or from ambient air, e.g. via a temperature–vacuum swing desorption/adsorption process.[7] The carbon material cycle can be closed by returning the captured CO_2 to the solar fuels process, as shown schematically in Fig. 2.

Solar process heat can also be used for the production of energy-intensive commodities. Pure metals can be extracted from metal ores in the thermal and carbothermal reduction processes described in the context of solar fuels. Thermally reduced redox materials can also be used as chemical or combined chemical and sensible heat storage media to provide high-temperature process heat on-demand, for example to a thermal power cycle. In this case, the reduced redox material is re-oxidized with an oxygen-rich gas, for example ambient air. Alternatively, metal oxides can be reduced carbothermally in nitrogen to form metal nitrides. These ceramic materials can either be used as high-temperature materials, or utilized in cyclic metal oxide/metal nitride processes to produce ammonia and syngas.[9] The first step of the $CaCO_3$-based calcination–carbonation process,

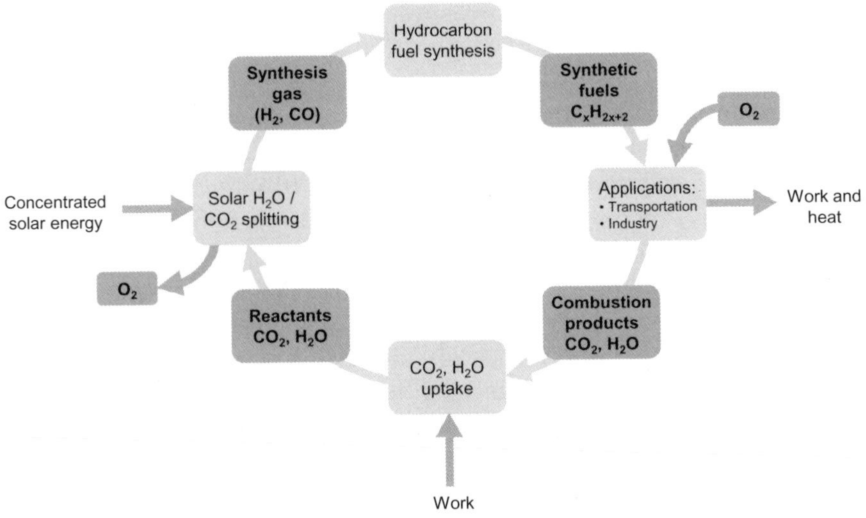

Fig. 2. CO_2 material cycle in a CO_2 neutral energy system based on solar fuels.

the calcination of limestone ($CaCO_3$), yields lime (CaO), the main ingredient of cement. In addition, the CO_2 evolving from this process could be used as a feedstock in solar fuels processes.

Before we enter the discussion of different solar processes, chemical reactions, and solar reactor technologies, we establish the thermodynamic basis that supports the idea of utilizing concentrated solar energy in thermochemical processes.

1.2 *Thermodynamic Limits*

For a chemical reaction operated at constant temperature and pressure, the total amount of energy to be supplied for the reaction to proceed is given by:

$$\Delta H_{\mathrm{rxn}} = \Delta G_{\mathrm{rxn}} + T\Delta S_{\mathrm{rxn}}, \tag{1}$$

By definition, ΔH_{rxn}, the enthalpy of reaction, is positive for endothermic reactions and negative for exothermic reactions. For most reactions, ΔH_{rxn} is only a weak function of temperature. The first term on the right-hand side of Eq. (1), ΔG_{rxn}, is the Gibbs free energy of reaction. It corresponds to the amount of energy that needs to be supplied in the form of work for the chemical reaction to proceed. A negative value of ΔG_{rxn} indicates that a chemical reaction proceeds spontaneously without any work input. The second term, $T\Delta S_{\mathrm{rxn}}$, corresponds to the amount of energy to be supplied in the form of heat.

At room temperature and ambient pressure, the magnitude of ΔG_{rxn} is usually much larger than the magnitude of $T\Delta S_{\mathrm{rxn}}$, and $\Delta H_{\mathrm{rxn}} \approx \Delta G_{\mathrm{rxn}}$. Hence, endothermic reactions ($\Delta H_{\mathrm{rxn}} > 0$) normally do not occur spontaneously at room temperature and ambient pressure without work input. With increasing temperature T, the magnitude of the term $T\Delta S_{\mathrm{rxn}}$ increases. Hence, an endothermic reaction with $\Delta S_{\mathrm{rxn}} > 0$ proceeds without work input above the temperature $T_{\Delta G=0}$, at which $\Delta G_{\mathrm{rxn}} = \Delta H_{\mathrm{rxn}} - T_{\Delta G=0}\Delta S_{\mathrm{rxn}} = 0$.[a] These trends of ΔH_{rxn}, $T\Delta S_{\mathrm{rxn}}$, and ΔG_{rxn} for an endothermic reaction are shown qualitatively in Fig. 3. Therefore, conducting an endothermic chemical reaction with thermal energy as the only source of energy requires heat input at elevated temperatures.

For example: For the dissociation of H_2O(l) into H_2(g) and $1/2O_2$(g), $\Delta G_{\mathrm{rxn}} = 237$ kJ mol^{-1} at room temperature and 1 bar pressure, and $\Delta G_{\mathrm{rxn}} = 0$ occurs at approximately 4330 K.[5,b]

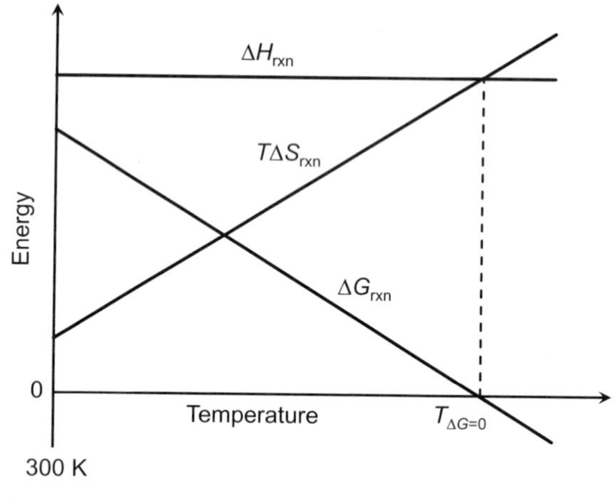

Fig. 3. Qualitative trends of the thermodynamic quantities enthalpy of reaction, ΔH_{rxn}, Gibbs free energy of reaction, ΔG_{rxn}, and $T\Delta S_{\mathrm{rxn}}$, for an endothermic reaction with positive ΔS_{rxn}. $T_{\Delta G=0}$ indicates the temperature above which the reaction proceeds spontaneously. Reprinted from Ref. 4, Copyright (2001), with permission from Elsevier.

[a]Note that $T_{\Delta G=0}$ is only indicative. The reaction starts to proceed at $T < T_{\Delta G=0}$ and complete reaction occurs at $T > T_{\Delta G=0}$.

[b]In practice, H, O, and OH will also be present at this temperature.

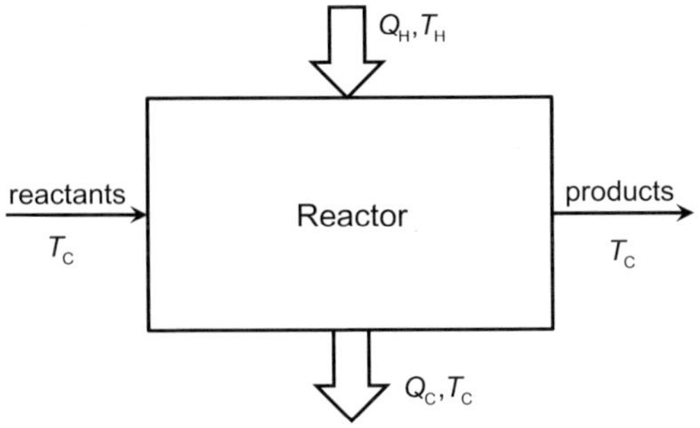

Fig. 4. Schematic of a generic thermochemical reactor.

Next, consider the thermochemical reactor shown schematically in Fig. 4. The reactor is assumed to host an endothermic reaction. It is further assumed to operate at steady-state and at ambient pressure. Heat is supplied at temperature T_{H} and rejected to the environment at ambient temperature, T_{C}. Reactants and products cross the reactor boundary at temperature, T_{C}. The first and second laws of thermodynamics yield:

$$0 = Q_{\mathrm{H}} - Q_{\mathrm{C}} - \Delta H_{\mathrm{rxn}}(T_{\mathrm{C}}), \tag{2}$$

$$0 = \frac{Q_{\mathrm{H}}}{T_{\mathrm{H}}} - \frac{Q_{\mathrm{C}}}{T_{\mathrm{C}}} - \Delta S_{\mathrm{rxn}}(T_{\mathrm{C}}) + S_{\mathrm{irr}}. \tag{3}$$

In Eq. (3), S_{irr} denotes the entropy generated within the system due to internal irreversibilities. Combining Eqs. (2) and (3) and using Eq. (1) yields:

$$Q_{\mathrm{H}} = \left(\frac{T_{\mathrm{H}}}{T_{\mathrm{H}} - T_{\mathrm{C}}}\right)\left[\Delta G_{\mathrm{rxn}}(T_{\mathrm{C}}) + T_{\mathrm{C}} S_{\mathrm{irr}}\right], \tag{4}$$

The energy efficiency of the process to store thermal energy in chemical products is defined as:

$$\eta_{\mathrm{energy}} = \frac{\Delta H_{\mathrm{rxn}}(T_{\mathrm{C}})}{Q_{\mathrm{H}}}, \tag{5}$$

The maximum efficiency is reached for an internally reversible process, $S_{\mathrm{irr}} = 0$. Inserting Eq. (4) into Eq. (5) yields:

$$\eta_{\mathrm{energy,ideal}} = \frac{\Delta H_{\mathrm{rxn}}(T_{\mathrm{C}})}{\Delta G_{\mathrm{rxn}}(T_{\mathrm{C}})}\left(1 - \frac{T_{\mathrm{C}}}{T_{\mathrm{H}}}\right). \tag{6}$$

For example: The maximum energy efficiency $\eta_{\text{energy,ideal}}$ of the dissociation of $H_2O(l)$ into $H_2(g)$ and $1/2O_2(g)$ in a reactor according to Fig. 4 with $T_H = 1000°C$ and $T_C = 298\,K$ is $\eta_{\text{energy,ideal}} = 92.3\%$.

If the products are used to derive work in a closed material system, they are returned to their initial form as they react exothermically to produce work. The maximum work that can be derived from a chemical energy carrier is that generated in a reversible fuel cell operating at ambient temperature T_C:

$$\eta_{\text{FC,ideal}} = \frac{\Delta G_{\text{rxn}}(T_C)}{\Delta H_{\text{rxn}}(T_C)}, \tag{7}$$

Hence, the maximum thermal efficiency to convert the heat input Q_H to work via a thermally-driven endothermic reaction and a reversible fuel cell in a closed material cycle equals the Carnot efficiency:

$$\eta_{\text{th,ideal}} = \eta_{\text{energy,ideal}}\eta_{\text{FC,ideal}} = 1 - \frac{T_C}{T_H}, \tag{8}$$

For example: The maximum thermal efficiency $\eta_{\text{th,ideal}}$ of the dissociation of $H_2O(l)$ into $H_2(g)$ and $1/2O_2(g)$ in a reactor according to Fig. 4 with $T_H = 1000°C$ and $T_C = 298\,K$, followed by the recombination of the products in a reversible fuel cell operating at $T_C = 298\,K$ is $\eta_{\text{th,ideal}} = 76.6\%$.

For an open material system in which the products of the work generation process are not equal to the reactants of the thermochemical fuel production process, the maximum efficiency of converting solar energy to work is:

$$\eta_{\text{th,ideal}} = \eta_{\text{energy,ideal}}\frac{\Delta G_{\text{FC}}(T_C)}{\Delta H_{\text{FC}}(T_C)}, \tag{9}$$

with $\eta_{\text{energy,ideal}}$ given in Eq. (6) and $\Delta G_{\text{FC}}(T_C)$ and $\Delta H_{\text{FC}}(T_C)$ denoting the Gibbs free energy and enthalpy of reaction, respectively, of the work-generating step occurring in the fuel cell.

To provide solar process heat, solar energy has to be captured with a solar receiver. Hence, the absorption efficiency of the solar receiver has to be taken into account. Neglecting convection and conduction losses, the absorption efficiency of an ideal blackbody absorber (absorptivity and emissivity equal to 1) at temperature T_H is:[c]

$$\eta_{\text{abs,ideal}} = 1 - \frac{\sigma T_H^4}{CI_{\text{sun}}}, \tag{10}$$

[c]The feasibility and benefit of a solar-selective absorber diminishes with increasing temperature of the absorber. Therefore, a blackbody absorber is considered as the ideal absorber in the present context.

where σ is the Stefan–Boltzmann constant, $I_{\text{sun}} = 1\,\text{kW}\,\text{m}^{-2}$ is the nominal direct normal solar irradiance, and C is the solar concentration ratio, defined as the ratio of the average radiative flux on the absorber to I_{sun}. Hence, for a solar-driven thermal process, the maximum thermal efficiency is:

$$\eta_{\text{solar,ideal}} = \eta_{\text{abs,ideal}}\eta_{\text{energy,ideal}}\eta_{\text{FC,ideal}}. \tag{11}$$

Figure 5 shows $\eta_{\text{solar,ideal}}$ as a function of T_{H} for a closed material system, $\eta_{\text{abs,ideal}}$, and the Carnot efficiency. With increasing solar concentration ratio, C, the maximum of $\eta_{\text{solar,ideal}}$ increases and shifts towards higher T_{H}. It can be shown with a radiation heat transfer analysis that the theoretical maximum solar concentration ratio is $C_{\text{ideal}} = 1/\sin^2\theta_{\text{sun}} \approx 46{,}000$,[d] where $\theta_{\text{sun}} = 4.65$ mrad is the cone half-angle subtended by the sun.[9] Hence, the maximum $\eta_{\text{solar,ideal}}$ that can be theoretically achieved is $\eta_{\text{solar,ideal,max}}(C_{\text{ideal}} = 46{,}000) = 84\%$ and occurs at $T_{\text{H,max}}(C_{\text{ideal}} = 46{,}000) = 2318\,\text{K}$. For existing three-dimensional solar concentrators (e.g. solar tower, solar dish), the solar concentration ratio is typically in the range $C = 1{,}000$–$10{,}000$ and the maximum of $\eta_{\text{solar,ideal}}$ is in

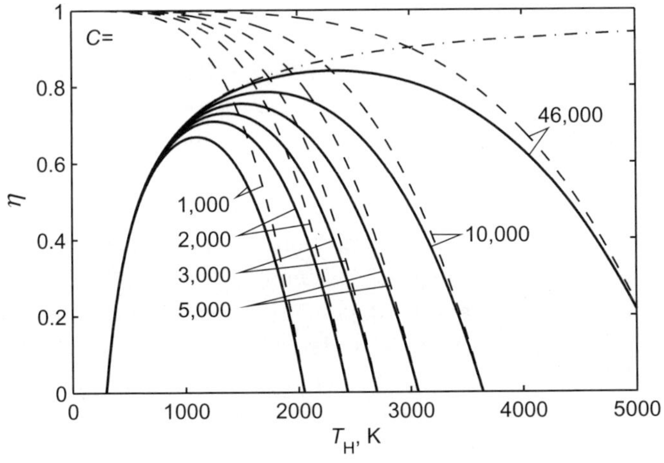

Fig. 5. Solid lines: theoretical maximum thermal efficiency, $\eta_{\text{solar,ideal}}$, of a solar-driven closed-cycle thermal process operating between the temperature reservoirs T_{H} and $T_{\text{C}} = 298\,\text{K}$. Dashed lines — absorption efficiency of an ideal blackbody receiver, $\eta_{\text{abs,ideal}}$; dash-dotted line — Carnot efficiency.

[d]This result is for an ideal three-dimensional concentrator containing air. In general, if the concentrator contains a medium with refractive index n, $C_{\text{ideal}} = n^2/\sin^2\theta_{\text{sun}}$.

the range of 67–79%, occurring between 1105 K and 1721 K. In practice, the peak efficiency will occur at a lower temperature due to additional heat losses by convection and conduction.

From the above, it follows that storing solar energy via endothermic thermochemical reactions (*i*) requires elevated temperatures (Fig. 3) and (*ii*) is most efficient at temperatures between approximately 1000–2000 K (Fig. 5).

2 Thermolysis

Direct thermal dissociation of H_2O and CO_2 according to:

$$H_2O(l) \rightarrow H_2 + \frac{1}{2}O_2, \quad \Delta H^\circ_{298\,\mathrm{K}} = 286\,\mathrm{kJ\,mol^{-1}}, \tag{12}$$

$$CO_2 \rightarrow CO + \frac{1}{2}O_2, \quad \Delta H^\circ_{298\,\mathrm{K}} = 283\,\mathrm{kJ\,mol^{-1}}, \tag{13}$$

is conceptually the simplest way to produce syngas from H_2O and CO_2. However, at ambient pressure, these reactions occur only at extremely high temperatures. For example, 20% dissociation of H_2O and CO_2 requires approximately 2800 K and 2600 K, respectively.[10,11] These temperatures are significantly higher than the optimum temperatures for solar thermal processes operated with state-of-the-art solar concentrators (Fig. 5). In addition, unwanted gaseous species (H, O, OH) are co-produced at these high temperatures.[10] Furthermore, in order to avoid the recombination of the product gases upon cooling, they must be separated at the extremely high dissociation temperatures required by the process. Potential separation techniques include effusional separation (the separation of gas species with differing molecular weights with porous membranes),[10,12] electrolytic separation,[13,14] and quenching.[4,15,16] Finally, the availability of materials of construction that can be operated at temperatures in excess of 2500 K over extended periods of time is limited. These drawbacks and challenges with the direct thermal dissociation of H_2O and CO_2 have been the motivation for the development of thermochemical cycles.

3 Thermochemical Cycles

The main advantages of thermochemical cycles compared to the direct thermolysis of H_2O and CO_2 are: (i) the need for high-temperature separation of H_2/CO and O_2 is avoided because the gases are produced in separate process steps, and (ii) the maximum process operating temperature is generally lower. Numerous candidate cycles have been scanned for their suitability to split H_2O using solar energy,[17–21] but only a handful of them have been investigated in-depth so far. While cycles with three or more steps have been proposed, the majority of today's research focuses on two-step cycles. Of the cycles that have been studied so far, the cycles comprising more than two steps tend to be operable at lower temperatures than the two-step cycles, which makes the reactor design and material selection less challenging and reduces the required solar concentration ratio. On the other hand, increasing the number of cycle steps increases irreversibilities in the process due to internal heat transfer, heat losses to the environment, and the number of transport and separation processes of reactants and products, leading to an increase in the overall complexity and costs of the process.[17] Important aspects that should be considered when selecting suitable thermochemical cycles for H_2O and CO_2 splitting include the required cycle operating conditions such as the temperatures and gaseous environments needed to conduct the reaction steps involved, the achievable product yield per mole of oxygen exchange material and cycle, the predicted energy and exergy efficiency of the overall process, the conversion of the reactants to products, the rates of the chemical reactions involved, the possibility of undesired side-reactions, the number, cost, availability, toxicity, and environmental safety of the chemical species involved, the ease of separation of the chemical species, their compatibility with reactor materials, and suitable reactor designs to conduct the process steps.[17] In the following sections we describe selected work that has been done to address some of the above research questions for the most prominent cycles.

3.1 *Two-Step Redox Cycles*

The chemical reactions in two-step redox cycles based on metal oxide redox pairs can be represented by:

Thermal reduction step (endothermic):

$$x\mathrm{MO_{ox}} \rightarrow x\mathrm{MO_{red}} + \frac{1}{2}\mathrm{O_2}. \tag{14}$$

Fuel production step (exothermic):

$$\begin{aligned} &x\mathrm{MO_{red}} + y\mathrm{H_2O} + (1-y)\mathrm{CO_2} \\ &\quad \rightarrow x\mathrm{MO_{ox}} + y\mathrm{H_2} + (1-y)\mathrm{CO}, \quad 0 \leqslant y \leqslant 1. \end{aligned} \tag{15}$$

The temperature of the reduction step can be lowered through the use of a reducing agent (e.g. carbon particles or methane). The overall net reaction is:

$$yH_2O + (1-y)CO_2 \rightarrow yH_2 + (1-y)CO + \frac{1}{2}O_2. \tag{16}$$

Metal oxide materials that have been considered for two-step redox cycles can be grouped into volatile and non-volatile materials, depending on whether the material evaporates or remains solid during the thermal reduction step. Non-volatile materials can be further grouped into stoichiometric and non-stoichiometric material systems, depending on the stoichiometry of the redox material in the cycle. An overview of the different categories, names, and common examples of two-step redox material pairs is given in Table 1.[19–21]

A benefit of the volatile candidate materials is that they undergo a stoichiometric change in their oxygen content, resulting in a relatively high theoretical fuel productivity per mole of redox material and cycle. Non-volatile materials typically undergo smaller changes in their oxygen content. A drawback of volatile processes is the need to separate the product gases of the reduction step, for example via quenching, to avoid their recombination. This challenge is avoided with non-volatile materials.

Not all metal oxide redox material pairs that have been considered were found to be useful. Many candidate metal oxides reduce at excessively high temperatures. For example, for TiO_2, Al_2O_3, SiO_2, and CaO,

Table 1: Main categories and examples of metal oxide redox material pairs used in two-step thermochemical H_2O/CO_2-splitting cycles. Adopted from Ref. 19 with modifications,[20,21] Copyright (2014), with permission from Elsevier.

Category	Name	Redox pair
Volatile, stoichiometric	Zinc oxide	ZnO(s)/Zn(g)
	Tin oxide	SnO_2(s)/SnO(g)
	Germanium oxide	GeO_2(s)/GeO(g)
Non-volatile, stoichiometric	Ferrite	$M_xFe_{3-x}O_4/xMO+(3-x)FeO$; M: Co, Ni, Zn, Mg, Mn
	Hercynite	$M_xFe_{3-x}O_4+3Al_2O_3/xMAl_2O_4$ $+(3-x)FeAl_2O_4$; M: Co
Non-volatile, non-stoichiometric	Ceria	$M_xCe_{1-x}O_2/M_xCe_{1-x}O_{2-\delta}$, M: 1+: Li; 2+: Mg, Ca, Sr; 3+: Sc, Y, Pr, Sm, Gd, Dy, La; 4+: Zr, Hf; 5+: Ta; 6+: Cr
	Perovskite	$ABO_3/ABO_{3-\delta}$, A: Sr, Ba, La; B: Al, Mn, Fe, Co

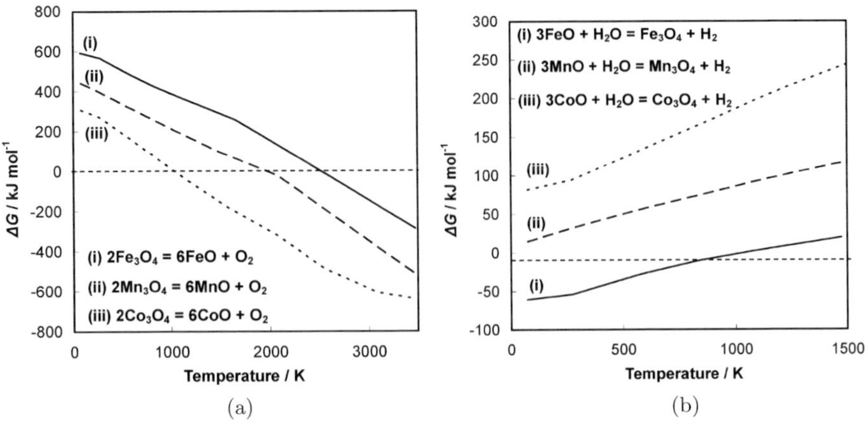

Fig. 6. Gibbs free energy for selected thermal reduction (a) and fuel production (b) reactions as functions of temperature at atmospheric pressure. Reprinted from Ref. 1, Copyright (2007), with permission from the American Chemical Society.

$T_{\Delta G=0} > 4000$ K.[4] Other materials were found to be unsuitable for water splitting. This is illustrated for the three candidate metal oxide redox pairs Fe_3O_4/FeO, Mn_3O_4/MnO, and Co_3O_4/CoO. As can be seen from Fig. 6(a), all three metal oxides are reducible in oxygen at temperatures above approximately 1000 K, 1900 K, and 2500 K, respectively. However, Fig. 6(b) shows that only FeO is able to react with water to form H_2. For the two other metal oxides, MnO and CoO, ΔG_{rxn} of the fuel production reaction is positive at all temperatures.[1]

In the remainder of this section, our goal is to discuss some of the key research and engineering work that has been done in the area of two-step thermochemical H_2O and CO_2 splitting, focusing on the most widely studied cycles. In view of the large amount of research that has been published over the last 15 years, a complete treatment of the literature in this area is beyond the scope of this text. References are provided to refer the interested reader to the primary literature.

3.1.1 *Zinc oxide cycle*

The zinc oxide cycle proceeds according to the following steps:[22]

Thermal reduction:

$$\text{ZnO(s)} \rightarrow \text{Zn(g)} + \frac{1}{2}\text{O}_2, \quad \Delta H_{298\text{K}} = 350.5 \text{ kJ mol}^{-1}, \tag{17}$$

Fuel production:

$$\mathrm{Zn(s)} + y\mathrm{H_2O} + (1-y)\mathrm{CO_2} \rightarrow \mathrm{ZnO(s)} + y\mathrm{H_2} + (1-y)\mathrm{CO}, \quad 0 \leqslant y \leqslant 1,$$
$$-67.5\,\mathrm{kJ\,mol^{-1}} < \Delta H_{298\,\mathrm{K}} < -64.7\,\mathrm{kJ\,mol^{-1}}. \tag{18}$$

For the thermal reduction step at ambient pressure, $T_{\Delta G=0} = 2235$ K,[4] whereas the boiling point of zinc is 1179 K. Hence, a gaseous mixture of zinc and oxygen is obtained. Somewhat lower reduction temperatures can be achieved if the product gases are diluted with inert gas to shift the equilibrium of reaction (17) to the right.[22] The fuel production step is thermodynamically favorable at temperatures below 1400 K at 1 bar,[23] and is usually carried out in the temperature range of 573–1273 K and at ambient pressure, i.e. below or slightly above the boiling point of zinc.

A second-law analysis of the zinc oxide cycle has been conducted by Loutzenhiser *et al.*[22] according to the process model shown in Fig. 7, to determine the theoretical thermal efficiency of the cycle. Thermal reduction is carried out in a solar reactor in the temperature range

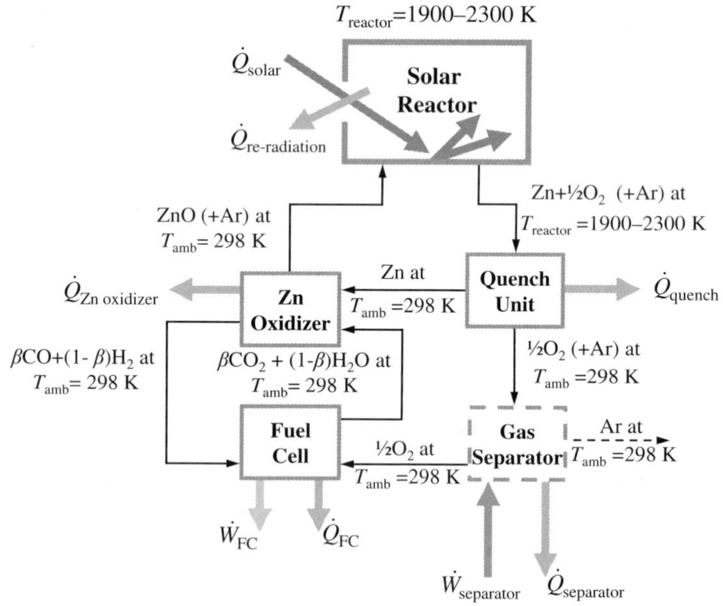

Fig. 7. Process model for the thermodynamic analysis of the zinc oxide cycle; β denotes the molar fractions $\dot{n}_{\mathrm{CO_2}}/(\dot{n}_{\mathrm{CO_2}} + \dot{n}_{\mathrm{H_2O}})$ at the inlet and $\dot{n}_{\mathrm{CO}}/(\dot{n}_{\mathrm{CO}} + \dot{n}_{\mathrm{H_2}})$ at the outlet of the zinc oxidizer. Reprinted from Ref. 22, Copyright (2011), with permission from Elsevier.

$1900\,K \leqslant T_{reactor} \leqslant 2300\,K$ and at ambient pressure. The solar reactor is modelled as a blackbody cavity receiver with a solar concentration ratio across the aperture of 5000. Convection and conduction losses are neglected. The product gas mixture of the thermal reduction step is separated via quenching to ambient temperature, $T_{amb} = 298\,K$. At temperatures below $T_{\Delta G=0}$, argon is used as inert gas in the solar reactor to shift the equilibrium of reaction (17) to the right-hand side. This increases the heat input required to heat the reactants from T_{amb} to $T_{reactor}$ and introduces a work requirement to separate Ar and O_2. Zinc is reacted with a variable mixture of CO_2 and H_2O in the zinc oxidizer, with reactants and products entering and exiting at T_{amb}. The produced fuel, a mixture of CO and H_2, and O_2 from the thermal reduction step are recombined in an ideal fuel cell operating at T_{amb}, generating a work output equivalent to the Gibbs free energy of the recombination reaction. The thermal efficiency of the cycle is defined as:

$$\eta_{th} = \frac{\dot{W}_{net}}{\dot{Q}_{solar}}, \tag{19}$$

where $\dot{W}_{net} = \dot{W}_{FC} - \dot{W}_{separator}$ and $\dot{Q}_{solar}$ are the net mechanical power output and the solar heat rate supplied to the cycle, respectively.

The thermal efficiency is shown in Fig. 8. In the absence of heat recovery, the maximum thermal efficiency is between 30–34% and peaks at $T_{\Delta G=0}$. Somewhat higher efficiencies are obtained with CO_2 than with H_2O, due to the lower entropy generation in the fuel cell with CO than with H_2 fuel. Decreasing $T_{reactor}$ below $T_{\Delta G=0}$ leads to increasing inert gas flow rate through the solar reactor which leads to increasing heat rejection in the quench unit, resulting in decreasing efficiency. Above $T_{\Delta G=0}$, inert gas is not used, but reradiation losses and the heat rejection by the quench unit increase with increasing $T_{reactor}$, resulting in decreasing efficiency. If the heat rejected by the quench unit and the zinc oxidizer is completely recuperated to preheat the reactants to the solar reactor, the energy penalty associated with the use of inert gas is reduced. In this case, the thermal efficiency reaches 50–55% and peaks below $T_{\Delta G=0}$, at 2070 K.

Schunk *et al.*[24] conducted the thermal reduction step (17) in a solar-driven thermogravimeter to measure the thermal dissociation rate under heat and mass transfer conditions closely approximating those in a directly irradiated solar reactor. Experiments were carried out in a windowed solar cavity receiver, purged with argon sweep gas. Samples of 12 mm-thick packed beds of micrometer-sized ZnO particles (sample mass 6.2 g) were

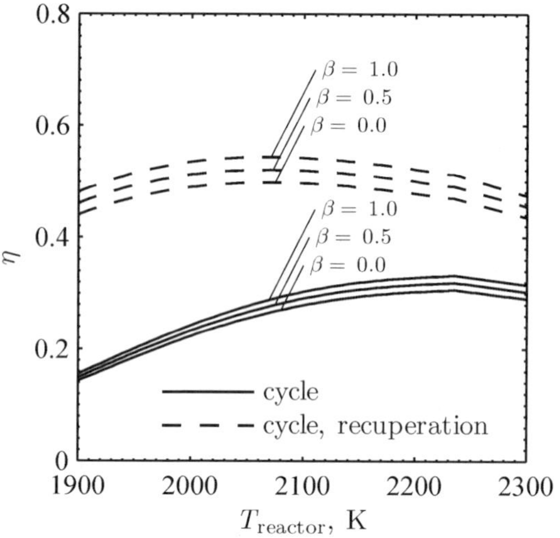

Fig. 8. Thermal efficiency of the cycle shown in Fig. 7, with (dashed lines) and without (solid lines) heat recovery from the quench unit and the zinc oxidizer. Reprinted from Ref. 22, Copyright (2011), with permission from Elsevier.

heated to different dissociation temperatures in the range of 1834–2109 K by direct exposure to concentrated solar radiation in excess of $2\,\mathrm{MW\,m^{-2}}$, supplied by a solar furnace. Samples were heated at a rapid rate of $150\,\mathrm{K\,s^{-1}}$ to 1300 K and at $20\,\mathrm{K\,s^{-1}}$ above 1300 K. The mass change of the sample was measured on-line with 0.01 g accuracy. The temperature of the irradiated surface of the sample was measured with a solar-blind pyrometer (operational wavelength 1.39 μm). A porosity change from 70% to 40% was observed due to sintering during the heating period up to 1800 K, creating a thin dense layer on the irradiated surface, but no appreciable weight change of the sample. At all dissociation temperatures, a linear weight loss was observed, characteristic of an ablation regime: The endothermic reaction occurred near the irradiated surface, because heat transfer to the surface, predominantly by radiation, proceeded faster than heat conduction into the packed bed. Hence, a zero-order Arrhenius rate equation was used. The apparent activation energy and frequency factor were calculated to $E_\mathrm{a} = 361 \pm 53\,\mathrm{kJ\,mol^{-1}K^{-1}}$ and $k_0 = 14.03 \times 10^6 \pm 2.73 \times 10^6\,\mathrm{kg\,m^{-2}}$ (at 95% confidence), consistent with previous work. As already observed in an earlier thermogravimetric study,[25] the reaction rate depends on gas-phase mass transport. Therefore, L'vov's theory was applied to predict the frequency factor as a function of the convective flow conditions.

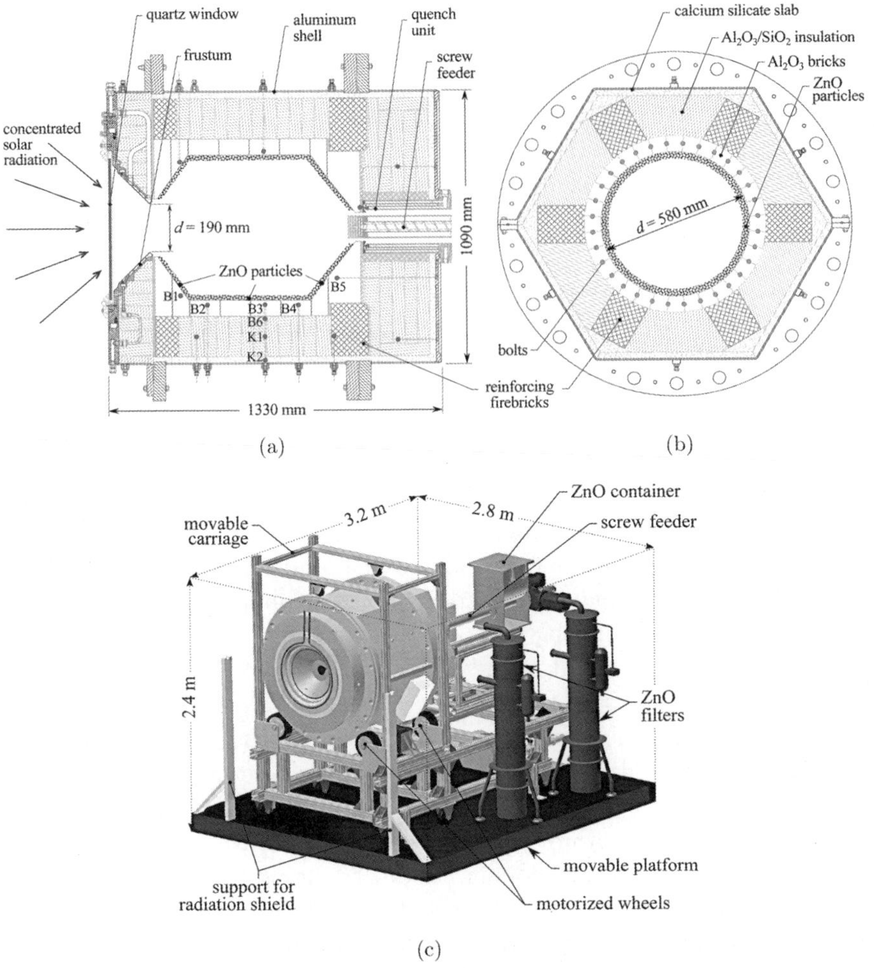

Fig. 9. A 100 kW pilot-scale solar reactor for the thermal dissociation of ZnO particles under direct concentrated solar irradiation: (a) Lateral cross-section, (b) frontal cross-section, and (c) 3-D rendering of the experimental setup. Reprinted from Ref. 29, Copyright (2014), with permission from the American Society of Mechanical Engineers.

A pilot-scale solar reactor, shown schematically in Fig. 9, has been developed to conduct the thermal dissociation of ZnO particles.[26–29] The reactor consists of a rotating cylindrical cavity receiver with a windowed aperture to control the gas atmosphere inside the reactor. Particles are fed into the reactor by a screw feeder and are distributed over the cylindrical wall of the cavity by its rotating motion, where the particles are directly exposed to concentrated solar radiation. The reactor front, the window, and the frustum are water-cooled. Argon sweep gas is injected through radial and tangential nozzles near the window and around the frustum, generating a vortex flow that protects the window from zinc deposits and carries the gaseous products out of the reactor. Product gases exit the reactor through an annular outlet at the back of the reactor that forms a water-cooled quench unit in which up to $1500\,\mathrm{L\,min^{-1}}$ of argon are injected to quench the products to avoid their recombination. The cavity wall comprises several layers. The innermost layer consists of 64-mm-thick 16%-porous alumina (corundum) bricks, bonded with high-temperature adhesive mortar. Maximum allowable temperature on the cavity inner wall is limited to about 2000 K, to avoid the formation of a eutectic phase of alumina and ZnO. The alumina cavity is lined with porous alumino-silicate insulation material, which is surrounded by a layer of 25-mm-thick fracture-tough calcium silicate. Mechanical rigidity of the assembly is enhanced with radially extending fire bricks connecting the alumina cavity and the calcium silicate plates. The entire reactor is encapsulated in a 5-mm-thick aluminium shell. The shell has a hexagonal cross-section to prevent slipping of the insulation assembly due to the rotating motion of the reactor.[29]

The reactor was tested in the 1 MW solar furnace in Odeillo, France. The solar power input to the reactor was in the range 115–140 kW, corresponding to mean solar concentration ratios over the reactor's aperture of $3682–4477\,\mathrm{kW\,m^{-2}}$. The solar power input was determined from a set of measured heat flux maps, obtained with an array of 5 Gardon-type heat flux gauges. Reactor temperatures were measured at 14 locations throughout the reactor, as indicated by the dots in Fig. 9(a). The pressure inside the reactor was maintained at 10 mbar above ambient pressure with a proportional-integral-derivative (PID) controller using a pressure transducer at the outlet of the reactor. Product gas composition was measured on-line with gas chromatography. Zinc particles were collected downstream of the quench unit with polytetrafluoroethylene (PTFE) membrane filters. Reactant and product particles were characterized with respect to their physicochemical properties. The particle size distribution was measured by laser scattering. Specific surface area, cumulative pore volume, and average pore diameter were determined with a BET surface area analyzer. The molar fraction of zinc in the product particles was determined by dissolving the particles in HCl and quantifying the evolved H_2 due to the reaction of zinc with HCl ($Zn + 2HCl \rightarrow ZnCl_2 + H_2$). The particle morphology was examined by scanning electron microscopy (SEM).

Table 2: Physicochemical properties of reactant and product particles from a typical experiment with the 100 kW pilot-scale reactor shown in Fig. 9. Reprinted from Ref. 29, Copyright (2014), with permission from the American Society of Mechanical Engineers.

	Reactant	Product
Mean particle diameter (μm)	5.6	25.8
BET specific surface area (m^2g^{-1})	4.0	27.1
Pore volume (cm^3g^{-1})	8	74
Pore diameter (nm)	8.1	11.1
Zn molar content in particle (%)	—	36.5

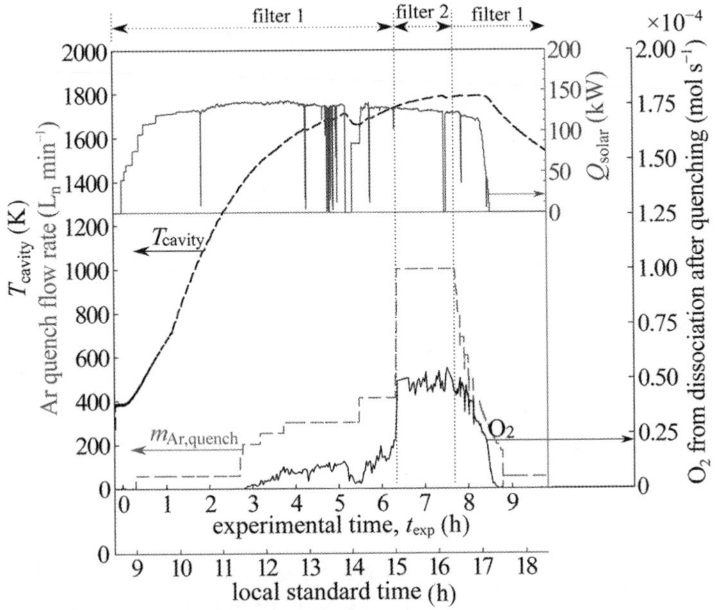

Fig. 10. Experimental results obtained with the 100 kW pilot-scale reactor shown in Fig. 9. T_{cavity} — cavity temperature measured at position B3 in Fig. 9(a); Q_{solar} — solar power input to the reactor; $m_{Ar,quench}$ — quench gas flow rate; "O_2" — molar rate of oxygen evolving from the dissociation of ZnO in the reaction. Reprinted from Ref. 29, Copyright (2014), with permission from the American Society of Mechanical Engineers.

The measured physicochemical properties of the particles from a typical experiment are listed in Table 2. The higher specific surface area of the product particles is due to their highly porous foam-like microstructure, as confirmed by SEM. Figure 10 shows temperature and product gas measurements from a typical experiment. The experimental run consisted of the 3 phases heat-up, experiment, and cool-down. The reactor was heated at a slow rate of approximately $4\,\mathrm{K\,min^{-1}}$ to a maximum cavity temperature of 1791 K by successively focusing additional heliostats, to prevent damage of the reactor due to thermal shock. Once the target reactor temperature was reached, a clean particle filter was put in place and the argon quench gas flow rate was rapidly increased to $1070\,\mathrm{L\,min^{-1}}$, resulting in a rapid increase of the oxygen content in the product gas stream. The experiment was continued for as long as the weather conditions were suitable. The experiment was terminated by defocusing the heliostats and replacing the sample filter. The total amount of dissociated ZnO in the experiment illustrated in Fig. 10 was 157 g.

Different reactors have been used to conduct the fuel production step (18). Berman *et al.*[30] performed the reaction by bubbling water vapor in argon carrier gas through liquid zinc at 450–500°C (melting point 420°C) and water partial pressures in the range of 65–560 mbar. They studied the reaction kinetics and found that the reaction rate increased with increasing water partial pressure, with a reaction order between 0 and 1 with respect to the water partial pressure. They further found that with increasing water partial pressure, the thickness of the ZnO layer forming around the water bubbles increases and at high water partial pressures diffusion through this layer becomes the rate limiting step. The continuous removal of solid ZnO product may be another challenge with this reactor type.

Hydrolysis of zinc particles was found to be characterized by an initial fast reaction rate followed by a slow reaction regime. In the initial phase, fast reaction occurs at the surface of the zinc particles. This results in the formation of a dense ZnO passivation layer surrounding the particle that inhibits further reaction of the zinc below.[31] Due to this reaction mechanism, small zinc particles with high surface-to-volume ratio are advantageous. An aerosol flow reactor was developed for the *in situ* formation and oxidation of zinc nanoparticles with H_2O and CO_2.[23,32–35] A schematic of the reactor is shown in Fig. 11. Zinc is evaporated from an alumina crucible contained in an electrically heated furnace and resting on a balance for on-line measuring of the evaporation rate. Zinc vapour is carried by a N_2 gas flow into a narrow 25-cm long quench unit where it is mixed with a preheated mixture of H_2O and N_2 to form nanoparticles. After the quench zone, the flow is expanded into a 50-cm long temperature-controlled stainless steel tube, which serves as the reaction zone for the oxidation of the nanoparticles. Product gases are analysed with a gas chromatograph. Product particles are collected with a glass-fiber filter and analyzed with

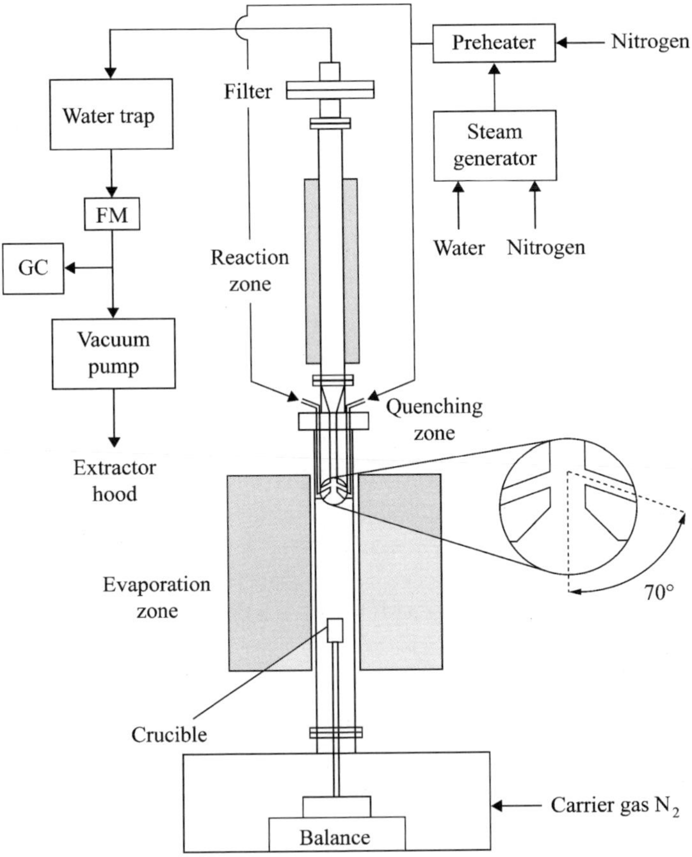

Fig. 11. Aerosol flow reactor for the *in situ* formation and oxidation of zinc nanoparticles. Reprinted from Ref. 34, Copyright (2009), with permission from Elsevier.

XRD, BET, and SEM. The reactor performance was measured in terms of the overall chemical conversion of the evaporated zinc, the fraction of the evaporated zinc ending up on the filter (particle yield), and the ZnO mole fraction in the particles collected at different locations inside the reactor. At the baseline conditions (evaporation zone temperature 1223 K, reaction zone temperature 673 K, quench gas flow rate $20\,l_n\,min^{-1}$, steam flow rate $20\,gh^{-1}$, N_2 carrier gas flow $2\,l_n\,min^{-1}$), the chemical conversion was 70%, the particle yield was 41%, and the ZnO mole fraction in particles collected on the filter was 22%. Hence, the majority of H_2 derived from hydrolysis of zinc deposits on the reactor walls. Decreasing the quench gas flow rate to $1\,l_n\,min^{-1}$ while maintaining all other parameters at the baseline values increased the chemical conversion to 95% but reduced the particle yield and the ZnO mole fraction in the product particles to below 20%. When the reaction zone temperature was increased from 573 to 873 K (Zn melting point 693 K), while all other parameters remained at the baseline values, the overall chemical conversion increased from 42 to 66%, the ZnO mole fraction in the particles increased from 10% to 50%, but the particle yield decreased from 52% to 28%. In all experiments, less than 25% of the evaporated zinc was hydrolyzed via the formation of zinc particles. The majority of the H_2 was produced via hydrolysis of zinc deposits on the reactor walls.[34] Zinc deposition on the reactor walls was also the primary concern when CO_2 was used as the gaseous reactant.[35]

3.1.2 *Ferrite cycle*

H_2O and CO_2 splitting via two-step redox cycling with ferrite materials can be represented by:

Thermal reduction:

$$\mathrm{M}_x\mathrm{Fe}_{3-x}\mathrm{O}_4 \rightarrow x\mathrm{MO} + (3-x)\mathrm{FeO} + \frac{1}{2}\mathrm{O}_2, \tag{20}$$

Fuel production:

$$x\mathrm{MO}+(3-x)\mathrm{FeO}+y\mathrm{H}_2\mathrm{O}+(1-y)\mathrm{CO}_2 \rightarrow \mathrm{M}_x\mathrm{Fe}_{3-x}\mathrm{O}_4+y\mathrm{H}_2+(1-y)\mathrm{CO}, \tag{21}$$

where M_x denotes the dopant concentration, $x = 0\ldots1$. For pure iron oxide ($x = 0$) at 1 bar total pressure, the thermal reduction reaction (20) proceeds spontaneously above $T_{\Delta G=0} = 2500$ K, which is higher than the melting point of both Fe_3O_4 (1870 K) and FeO (1650 K).[1,36] If the oxygen partial pressure is reduced to 10^{-7} bar, the thermal reduction step proceeds at around 1623 K, i.e. below the melting point of Fe_3O_4 and FeO.[1,36,37] The fuel production reaction (21) with H_2O proceeds spontaneously below 650 K and the product H_2/H_2O ratio increases with decreasing temperature.[36]

A thermodynamic process analysis was performed by *Steinfeld et al.*[38] to evaluate the theoretical thermal efficiency of the ferrite cycle operated with pure iron oxide, analogous to that conducted by Loutzenhiser *et al.*[22] for the zinc oxide cycle. Chemical equilibrium compositions are taken into account, assuming the condensed phase to be an ideal solution. Thermal reduction is conducted at temperatures between 1900 and 2500 K in air, with air/Fe_3O_4 ratios of 0, 1, and 10. The fuel production step is conducted at ambient temperature. The system is at a uniform total pressure of 1 bar.

The thermal efficiency strongly depends on whether the heat rejected in the quench unit is recovered. Without heat recovery, the predicted thermal efficiency is below 8% (energy efficiency <10%[1]). If the heat rejected in the quench unit is converted to work with an ideal Carnot engine, predicted efficiencies are up to 61%. In another thermodynamic analysis, Diver *et al.*[39] estimated energy efficiencies of 36% and 76% without and with heat recovery, respectively. However, these results were obtained by assuming complete conversion of the iron oxide in both process steps and neglecting reradiation losses from the high-temperature solar reactor.

In practice, continuous H_2O/CO_2 splitting using undoped ferrite materials is inhibited by material sintering and melting during heating to the thermal reduction temperature, followed by solidification during cooling of the reduced iron oxide to the fuel production temperature. This results in a loss of the material's reactivity and complicates the handling of the material, as the solid reduced iron oxide needs to be ground into small particles to effect the fuel production step.[40,41] Lowering the oxygen partial pressure to 10^{-7} bar to operate the thermal reduction step at temperatures below the melting points of Fe_3O_4 and FeO can theoretically avoid the sintering problem and reduces the energy input required to heat the reactants to the thermal reduction temperature, but requires either a high flow rate of sweep gas in the solar reactor or vacuum pumping, which increases the auxiliary energy consumption and the complexity of the process.[1]

Allendorf *et al.*[42] calculated the equilibrium compositions via Gibbs free energy minimization for the metal ferrites $M_xFe_{3-x}O_4$ with M = Fe, Co, Ni, and Zn. Each ferrite exhibits a distinct temperature, above which the ferrite decomposes (Fig. 12(a)), and a narrow temperature band within which a transition takes place between the solid MeO solution and liquid slag phases (Fig. 12(b)). For undoped ferrite, complete dissociation of the ferrite phase occurs above 1730 K, whereas liquid slag phases are formed above 1700 K. Therefore, it is not possible to completely reduce Fe_3O_4 without sintering. In contrast, for both $CoFe_2O_4$ and $NiFe_2O_4$, dissociation of the ferrite occurs at a lower temperature than the formation of slag phases. $CoFe_2O_4$ dissociates completely above 1690 K, while slag phases occur only above 1815 K. For $NiFe_2O_4$, the corresponding temperatures are 1700 K and 1795 K. Hence, $CoFe_2O_4$ and $NiFe_2O_4$ offer a temperature window of approximately 100 K in which the ferrite completely dissociates without the formation of slag phases and hence without sintering. This makes these ferrites more promising candidates for redox cycling. The dissociation of $ZnFe_2O_4$ differs from the other ferrites shown. Heating $ZnFe_2O_4$ leads to

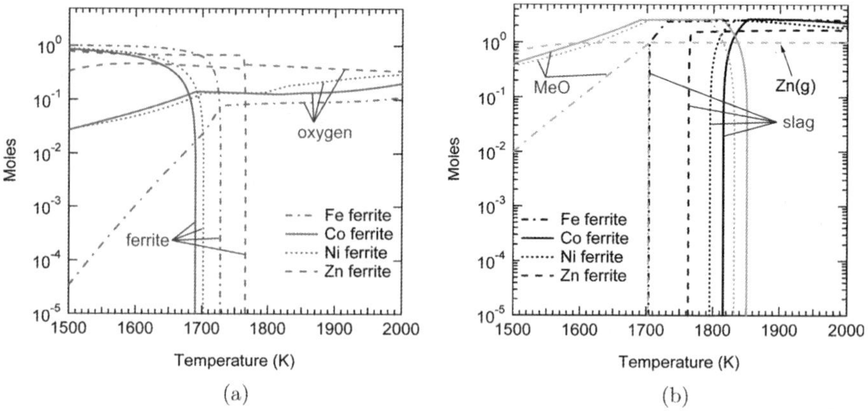

Fig. 12. Equilibrium composition as a function of temperature for the metal ferrites $M_xFe_{3-x}O_4$, with M = Fe, Co, Ni and Zn: (a) solid ferrite and gaseous oxygen; (b) MeO solid solution, Zn, and liquid slag phase. Dash-dotted lines — undoped ferrite, solid lines — $CoFe_2O_4$, and dotted lines — $NiFe_2O_4$. Inputs: 1 mol $M_xFe_{3-x}O_4$, 10^4 mol Ar. Reprinted from Ref. 42, Copyright (2008), with permission from the American Chemical Society.

the release of gaseous zinc and zinc-depleted ferrite at temperatures below 1760 K. Above 1760 K, the zinc-depleted ferrite melts to form a Zn–Fe–O slag. In practice, inert sweep gas flow or vacuum pumping is used to lower the oxygen partial pressure in the gas phase. Hence, the high zinc vapour concentration would result in rapid depletion of the zinc contained in the system, resulting in the formation of pure iron ferrite. Therefore, $ZnFe_2O_4$ is not a suitable candidate material for H_2O/CO_2 splitting via reactions (20) and (21).[42]

Mixed iron oxides, $M_xFe_{3-x}O_4$, with M = Mg, Mn, Co, Ni, Zn, and $x = 0 \ldots 1$, deposited on monoclinic zirconia (m-ZrO_2) or yttrium-stabilized zirconia (YSZ) support structures (particles, ceramic foams), have been studied experimentally to increase the fuel productivity, lower the thermal reduction temperature, and reduce material sintering.[43] Kodama *et al.*[1] found that $NiFe_2O_4$ supported on m-ZrO_2 resulted in a doubling of the hydrogen productivity compared to undoped ferrite on m-ZrO_2. They further reported that supporting $NiFe_2O_4$ on m-ZrO_2 resulted in significantly higher hydrogen production rate than unsupported $NiFe_2O_4$ per unit mass, despite a mass loading of $NiFe_2O_4$ on m-ZrO_2 of only 19 wt%, which indicates that the support material promotes the reaction kinetics.

Different reactors have been designed to execute reactions (20)–(21). A 100 kW pilot-scale solar reactor has been developed to conduct both

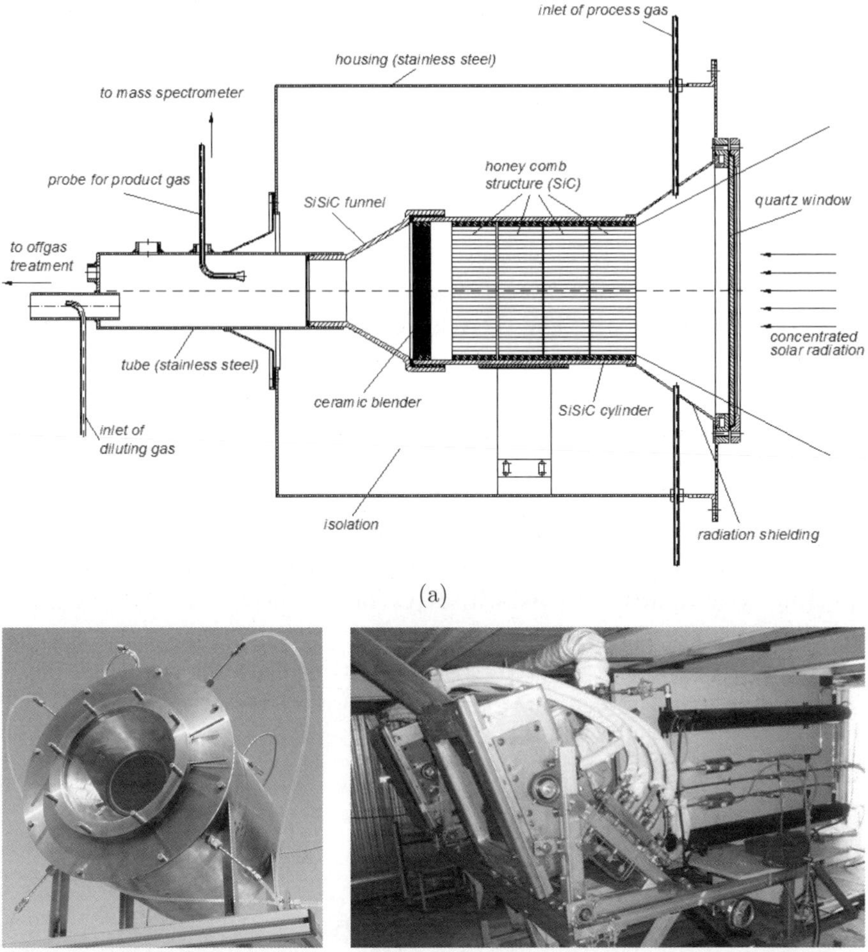

Fig. 13. Solar receiver-reactor based on honeycomb absorber structure coated with ferrite-based redox materials: (a) schematic of the reactor setup (reprinted from Ref. 44, Copyright (2006), with permission from the American Society of Mechanical Engineers.); (b) laboratory-scale prototype reactor (reprinted from Ref. 45. Copyright (2005), with permission from Elsevier.); and (c) dual-chamber pilot-scale reactor (reprinted from Ref. 46, Copyright (2011), with permission from Elsevier).

process steps sequentially in one reactor (Fig. 13).[44–46] The redox material is coated on a monolithic silicon carbide honeycomb structure contained in a solar receiver and exposed to direct concentrated solar radiation. The aperture of the receiver is covered with a quartz glass window. Gases enter through inlet ports in the frustum of the receiver, flow through the axial channels of the honeycomb structure, and are collected in a funnel at the back of the reactor. Different Mn-, Ni-, and Zn-iron oxides were synthesized via solid-state synthesis, self-propagating high-temperature synthesis, and aerosol spray pyrolysis (ASP).[44] The material reactivity was enhanced by incorporating oxygen vacancies in the material during the synthesis process. Highest hydrogen productivities of >3 mmol per gram of metal oxide were observed for ASP-synthesized Zn- and NiZn-ferrites. The 100 kW pilot-scale reactor shown in Fig. 13(c) was tested on a 1.5 MW_{th} solar tower facility at the Plataforma Solar de Almería in Spain. The reactor consists of two separate reaction chambers. The thermal reduction and fuel production steps were conducted in parallel in the two reaction chambers. Each cycle step was conducted for 20–30 min. Target reactor temperatures during the thermal reduction and fuel production steps were 1200°C and 800°C, respectively. Temperatures were controlled via heliostat focusing/defocusing. To switch between thermal reduction and fuel production steps, a number of the heliostats was refocused from the reaction chamber undergoing thermal reduction to the reaction chamber undergoing oxidation. Two full H_2O splitting cycles were conducted with H_2O conversion of up to 30%, resulting in approximately 35 g of H_2 produced.

Other reactors tested for H_2O/CO_2 splitting with ferrite-based materials include fluidized bed reactors,[47,48] and rotary-type reactors.[39,49]

3.1.3 *Ceria cycle*

The ceria cycle can be represented by:

Thermal reduction:

$$\mathrm{M}_x\mathrm{Ce}_{1-x}\mathrm{O}_{2-\delta_{\mathrm{ox}}}\,(\mathrm{s}) \rightarrow \mathrm{M}_x\mathrm{Ce}_{1-x}\mathrm{O}_{2-\delta_{\mathrm{red}}}\,(\mathrm{s}) + \frac{\Delta\delta}{2}\mathrm{O}_2, \tag{22}$$

Fuel production:

$$\begin{aligned}&\mathrm{M}_x\mathrm{Ce}_{1-x}\mathrm{O}_{2-\delta_{\mathrm{red}}}(\mathrm{s}) + \Delta\delta(y\mathrm{H}_2\mathrm{O} + (1-y)\mathrm{CO}_2)\\ &\quad \rightarrow \mathrm{M}_x\mathrm{Ce}_{1-x}\mathrm{O}_{2-\delta_{\mathrm{ox}}}(\mathrm{s}) + \Delta\delta(y\mathrm{H}_2 + (1-y)\mathrm{CO}).\end{aligned} \tag{23}$$

M_x denotes the dopant concentration; $x = 0$ for undoped ceria. δ_{red} and δ_{ox} denote the oxygen deficiency in the ceria lattice at the end of the reduction and fuel production step, respectively, and $\Delta\delta = \delta_{\mathrm{red}} - \delta_{\mathrm{ox}}$.

Initially, a stoichiometric reaction scheme with the redox pair CeO_2/Ce_2O_3 was proposed, with the reduction step operated at 2000°C and a slightly reduced total pressure of 100–200 mbar ($T_{\Delta G=0} = 2350°\mathrm{C}$[50]).[51]

Since the melting points of both metal oxides are well above 2000°C (CeO_2: 2397°C; Ce_2O_3: 2177°C[52]), ceria would be expected to remain solid during the process. However, intermediate phases with lower melting point and higher volatility form during the reduction step, which resulted in melting at 1950°C and excessive sublimation of ceria at temperatures above 1950°C. To solve the problem of material instability, Chueh *et al.*[53] elucidated the feasibility of a non-stoichiometric cycle in which solid ceria undergoes changes in its oxygen content while remaining in the cubic fluorite phase, i.e. without undergoing any phase change.

The lower oxygen exchange capacity of a non-stoichiometric cycle raised the question of its theoretical efficiency, as less fuel is produced per mole of redox material and cycle compared to stoichiometric cycles. Lapp *et al.*[54] conducted a first-law analysis to calculate the theoretical energy and exergy efficiencies for the model system shown in Fig. 14. A continuous stream of undoped ceria (e.g. in the form of particles or a rotating cylinder) is cycled between a reduction zone and a fuel production (oxidation) zone. The reduction step is carried out in a flow of 99 mol% nitrogen and 1 mol% oxygen at temperatures in the range of $T_{\text{red}} = 1400-2200\,\text{K}$. The fuel production step is conducted with a pure stream of H_2O/CO_2 at $T_{\text{ox}} = 1073\,\text{K}$, resulting in $\delta_{\text{ox}} \approx 0$. Inlet gas streams to the two reactor zones are preheated with

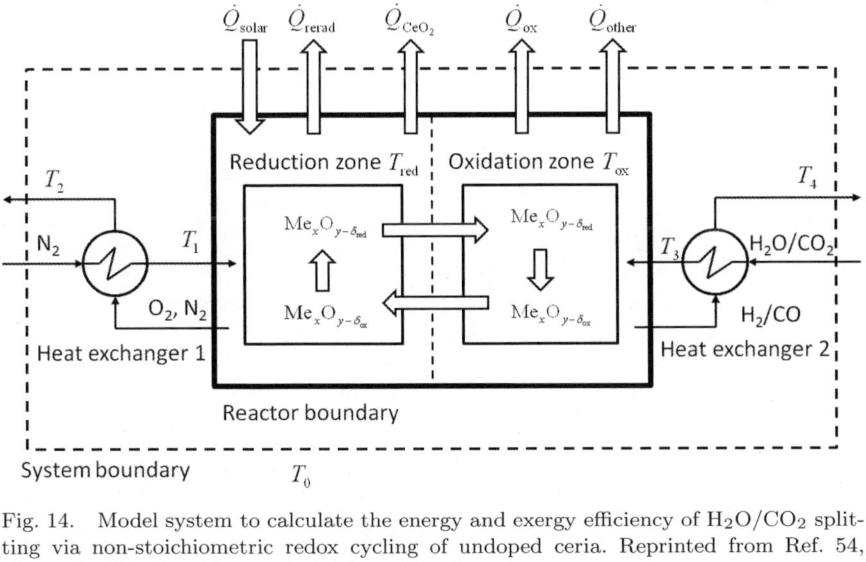

Fig. 14. Model system to calculate the energy and exergy efficiency of H_2O/CO_2 splitting via non-stoichiometric redox cycling of undoped ceria. Reprinted from Ref. 54, Copyright (2012), with permission from Elsevier.

the corresponding product gas streams in counterflow heat exchangers. In addition, solid phase heat recovery between the two streams of ceria passing between the reaction zones is taken into account. The energy balance further takes into account reradiation losses from a blackbody receiver at $T_{\rm red}$ and a lumped heat loss term accounting for convective and radiative heat losses from the hot outer surfaces of the reactor. The total pressure in the reactor is 1 atm and pressure drop in the gas flow is omitted. Thermal and chemical equilibrium between gas and solid are assumed in both steps. The energy efficiency is calculated according to:

$$\eta_{\rm energy} = \frac{\dot{n}_{\rm H_2/CO}|\Delta H_{\rm c}^{\circ}|}{\dot{Q}_{\rm solar}}, \tag{24}$$

where $|\Delta H_{\rm c}^{\circ}|$ is the higher heating value of the produced fuel.

Selected results are shown in Fig. 15. Results are valid for both H_2O and CO_2 splitting. As shown in Fig. 15(a), high efficiencies exceeding 50% are theoretically possible with 100% heat recuperation from both the solid and gas streams. In contrast, in the absence of any heat recovery, the theoretical efficiency is below 6%. Figure 15(b) shows the influence of solid-phase heat recovery on the efficiency for a realistic scenario with $C = 3000$ and 90% gas-phase heat recovery. Without solid-phase heat recovery, the maximum efficiency of the cycle is limited to less than 10% — hardly enough for

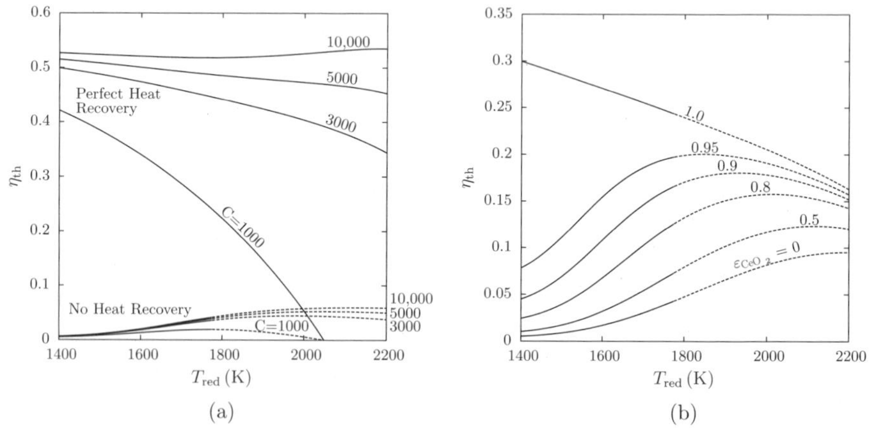

Fig. 15. Energy efficiency calculated for the model system shown in Fig. 14: (a) limiting cases with 0% and 100% gas- and solid-phase heat recovery; (b) gas-phase heat recovery of 90% and variable solid-phase heat recovery, $C = 3000$. Dashed lines represent results obtained with extrapolated non-stoichiometry data for ceria. Reprinted from Ref. 54, Copyright (2012), with permission from Elsevier.

cost-effective solar fuel production. Only with solid-phase heat recuperation of 95% or more can efficiencies beyond 20% be achieved. These results indicate that efficient splitting of H_2O/CO_2 with undoped non-stoichiometric ceria in a temperature-swing cycle requires very high degrees of both gas- and solid-phase heat recovery, which is technically challenging to realize.

To enhance the efficiency of non-stoichiometric redox cycles, the effect of incorporating dopants into the cubic fluorite structure of ceria on fuel productivity and fuel production rate has been examined. At a fixed dopant concentration of 10 mol%, a trend of increasing reducibility with decreasing ionic radius of the dopant has been observed for the dopants Dy, Y, Mg, Sc, Zr, and Hf.[55] For Zr-doped ceria, it has been shown experimentally that both the oxygen yield and the fuel (H_2) productivity increase significantly with increasing dopant concentration up to 54 mol%.[56,57] In the experiments by Le Gal *et al.*[56] with reduction and H_2 production steps operated at 1400°C and 1050°C, respectively, the oxygen release increased from 123 to 236 μ mol g^{-1} and the fuel productivity increased from 234 to 468 μ mol g^{-1} as the dopant concentration was increased from 10 to 50 mol%. These results illustrate that the higher oxygen deficiencies achievable with higher Zr concentrations can be effectively used to split additional H_2O. However, it was observed that the amount of oxygen released starts to level off at Zr concentrations above 25 mol%. This has been attributed to the increasing fraction of non-reducible Zr cations in the ceria lattice.[57] In addition, at Zr concentrations above 20 mol%, the formation of the tetragonal phase may occur, which is expected to have lower oxygen storage capacity than the cubic fluorite structure.[58] Furthermore, it has been observed both theoretically and experimentally that Zr-doped ceria has lower reducing power to split H_2O than undoped ceria, thus requiring higher amounts of excess steam and lower temperatures during the fuel production step.[19,58] These factors can have an impact on the energy efficiency of the cycle. In addition, the results by Le Gal *et al.*[56,59] suggest that Zr-doped ceria samples sinter upon repeated heating to 1400°C, which affects the cyclability of the material.

In contrast to Zr-doping, the incorporation of trivalent rare earth metals (Y, La, Pr, Sm, Gd) into the ceria lattice did not improve the reducibility or the H_2 productivity compared to undoped ceria.[57,59] However, doping ceria with 10 mol% of La, Sm, and Gd resulted in redox materials with good thermal stability upon repeated redox cycling.[59] In addition, doping with trivalent rare earth metals (Sc, Y, La, Pr, Sm, Gd) and Mg has been shown to increase the ionic conductivity in the redox material.[53,55,57] In order to combine the enhanced fuel productivity of Zr-doped ceria with the enhanced thermal stability and ionic conductivity obtained with rare earth metal dopants, ternary oxides based on 25 mol% Zr-doped ceria, containing minor concentrations of Y, La, or Gd, were tested. Like Zr-doped ceria, these materials lead to significantly enhanced H_2 productivity, but suffer from a loss in reactivity upon repeated cycling, due to sintering.[59]

In order to eliminate the need for solid-phase heat recovery, Hao *et al.*[60] demonstrated experimentally that the non-stoichiometric ceria cycle can also be operated isothermally. This is possible if ceria is reduced at an oxygen partial pressure lower than that during the fuel production step. This was the motivation for another first-law analysis, similar to that by Lapp *et al.*[54] for isothermal and near-isothermal cycle operation.[61] The main difference between the two models lies in the way the reaction zones are treated. Lapp *et al.*[54] assumed that the oxygen partial pressure in the reduction zone is uniform and equal to that at the gas outlet, whereas Bader *et al.*[61] assumed the gas and solid streams to be in a counterflow arrangement in both reaction zones. This arrangement minimizes the sweep gas and H_2O/CO_2 gas flow rates, which reduces the heat losses via unrecovered thermal energy in the product gas streams. However, even in this ideal arrangement, high sweep gas flow rates are required to operate the process isothermally, and hence the cycle efficiency strongly depends on high extents of gas-phase heat recovery. For example, with $T_{red} = T_{ox} = 1773\,K$, $C = 3000$, and 50% gas-phase heat recovery, the theoretical energy efficiency is 1.1% with H_2O as the reactant and 2.6% with CO_2. An energy efficiency of 10% for H_2O splitting (18% for CO_2) requires 95.5% gas-phase heat recovery. It was further shown that the efficiency can theoretically be boosted to more than 30% if a small temperature difference $T_{red} - T_{ox}$ of up to 150 K is introduced, mainly due to a reduction of the required sweep gas flow rate.

Different reactors have been designed for both temperature-swing and isothermal cycle operation. A solar cavity receiver-reactor has been developed to effect both cycle steps consecutively in a single reactor (Fig. 16).[62] The reactor consists of an insulated cylindrical cavity containing a porous ceria cylinder. Solar radiation enters the reactor through a conical frustum and a compound parabolic secondary concentrator. Gases enter the reactor through four radial inlet ports. The gases flow through the ceria structure into the plenum of the reactor and exit through a central outlet port at the bottom of the reactor. The reactor was tested in a high-flux solar simulator, consisting of seven xenon short-arc lamps close-coupled to ellipsoidal reflectors. Thermal reduction is conducted in argon purge gas at temperatures

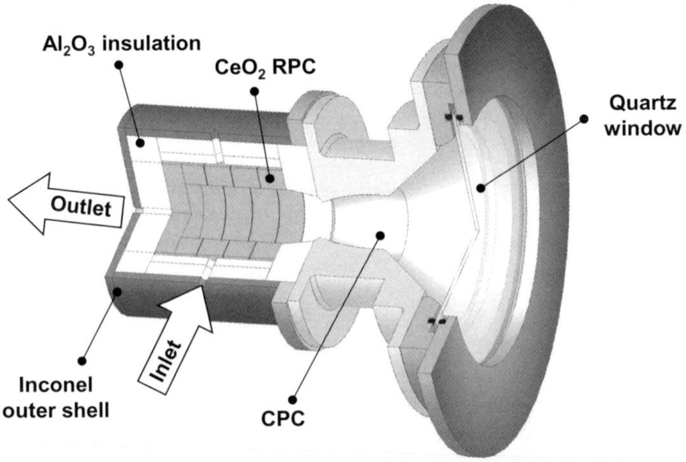

Fig. 16. Solar reactor to conduct the ceria cycle. Reprinted from Ref. 64, Copyright (2012), with permission from the American Chemical Society.

between 1400 and 1650°C, radiative power inputs of approximately 3–4 kW, and solar concentration ratio of approximately 3000. The fuel production step is conducted at 500–1000°C with a radiative power input of 0.8 kW and a flow of H_2O/CO_2 diluted in argon carrier gas. Four consecutive fuel production cycles were reported, resulting in relatively stable fuel production. Experimentally determined energy efficiency of the reactor was in the order of 1–2%. Energy losses are attributed mainly to thermal losses by conduction and radiation.[62–65] Figure 15 suggests that, in addition to reducing thermal losses, a substantial improvement of the reactor efficiency also requires heat recovery from the gas and solid-phases.

Another reactor concept for temperature-swing redox cycling of ceria particles was proposed by Ermanoski *et al.*[66] (Fig. 17). This reactor consists of separate reaction chambers for the thermal reduction and fuel production steps. In the thermal reduction reactor, a packed bed of particles is fed with a conveyor auger to the sun-lit reduction zone. Hot reduced particles moving down the center of the conveyor auger exchange heat with the oxidized particles that move up along the screw blades. The reaction chamber is sealed and evacuated with a vacuum pump to lower the oxygen partial pressure. The fuel production step is conducted off-sun at atmospheric pressure with a flow of oxidant gas through the particle bed.

A solar reactor to realize near-isothermal cycles with gas-phase heat recovery was proposed by Bader *et al.*[67] and is shown in Fig. 18. This

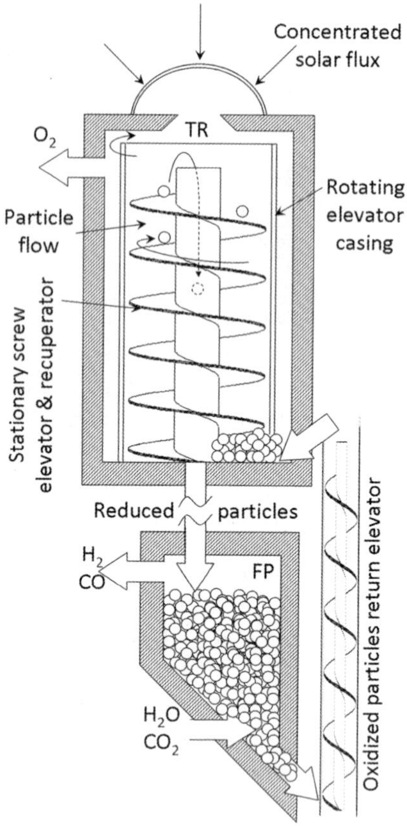

Fig. 17. Reactor for temperature-swing cycling of redox particles. Reprinted from Ref. 66, Copyright (2013), with permission from the American Society of Mechanical Engineers.

reactor consists of an insulated cylindrical cavity receiver lined with tubular reactive elements. Each reactive element consists of two concentric alumina tubes (Fig. 18(b)). A packed bed of mm-sized porous ceria particles is contained in the annular space between the tubes. The redox cycle is conducted by alternating the gas flowing through the reactive elements between high-purity nitrogen sweep gas to reduce the ceria and pure CO_2 reactant gas. Gas enters the reactive element through the inner tube. The flow direction of the gas is reversed at the domed end of the outer tube where the preheated gas enters the packed bed of ceria particles. After passing through the packed bed, the gas leaves the reactor through the annular gap between the inner and outer alumina tubes. The reactive element

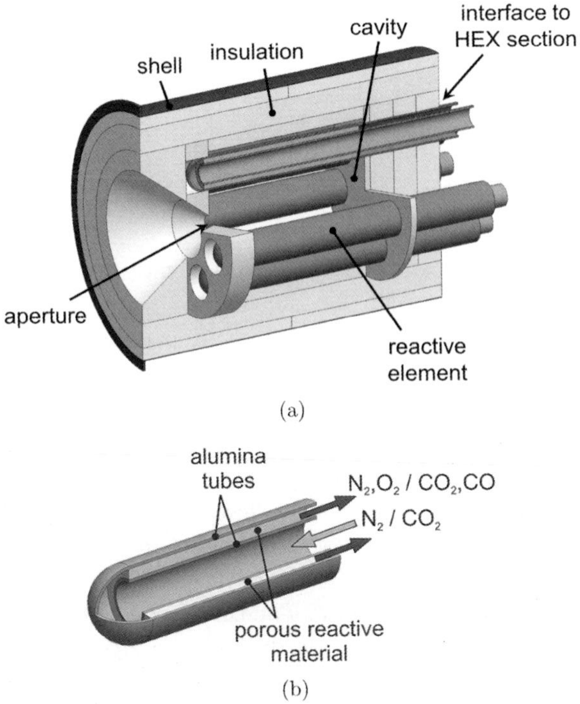

Fig. 18. Solar reactor for near-isothermal redox cycling of ceria: (a) reactor assembly, (b) enlarged view of a single reactive element. Reprinted from Ref. 67, Copyright (2015), with permission from the American Society of Mechanical Engineers.

tubes extend beyond the back of the reactor to form a ceramic foam-filled tube-in-tube counterflow heat exchanger that implements gas-phase heat recovery.[68] This reactor can be operated either in batch mode or in quasi-continuous mode. In the latter case, each reactive element is at a different stage of the periodic two-step process, but the numbers of reactive elements that are simultaneously undergoing reduction and oxidation, respectively, are constant over time. This allows for a constant solar power input to the reactor while maintaining a nearly constant reactor temperature.

3.1.4 *Perovskite cycle*

Recently, redox materials with the perovskite structure (ABO_3) have been considered for solar thermochemical H_2O/CO_2-splitting. This class of metal oxide materials has long been used as electrolyte material in solid oxide fuel cells due to their high oxygen exchange capacity. The perovskite cycle can be represented by:

$$ABO_{3-\delta_{ox}} \rightarrow ABO_{3-\delta_{red}} + \frac{\Delta\delta}{2}O_2, \quad (25)$$

$$ABO_{3-\delta_{red}}(s) + \Delta\delta(yH_2O + (1-y)CO_2) \rightarrow ABO_{3-\delta_{ox}}(s) + \Delta\delta(yH_2 + (1-y)CO), \quad (26)$$

where $\Delta\delta = \delta_{red} - \delta_{ox}$. Similar to ceria, the A- and B-site metals of the perovskite lattice can be substituted with dopants. McDaniel *et al.*[69] tested strontium- and manganese-doped lanthanum aluminates ($Sr_xLa_{1-x}Mn_yAl_{1-y}O_{3-\delta}$) with three different compositions (x, y) with respect to their oxygen release and uptake capacity and their potential to split H_2O and CO_2 and compared the results with those for undoped ceria. The perovskites were shown to release several times more oxygen under the same conditions (1350°C, He sweep gas flow with O_2 partial pressure of 0.2 mbar) compared to ceria. Subsequent re-oxidation of the perovskites with 40 vol% H_2O and CO_2 at 1000°C yielded up to 9× and 6× more H_2 and CO, respectively, compared to ceria. The time to complete the re-oxidation step was comparable for the two material classes. Redox cycling between 1350°C and 1000°C over 80 cycles resulted in constant CO production of 125–150 μmol g^{-1} per cycle.

Scheffe *et al.*[70] conducted a thermodynamic analysis of strontium-doped lanthanum manganites ($Sr_xLa_{1-x}MnO_{3-\delta}$). These perovskites exhibit higher oxygen non-stoichiometry than ceria at identical conditions (1773 K, oxygen partial pressure of 1 Pa), but lower reactivity with H_2O/CO_2. Nevertheless, higher fuel productivity was obtained compared to ceria.

Several additional perovskite materials with the compositions La_xSr_{1-x} (Mn, Co, Fe)O_3, Ba_xSr_{1-x}(Co, Fe)O_3, $LaSrCoO_4$, and $LaSrFeO_4$ have been studied by Demont *et al.*[71] Most of these materials were found to exhibit higher oxygen exchange capacity compared to ceria at temperatures in the range of 1000–1400°C and noticeable access to metallic valence transitions during re-oxidation in steam.[71]

3.2 *Multi-Step Cycles*

A thermodynamic analysis has shown that H_2O-splitting redox cycles operating below 723°C require at least three process steps.[72] In fact, there are no known two-step cycles that operate at temperatures below 1100°C.[17,73]

Hence, H_2O/CO_2 splitting at temperatures below 1100°C requires thermochemical cycles with at least three process steps. Since state-of-the-art solar concentrating systems can readily produce temperatures >1100°C, two-step cycles have been the primary focus of research in recent years, due to their inherent potential to reach higher energy efficiencies and their lower complexity. Therefore, thermochemical cycles with three or more cycle steps are not discussed here. A review and discussion of multistep cycles can be found in Kodama *et al.*[1]

4 Conversion of Carbonaceous Feedstocks

High-temperature solar thermal processing of carbonaceous feedstocks has been proposed as a hybrid solar–fossil and near-term route to production of solar fuels and carbonaceous commodity materials. Solar energy is used as the exclusive source of process heat, and the carbonaceous feedstocks are the source of carbon and hydrogen. The processes include gasification, cracking and reforming.[74] Chemical reaction equations presented in this section are written for carbonaceous feedstocks containing exclusively carbon, hydrogen and oxygen. Impurities and other elements are typically found in fossil fuels and biomass, e.g. sulfur and nitrogen, and lead to formation of additional product compounds as compared to those reported in this section.

4.1 *Gasification*

Solar thermal gasification has been studied for a variety of carbonaceous feedstocks including coal, coke, biomass, bitumen and carbonaceous wastes.[75] While the use of biomass as the feedstock promises a carbon-neutral fuel, the availability of biomass for energetic uses is limited. Coal remains the most abundant fossil fuel resource in the world, with the total recoverable reserves estimated in year 2008 to be 909 billion tons. The United States share the largest portion of these reserves, which are 262.7 billion tons (29%). The main approach to coal utilization is the direct combustion for electricity generation, which produces approximately 0.92 kg of CO_2 per 1 kWh_e of electricity generated. In spite of environmental concerns, coal is likely to remain one of the leading energy sources in future. Coal gasification technologies can substantially reduce greenhouse gas emissions through increased conversion efficiency. Further emission reductions are possible, if the process heat required to drive the gasification process is obtained from a CO_2-free energy source such as concentrated solar energy, eliminating the need for combustion of a significant portion of the feedstock. In addition, elimination of internal combustion for supplying the process heat mitigates contamination of the synthesis gas with the combustion byproducts.[5,74] The produced synthesis gas can then be used for power generation via power cycles or for synthesis of hydrocarbon fuels.

Thermal gasification of carbonaceous materials is a complex chemical process that involves several different intermediate reactions. Steam and carbon dioxide are the common oxidizing agents. A simplified net reaction of steam gasification was proposed by Zedtwitz and Steinfeld[76]:

$$CH_xO_y + (1-y)H_2O = \left(\frac{x}{2} + 1 - y\right) H_2 + CO, \tag{27}$$

where x and y are the molar ratio of hydrogen to carbon (H/C) and oxygen to carbon (O/C), respectively. Analogously, a simplified net reaction of dry gasification can be written as[77]:

$$CH_xO_y + (1-y)CO_2 = \frac{x}{2}H_2 + (2-y)CO, \tag{28}$$

Reactions (27) and (28) proceed endothermically under atmospheric pressure at temperatures above 850 K and 900 K, respectively. Intermediate reactions occurring in the simplified reactions given by Eqs. (27) and (28) are:

Bouduard reaction:

$$C + CO_2 = 2CO, \tag{29}$$

Methane cracking–hydrogasification:

$$C + 2H_2 = CH_4, \tag{30}$$

Reforming–methanation:

$$CH_4 + H_2O = CO + 3H_2, \tag{31}$$

Water-gas shift:

$$CO + H_2O = CO_2 + H_2. \tag{32}$$

Reactions (29)–(32) are strongly dependent on temperature, pressure and the C/O ratio. The composition of syngas, and thus its suitability for production of synthetic fuels is highly dependent on the outcome of reactions (29)–(32). Thermodynamic analyses of solar thermal gasification of four types of coal, anthracite, bituminous, lignite and peat, were studied by Zedtwitz and Steinfeld,[76] and by Ng and Lipiński.[77] Figures 19(a) and 19(b)

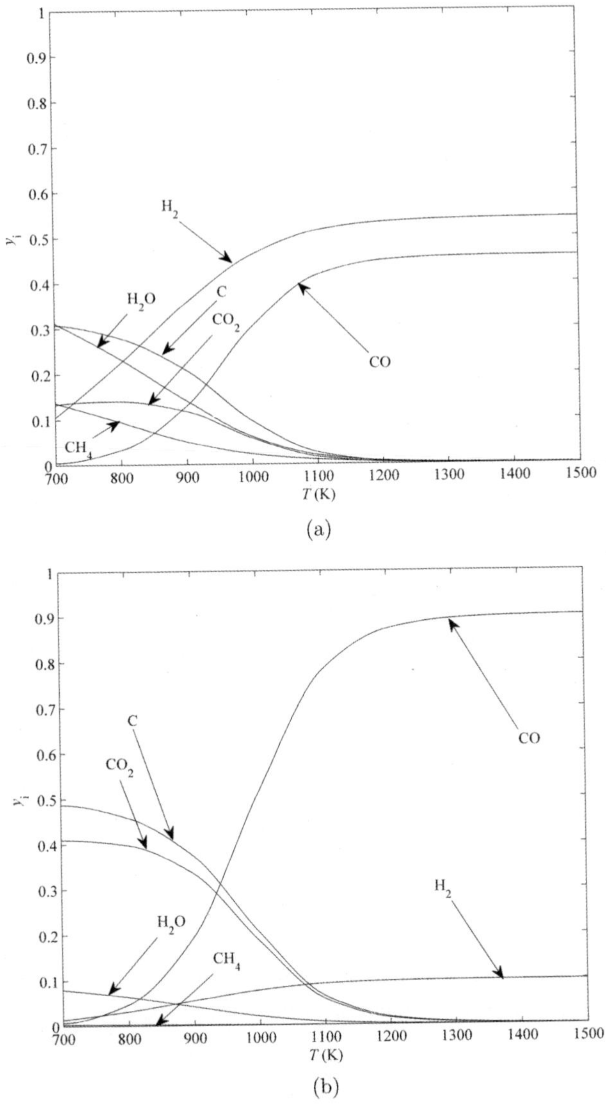

Fig. 19. Equilibrium composition for (a) steam and (b) dry gasification of anthracite as a function of temperature. Reprinted from Ref. 77, Copyright (2012), with permission from Elsevier.

show the equilibrium mole fractions of the main species for the steam and dry gasification, respectively, of anthracite over the temperature range 700–1500 K. For temperatures below 700 K, the intermediate reactions do not take place. Thus, components such as C, CH_4, CO_2, H_2O are thermodynamically stable at temperatures below 700 K. As the temperature increases from 700 K to approximately 1200 K, rates of the intermediate reactions increase and the fractions of C, CH_4, CO_2, and H_2O decrease. At temperatures higher than 1200 K, both steam and dry gasification produce syngas that only consists of H_2 and CO with a constant H_2/CO ratio.

Kinetic models of solar thermal gasification processes have been proposed in several studies.[78–80] The models are typically based on the Langmuir–Hinshelwood type of kinetic rate expressions for production and consumption of each gas species.[75] A detailed discussion of the kinetic analyses and models is omitted from this work, and the reader is referred to the cited literature references.

Solar reactors for gasification of carbonaceous feedstocks are among the most mature reactor concepts in solar thermochemistry. Both directly- and indirectly-irradiated reactors have been developed.[75,81] The configurations of the reacting two-phase solid–gas media include fluidized beds,[82] entrained particle flows,[83] vortex flows,[84] and packed bed reactors.[85] Example solar reactors containing reacting media of such configurations are depicted in Fig. 20. More recently, reactors implementing more complex approaches to solar thermal gasification have been proposed, which include gasification of biomass in molten salts[86] and with supercritical water.[87] The solar reactor from the latter study is shown in Fig. 21.

4.2 *Cracking and Reforming*

Solar-driven cracking of hydrocarbons to co-produce hydrogen and carbon black can be represented by the simplified reaction:

$$C_xH_y = x\mathrm{C(gr)} + \frac{y}{2}\mathrm{H}_2, \tag{33}$$

The process is typically realized in the form of a directly-irradiated gas flow laden with carbon particles,[88–90] or an indirectly-irradiated flow in tubes.[91–94] The particles serve as radiant absorbers and provide surface to the heterogeneous reaction. Figure 22 shows an example directly-irradiated solar thermochemical reactor featuring a horizontal cylindrical reaction cavity containing the reactive mixture of carbon particles and methane.[88,89]

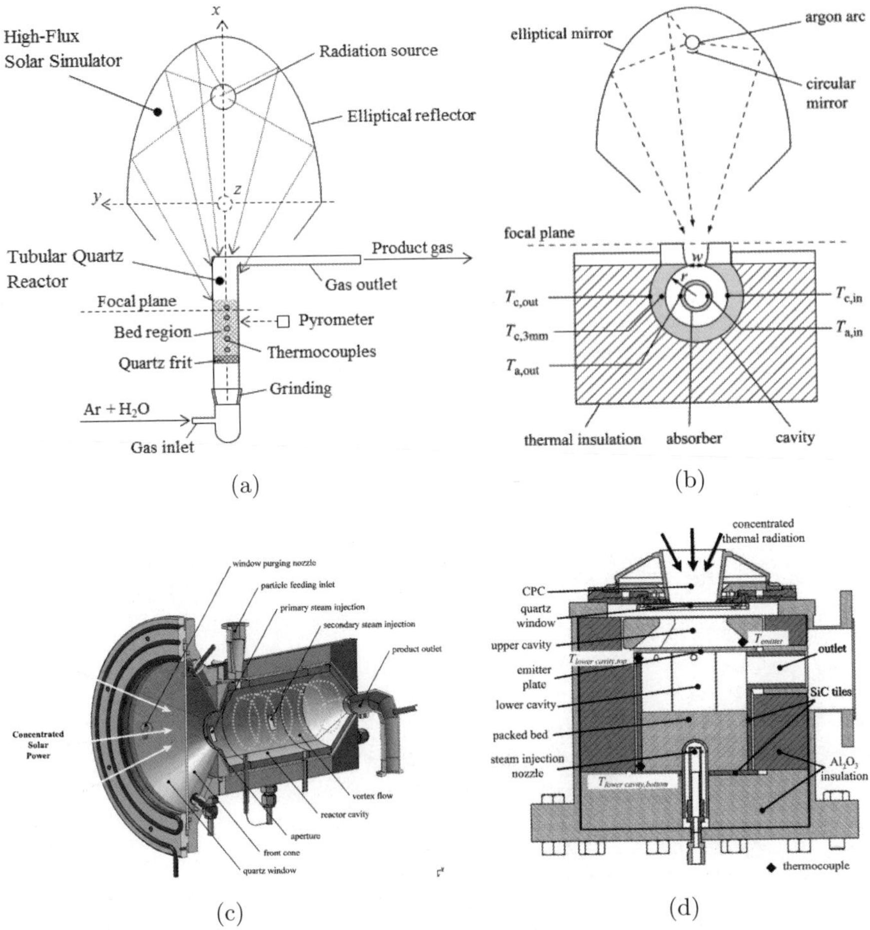

Fig. 20. Examples of solar thermochemical reactors for thermal gasification of carbonaceous materials: (a) a directly-irradiated fluidized bed reactor (reprinted from Ref. 82, Copyright (2007), with permission from Elsevier); (b) an indirectly-irradiated entrained particle flow reactor (reprinted from Ref. 83, Copyright (2009), with permission from Elsevier); (c) a directly-irradiated vortex-flow reactor (reprinted from Ref. 84, Copyright (2006), with permission from Elsevier); and (d) an indirectly-irradiated packed bed reactor (reprinted from Ref. 85, Copyright (2009), with permission from Elsevier).

Steam and dry reforming of hydrocarbons are given by the simplified reactions:

$$C_xH_y + xH_2O = \left(x + \frac{y}{2}\right) H_2 + xCO, \tag{34}$$

$$C_xH_y + xCO_2 = \frac{y}{2}H_2 + 2xCO. \tag{35}$$

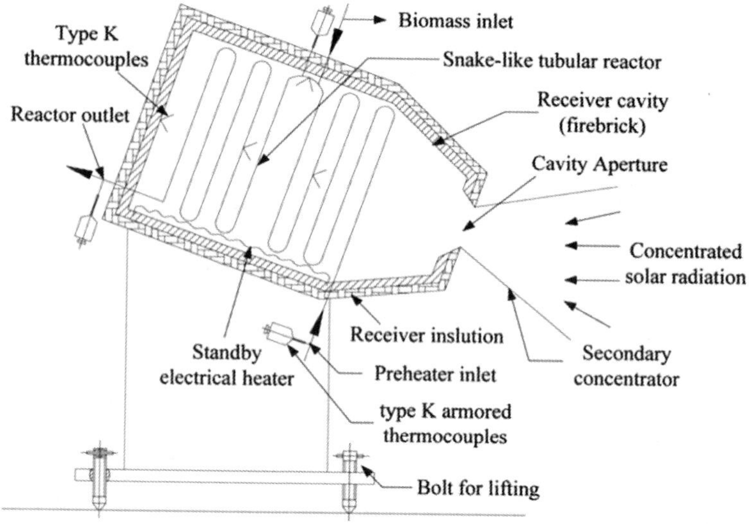

Fig. 21. Solar thermochemical reactors for supercritical water gasification of biomass. Reprinted from Ref. 87, Copyright (2010), with permission from Elsevier.

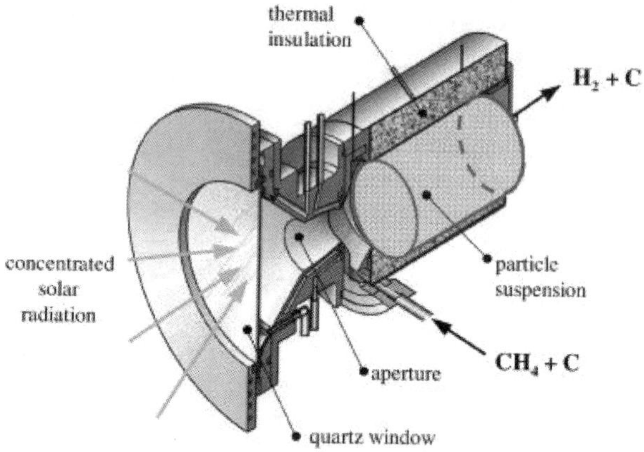

Fig. 22. Schematic of the solar reactor prototype, featuring a flow of methane laden with carbon particles and directly exposed to concentrated solar radiation. Reprinted from Ref. 88, Copyright (2009), with permission from Elsevier.

Figure 23 shows a directly-irradiated solar reactor for the solar reforming of methane.[95] The reactor uses a reticulate porous ceramics coated with rhodium-based catalyst. This reactor has been scaled up to power levels of 300 kW in a solar tower facility and used to conduct both steam- and

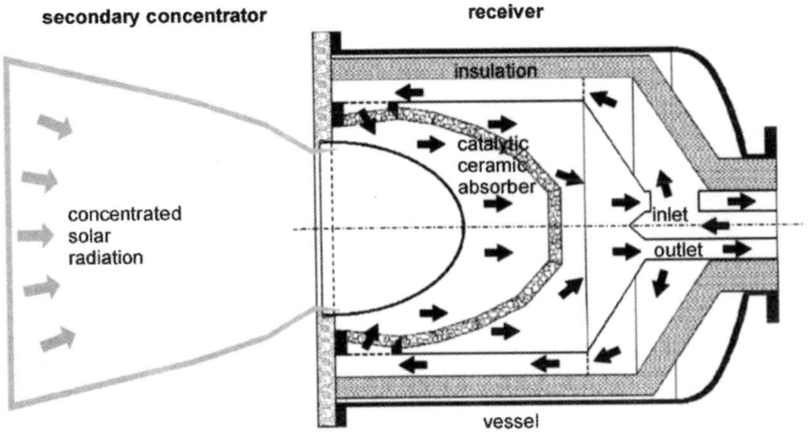

Fig. 23. A directly-irradiated solar reactor for the solar reforming of methane. Reprinted from Ref. 95, Copyright (2001), with permission from the American Society of Mechanical Engineers.

dry-reforming of methane.[95–97] Selected design aspects of a solar reactor for methane decomposition in both cracking and reforming processes based on thermodynamic and kinetic analyses are discussed in the work by Ozalp and Shilapuram.[98]

5 Production of Lime and Cement

The calcination of limestone ($CaCO_3$) to lime (CaO) according to:

$$CaCO_3(s) \rightarrow CaO(s) + CO_2, \quad \Delta H^\circ_{298\,\mathrm{K}} = 178\,\mathrm{kJ\,mol^{-1}}, \tag{36}$$

is the main endothermic step in the production of lime and cement, and is operated above 900°C.[8] In most lime kilns, process heat is supplied by combustion of fossil fuels. The CO_2 emissions from the lime industry are approximately 1% of the global anthropogenic CO_2 emissions.[99] The CO_2 emissions from the process can be mitigated by operating the process with high-temperature solar process heat and sequestering the CO_2 evolving from the calcination reaction (36). The CO_2 can be stored or used as a feedstock in the production of solar fuels.

A 10 kW prototype solar-heated rotary kiln has been developed and tested (Fig. 24).[8,99,100] It consists of a 225 × 252 mm indirectly irradiated

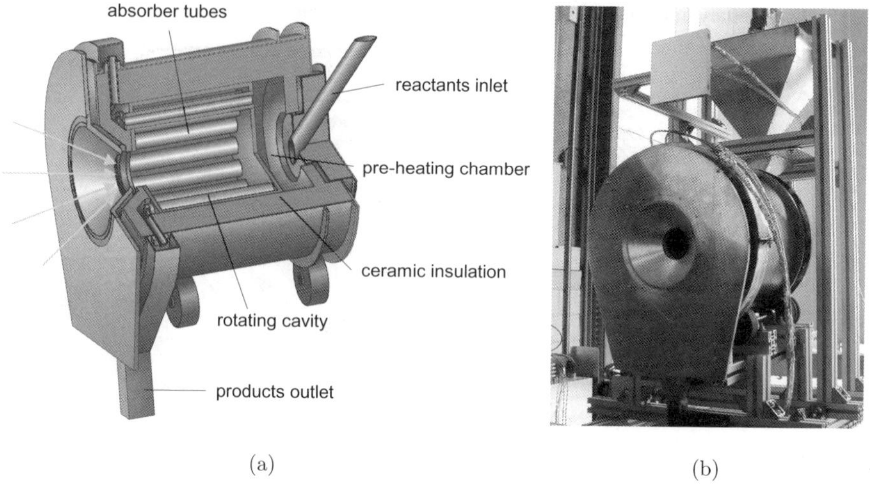

Fig. 24. 10 kW prototype solar rotary kiln for the calcination of limestone. Reprinted from Ref. 99, Copyright (2006), with permission from Elsevier.

cylindrical solar cavity receiver made from recrystallized silicon carbide (maximum operating temperature 1600°C). Solar radiation at a concentration ratio of approximately 2000, provided by a solar furnace, enters the reactor through a 9 cm-dia. aperture in the water-cooled aluminum front. 1–5 mm large limestone particles are fed into the pre-heating chamber at the back of the reactor. Due to the inclination and the rotation of the reactor, the particles are moved into the 16 silicon carbide absorber tubes that line the solar cavity. Calcined CaO particles are collected in a funnel at the front of the reactor. The reactor has been tested on-sun for over 100 h, producing up to 4 kg h^{-1} of lime at solar energy conversion efficiencies of up to 35%. The cost of lime produced in a 20 MW solar calcination plant has been estimated to amount to 131–158 \$ t^{-1}, corresponding to two to three times the current selling price.[101]

6 CO_2 Capture

Operated in a closed cycle, the calcination of limestone followed by the carbonation of lime can be utilized to capture CO_2 from CO_2-rich flue gas streams or from air according to:

$$\mathrm{CaCO_3(s)} \rightarrow \mathrm{CaO(s)} + \mathrm{CO_2} \quad \Delta H^\circ_{298\mathrm{K}} = 178\,\mathrm{kJmol^{-1}} \tag{37}$$

$$\mathrm{CaO(s)} + \mathrm{CO_2} \rightarrow \mathrm{CaCO_3(s)} \quad \Delta H^\circ_{298\mathrm{K}} = -178\,\mathrm{kJmol^{-1}} \tag{38}$$

To obtain a product stream of pure CO_2 during the calcination step, the reaction needs to be operated in a sweep gas flow of pure CO_2.

A thermodynamic process analysis has been conducted by Matthews *et al.*[102] to calculate the theoretical energy input required by the process as a function of CO_2 concentration in the flue gas and heat recovery from the hot products. The energy requirement of the process (37)–(38) strongly depends on the CO_2 concentration in the inlet gas and on gas-phase heat recovery, while solid-phase heat recovery is relatively unimportant. With 100% gas- and solid-phase heat recovery, 207 kJ of thermal energy is required per mole of captured CO_2, irrespective of the CO_2 concentration in the flue gas stream. In the absence of gas-phase and solid-phase heat recovery, the energy requirement is between 45 MJ and 283 kJ per mole of captured CO_2 for initial CO_2 concentrations spanning those encountered in air (~0.03%) and in flue gas (15%). For comparison, the theoretical minimum work to separate CO_2 from a binary gas mixture is between 20 and 30 kJ mol^{-1} of captured CO_2 at atmospheric CO_2 concentrations and 7 kJ mol^{-1} of captured CO_2 for an initial CO_2 concentration of 15%.

A packed-bed solar reactor has been designed to effect both process steps (Fig. 25).[103,104] The reactor has a beam-up configuration. Calcium

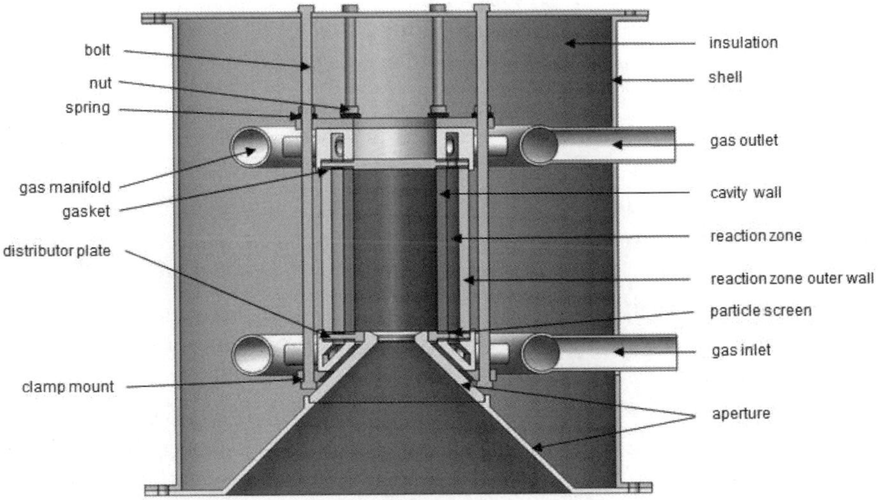

Fig. 25. Packed-bed solar reactor to capture CO_2 via process (37)–(38). Reprinted from Ref. 104, Copyright (2015), with permission from the American Society of Mechanical Engineers.

carbonate particles are contained in an annular space surrounding a downward-facing solar cavity receiver. The downward-facing orientation of the reactor reduces convective heat losses from the open receiver aperture. Calcium carbonate particles are calcined in a pure stream of CO_2 with solar heat input. During the carbonation step, flue gases or atmospheric air flow through the particle bed. Due to the exothermicity of the chemical reaction, solar heat input to the reactor is not required during the carbonation step. Hence, two or more reactors could be operated in parallel and the solar energy input be switched between the reactors, e.g. by changing the orientation of the heliostats of a solar tower plant like in Roeb *et al.*[46]

Other reactors that have been examined to effect one or both of the process steps (37)–(38) include rotary kilns (including the reactor shown in Fig. 24),[99,105] fluidized beds,[105,106] and cyclone and vortex-flow reactors.[107–109]

7 Other Processes

The high temperatures and heating rates that can be obtained with concentrating solar technologies facilitate development of further solar-driven processes for production of material commodities and high-temperature thermal energy storage. Ammonia production from nitrogen, water, and solar energy has been proposed using metal nitride-based thermochemical redox cycles driven by concentrated solar radiation.[110] Thermodynamic and economic aspects of solar production of ammonia are discussed by Gálvez *et al.*[111] and Michalsky *et al.*[112] Employing high-temperature solar thermal technologies in metallurgical processing promises a green transformation of one of the environmentally most challenging industrial sectors. The temperatures achievable with state-of-the-art solar concentrators allow for a reduction of the electricity input in electrolysis during extraction of some metals and semiconductors. Solar-driven carbothermal or/and vacuum processing can alternatively be considered for metal extraction at reduced temperatures. Solar production of metals, nitrides, and carbides via solar carbothermal reduction of metal oxides is discussed by Murray *et al.*[113] More recent thermodynamic analyses of carbothermal reduction of a range of metal oxides under vacuum were presented by Halmann *et al.*[114] The solar carbothermal production of silicon from silica under vacuum was studied by Loutzenhiser *et al.*[115] Solar production of iron, copper, nickel, chromium, aluminum, magnesium, cobalt, lithium, and other metals promises substantial reduction of fossil fuel use in industry due to large demand and the highly endothermic character of the metal-extractive processes. Solar metallurgy is a particularly interesting research and industrial field for geographical areas with co-located mineral and solar resources such as Australia.

8 Summary and Conclusions

Solar energy is a vast source of highly-exergetic energy. If harvested with state-of-the-art solar concentrators at temperatures of approximately 1000–2000 K, about 70–80% of the collected energy can theoretically be converted to work. Hence, solar energy can be used to efficiently produce energy-related products, such as electricity and fuels. In addition, solar energy can also be used as a source of high-temperature process heat to produce sustainable materials. Solar-driven processes and reactors have been developed for various purposes, including H_2O/CO_2 splitting, coal gasification, methane cracking and reforming, lime production, and CO_2 capture from air and flue gases. Additional potential applications of solar energy include fertilizer (ammonia, urea) production, metal extraction, seawater desalination, and more.

References

1. T. Kodama and N. Gokon, *Chem. Rev.* **107** (2007), pp. 4048–4077.
2. M. Romero and A. Steinfeld, *Energy Environ. Sci.* **5** (2012), pp. 9234–9245.
3. E.A. Fletcher, *Journal of Solar Energy Engineering* **123** (2001), pp. 63–74.
4. A. Steinfeld and R. Palumbo, *Encycl. Phys. Sci. Technol.* **15** (2001), pp. 237–256.
5. T. Kodama, *Progress in Energy and Combustion Science* **29** (2003), pp. 567–597.
6. A. Steinfeld, *Annual Review of Heat Transfer* (Begell House, 2012), pp. 255–275.
7. J.A. Wurzbacher, C. Gebald and A. Steinfeld, *Energy Environ. Sci.* **4** (2011), pp. 3584–3592.
8. A. Meier and A. Steinfeld, *Encycl. Sustain. Sci. Technol.* (2012), pp. 9588–9619.
9. A. Rabl, *Sol. Energy,* **18** (1976), pp. 93–111.
10. E.A. Fletcher and R.L. Moen, *Science* **197** (1977), pp. 1050–1056.
11. M.E. Gálvez, P.G. Loutzenhiser, I. Hischier and A. Steinfeld, *Energy & Fuels,* **22** (2008), pp. 3544–3550.
12. A. Kogan, *Int. J. Hydrogen Energy* **23** (1998), pp. 89–98.
13. S. Ihara, *Int. J. Hydrogen Energy* **5** (1980), pp. 527–534.
14. E.A. Fletcher, *Ind. Eng. Chem. Res.* **38** (1999), pp. 2275–2282.

15. R.B. Diver, S. Pederson, T. Kappauf and E.A. Fletcher, *Energy* **8** (1983), pp. 947–955.
16. J. Lede and J. Villermaux, *Int. J. Hydrogen Energy* **12** (1987), pp. 3–11.
17. S. Abanades, P. Charvin, G. Flamant and P. Neveu, *Energy* **31** (2006), pp. 2469–2486.
18. R. Perret, *Sandia Report SAND2011-3622* (2011).
19. J.R. Scheffe and A. Steinfeld, *Mater. Today* **17** (2014), pp. 341–348.
20. C.L. Muhich, B.W. Evanko, K.C. Weston, P. Lichty, X. Liang, J. Martinek, C.B. Musgrave and A.W. Weimer, *Science* **341** (2013), pp. 540–542.
21. C. Agrafiotis, M. Roeb and C. Sattler, *Renew. Sustain. Energy Rev.* **42** (2015), pp. 254–285.
22. P.G. Loutzenhiser and A. Steinfeld, *Int. J. Hydrogen Energy* **36** (2011), pp. 12141–12147.
23. K. Wegner, H.C. Ly, R.J. Weiss, S.E. Pratsinis and A. Steinfeld, *Int. J. Hydrogen Energy* **31** (2006), pp. 55–61.
24. L.O. Schunk and A. Steinfeld, *AIChE J.* **55** (2009), pp. 1497–1504.
25. A. Weidenkaff, A.W. Reller, A. Wokaun and A. Steinfeld, *Thermochimica Acta* **359** 69–75 (2000).
26. P. Haueter, S. Moeller, R. Palumbo and A. Steinfeld, *Sol. Energy* **67** (1999), pp. 161–167.
27. L.O. Schunk, P. Haeberling, S. Wepf, D. Wuillemin, A. Meier and A. Steinfeld, *J. Sol. Energy Eng.* **130** (2008), pp. 021009–1/6.
28. L.O. Schunk, W. Lipiński and A. Steinfeld, *Chem. Eng. J.* **150** (2009), pp. 502–508.
29. W.Villasmil, M. Brkic, D. Wuillemin, A. Meier and A. Steinfeld, *J. Sol. Energy Eng.* **136** (2014), pp. 011017–1/11.
30. A. Berman and M. Epstein, *Int. J. Hydrogen Energy* **25** (2000), pp. 957–967.
31. A. Stamatiou, P. G. Loutzenhiser and A. Steinfeld, *Energy & Fuels* **24** (2010), pp. 2716–2722.
32. R.J. Weiss, H.C. Ly, K. Wegner, S.E. Pratsinis and A. Steinfeld, *AIChE J.* **51** (2005), pp. 1966–1970.
33. F.O. Ernst, A. Tricoli, S.E. Pratsinis and A. Steinfeld, *AIChE J.* **52** (2006), pp. 3297–3303.
34. T. Melchior, N. Piatkowski and A. Steinfeld, *Chem. Eng. Sci.* **64** (2009), pp. 1095–1101.
35. P.G. Loutzenhiser, M.E. Gálvez, I. Hischier, A. Graf and A. Steinfeld, *Chem. Eng. Sci.* **65** (2010), pp. 1855–1864.
36. T. Nakamura, *Sol. Energy* **19** (1977), pp. 467–475.
37. L.S. Darken and R.W. Gurry, *J. Am. Chem. Soc.* **68** (1946), pp. 798–816.
38. A. Steinfeld, S. Sanders, and R. Palumbo, *Sol. Energy* **65** (1999), pp. 43–53.
39. R.B. Diver, J.E. Miller, M.D. Allendorf, N.P. Siegel and R.E. Hogan, *J. Sol. Energy Eng.* **130** (2008), pp. 041001–1/8.
40. T. Kodama, N. Gokon and R. Yamamoto, *Sol. Energy* **82** (2008), pp. 73–79.
41. P. Charvin, S. Abanades, G. Flamant and F. Lemort, *Energy* **32** (2007), pp. 1124–1133.

42. M.D. Allendorf, R.B. Diver, N.P. Siegel, and J.E. Miller, *Energy & Fuels* **22** (2008), pp. 4115–4124.
43. N. Gokon, H. Murayama, A. Nagasaki and T. Kodama, *Solar Energy* **83** (2009), pp. 527–537.
44. M. Roeb, C. Sattler, R. Klüser, N. Monnerie, L. de Oliveira, A.G. Konstandopoulos, C. Agrafiotis, V.T. Zaspalis, L. Nalbandian, A. Steele and P.J. Stobbe, *Sol. Energy Eng.* **28** (2006), pp. 125–133.
45. C. Agrafiotis, M. Roeb, A.G. Konstandopoulos, L. Nalbandian, V.T. Zaspalis, C. Sattler, P. Stobbe and A.M. Steele, *Solar Energy* **79** (2005), pp. 409–421.
46. M. Roeb, J.-P. Säck, P. Rietbrock, C. Prahl, H. Schreiber, M. Neises, L. de Oliveira, D. Graf, M. Ebert, W. Reinalter, M. Meyer-Grünefeldt, C. Sattler, A. Lopez, A. Vidal, A. Elsberg, P. Stobbe, D. Jones, A. Steele, S. Lorentzou, C. Pagkoura, A. Zygogianni, C. Agrafiotis, A.G. Konstandopoulos, *Solar Energy* **85** (2011), pp. 634–644.
47. N. Gokon, S. Takahashi, H. Yamamoto and T. Kodama, *Int. J. Hydrogen Energy* **33** (2008), pp. 2189–2199.
48. N. Gokon, T. Mataga, N. Kondo and T. Kodama, T. *Int. J. Hydrogen Energy* **36** (2011), pp. 4757–4767.
49. H. Kaneko, T. Miura, A. Fuse, H. Ishihara, S. Taku, H. Fukuzumi, Y. Naganuma and Y. Yamaura, *Energy & Fuels* **21** (2007), pp. 2287–2293.
50. J.E. Miller, M.D. Allendorf, R.B. Diver, L.R. Evans, N.P. Siegel J.N. Stuecker, *J. Mater. Sci.* **43** (2008), pp. 4714–4728.
51. S. Abanades and G. Flamant, *Solar Energy* **80** (2006), pp. 1611–1623.
52. M. Binnewies and E. Milke, *Thermochemical Data of Elements and Compounds* (Wiley-VCH Verlag GmbH, 2002).
53. W.C. Chueh and S.M. Haile, *Philos. Trans. R. Soc. A Math. Phys. Eng. Sci.* **368** (2010), pp. 3269–3294.
54. J. Lapp, J.H. Davidson and W. Lipiński, *Energy*, **37** (2012), pp. 591–600.
55. Q.-L. Meng, C. Lee, T. Ishihara, H. Kaneko and Y. Tamaura, *Int. J. Hydrogen Energy* **36** (2011), pp. 13435–13441.
56. A. Le Gal and S. Abanades, *Int. J. Hydrogen Energy* **36** (2011), pp. 4739–4748.
57. A. Le Gal, N. Bion, T. Mercier, Le and V. Harle, *Energy & Fuels* **27** (2013).
58. Y. Hao, C.-K. Yang and S.M. Haile, *Chem. Mater.* **26** (2014), pp. 6073–6082.
59. A. Le Gal and S. Abanades, *J. Phys. Chem. C* **116** (2012), pp. 13516–13523.
60. Y. Hao, C.-K. Yang and S.M. Haile, *Phys. Chem. Chem. Phys.* **15** (2013), pp. 17084–92.
61. R. Bader, L.J. Venstrom, J.H. Davidson, W. Lipiński, *Energy & Fuels* **27** (2013), pp. 5533–5544.
62. W.C. Chueh, C. Falter, M. Abbott, D. Scipio, P. Furler, S.M. Haile, and A. Steinfeld, *Science* **330** (2010), pp. 1797–1801.
63. P. Furler, J. Scheffe and A. Steinfeld, *Energy Environ. Sci.* **5** (2012), pp. 6098–6103.
64. P. Furler, J. Scheffe, M. Gorbar, L. Moes, U. Vogt and A. Steinfeld, *Energy & Fuels* **26** (2012), pp. 7051–7059.

65. P. Furler, J. Scheffe, D. Marxer, M. Gorbar, A. Bonk, U. Vogtam and A. Steinfeld, *Phys. Chem. Chem. Phys.* **16** (2014), pp. 10503–10511.
66. I. Ermanoski, N.P. Siegel and E.B. Stechel, *J. Sol. Energy Eng.* **135** (2013), pp. 031002–1/10.
67. R. Bader, R. Bala Chandran, S. Sedler, R. De Smith, L.J. Venstrom, A. Banerjee, T.R. Chase, J.H. Davidson and W.J. Lipiński, *Sol. Energy Eng.* **137** (2015), pp. 031007–1/10.
68. A. Banerjee, R. Bala Chandran and J.H. Davidson, *Appl. Therm. Eng.* **75** (2015), pp. 889–895.
69. A.H. McDaniel, E.C. Miller, D. Arifin, A. Ambrosini, E.N. Coker, R. O'Hayre, W.C. Chueh and J. Tong, *Energy Environ. Sci.* **6** (2013), p. 2424.
70. J.R. Scheffe, D. Weibel and A. Steinfeld, *Energy & Fuels* **27** (2013), pp. 4250–4257.
71. A. Demont, S. Abanades and E. Beche, *J. Phys. Chem. C* **118** (2014), pp. 12682–12692.
72. B.M. Abraham and F. Schreiner, *Ind. Eng. Chem. Fundam.* **13** (1974), pp. 305–310.
73. J.E. Funk and R.M. Reinstrom, *Ind. Eng. Chem. Process Des. Dev.* **5** (1966), pp. 336–342.
74. P. Von Zedtwitz, J. Petrasch, D. Trommer and A. Steinfeld, *Solar Energy* **80** (2006), pp. 1333–1337.
75. N. Piatkowski, C. Wieckert, A. W. Weimer and A. Steinfeld, *Energy Environ. Sci.* **4** (2011), pp. 73–82.
76. P. Von Zedtwitz and A. Steinfeld, *Energy* **28** (2003), pp. 441–456.
77. Y.C. Ng and W. Lipiński, *Energy* **44** (2012), pp. 720–731.
78. R. Müller, P. von Zedtwitz, A. Wokaun and A. Steinfeld, *Chem. Eng. Sci.* **58** (2003), pp. 5111–5119.
79. D. Trommer and A. Steinfeld, *Energy & Fuels* **20** (2006), pp. 1250–1258.
80. D. Trommer, F. Noembrini, M. Fasciana, D. Rodriguez, A. Morales, M. Romero and A. Steinfeld, *Int. J. Hydrogen Energy* **30** (2005), pp. 605–618.
81. W. Lipiński, J.H. Davidson, S. Haussener, J.F. Klausner, A.M. Mehdizadeh, J. Petrasch, A. Steinfeld and L. Venstrom, *J. Therm. Sci. Eng. Appl.* **5** (2013), 021005.
82. P. Von Zedtwitz, W. Lipiński and A. Steinfeld, *Chem. Eng. Sci.* **62** (2007), pp. 599–607.
83. T. Melchior, C. Perkins, P. Lichty, A.W. Weimer and A. Steinfeld, *Chem. Eng. Process.* **48** (2009), pp. 1279–1287.
84. A. Z'Graggen, P. Haueter, D. Trommer, M. Romero, J.C. de Jesus and A. Steinfeld, *International Journal of Hydrogen Energy*, **31**(6) (2006), pp. 797–811.
85. N. Piatkowski, C. Wieckert and A. Steinfeld, *Fuel Process. Technol.* **90** (2009), pp. 360–366.
86. B.J. Hathaway, J.H. Davidson and D.B. Kittelson, *J. Sol. Energy Eng.* **133** 021011 (2011).
87. J. Chen, Y. Lu, L. Guo, X. Zhang and P. Xiao, *Int. J. Hydrogen Energy* **35** (2010), pp. 7134–7141.

88. G. Maag, W. Lipiński and A. Steinfeld, *Int. J. Heat Mass Transf.* **52** (2009), pp. 4997–5004.
89. G. Maag, G. Zanganeh and A. Steinfeld, *Int. J. Hydrogen Energy* **34** (2009), pp. 7676–7685.
90. A. Kogan, M. Kogan and S. Barak, *Int. J. Hydrogen Energy* **30** (2005), pp. 35–43.
91. S. Abanades and G. Flamant, *Int. J. Hydrogen Energy* **32** (2007), pp. 1508–1515.
92. J.K. Dahl, K.J. Buechler, A.W. Weimer, A. Lewandowski and C. Bingham, *Int. J. Hydrogen Energy* **29** (2004), pp. 725–736.
93. J. Wyss, J. Martinek, M. Kerins, J.K. Dahl, A. Weimer, A. Lewandowski and C. Bingham, *Int. J. Chem. React. Eng.* **5** (2007), p. A69.
94. S. Rodat, S. Abanades and G. Flamant, *Int. J. Chem. React. Eng.* **8** (2010), p. A25.
95. R. Tamme, R. Buck, M. Epstein, U. Fisher and C. Sugarmen, Solar Upgrading of Fuels for Generation of Electricity. *Journal of Solar Energy Engineering* **123** (2001), pp. 160–163.
96. A. Wörner and R. Tamme, *Catal. Today* **46** (1998), pp. 165–174.
97. C. Agrafiotis, H. Von Storch, M. Roeb and C. Sattler, *Renew. Sustain. Energy Rev.* **29** (2014), pp. 656–682.
98. N. Ozalp and V. Shilapuram, *Int. J. Hydrogen Energy* **35** (2010), pp. 4484–4495.
99. A. Meier, N. Gremaud and A. Steinfeld, *Energy Convers. Manag.* **46** (2005), pp. 905–926.
100. A. Meier, E. Bonaldi, G.M. Cella, W. Lipinski, D. Wuillemin and R. Palumbo, *Energy* **29** (2004), pp. 811–821.
101. A. Meier, E. Bonaldi, G.M. Cella, W. Lipinski and D. Wuillemin, *Solar Energy* **80** (2006), pp. 1355–1362.
102. L. Matthews and W. Lipinski, *Energy* **45** (2012), pp. 900–907.
103. L. Reich, R. Bader, T. Simon and W. Lipiński, *Spec. Top. Rev. Porous Media* **6** (2015).
104. L. Reich, L. Melmoth, R. Gresham, T. Simon and W. Lipiński, *Proc. ASME 2015 Power & Energy Conf.* (2015), Paper No. PowerEnergy2015-49507.
105. G. Flamant, D. Hernandez and C. Bonet, *Solar Energy* **24** (1980), pp. 385–395.
106. V. Nikulshina, C. Gebald and A. Steinfeld, *Chem. Eng. J.* **146** (2009), pp. 244–248.
107. A. Imhof, *Sol. Energy Mater.* **24** (1991), pp. 733–741.
108. A. Steinfeld, A. Imhof and D. Mischler, *J. Sol. Energy Eng.* **114** (1992), pp. 171–174.
109. V. Nikulshina, M. Halmann and A. Steinfeld, *Energy & Fuels* **23** (2009), pp. 6207–6212.
110. R. Michalsky, P.H. Pfromm and A. Steinfeld, *Interface Focus* **5** (2015), pp. 20140084, 1–10.
111. M.E. Gálvez, M. Halmann and A. Steinfeld, *Ind. Eng. Chem. Res.* **46** (2007), pp. 2042–2046.

112. R. Michalsky, B.J. Parman, V. Amanor-Boadu and P.H. Pfromm, *Energy* **42** (2012), pp. 251–260.
113. J.P. Murray, A. Steinfeld and E.A. Fletcher, *Energy* **20** (1995), pp. 695–704.
114. M. Halmann, A. Frei and A. Steinfeld, *Miner. Process. Extr. Metall. Rev.* **32** (2011), pp. 247–266.
115. P.G. Loutzenhiser, O. Tuerk and A. Steinfeld, *JOM* **62** (2010), pp. 49–54.

Index

F

G

H

M

N

O

P

S

About the Contributors

Roman Bader received his B.Sc. (2007), M.Sc. (2008), and Dr. Sc. (2011) degrees in Mechanical Engineering from the Swiss Federal Institute of Technology (ETH Zurich). He has over 8 years of research experience with solar thermal technologies. From 2007 to 2012, he worked on the development of a novel parabolic trough solar field technology and a novel solar dish technology (www.airlightenergy.com). From 2012 to 2013, he was in charge of developing a prototype solar reactor for efficient solar thermochemical CO_2 and H_2O splitting at the University of Minnesota. Since 2013, he has been a Research Fellow in the Solar Thermal Group at the Australian National University. He has designed the ANU high-flux solar simulator. He is currently in charge of developing solar thermochemical processes and reactors for high-temperature energy storage and solar fuels production. He has authored over 30 scientific publications and over 40 scientific presentations. He has reviewed over 30 papers for over 10 journals, including *Solar Energy, Journal of Solar Energy Engineering*, and *Applied Energy*. He has acted as a session organizer at several international conferences. He is a member of ASME, AIChE, ISES, and the ANU Energy Change Institute.

Manfred Becker graduated from RWTH (Technical University) Aachen as Diplom-Ingenieur for Mechanical Engineering and obtained a Ph.D. there with an analysis of the transition regime conditions between continuum and free molecular flow. He had two leaves, both for one year in the United States at Arnold Engineering Development Center (AEDC), Tullahoma–Tennessee within an exchange program of the Departments of Defense of Germany and USA. The second was a post doc position at the University of California, Department of Mechanical Engineering. Dr. Becker also participated as a member of DLR in the project Small Solar Power Systems (SSPS) by the International Energy Agency including nine countries at Almeria, Spain. The objective was to define the potential of Concentrating Solar Power (CSP) by critically comparing, at a small scale, new concepts of tower and parabolic trough collectors. Professor Becker was appointed Head of the Energy Technology Division of DLR-Cologne. He acted in the Solar Power and Chemical Energy Systems (SolarPACES) project at first as operating agent for technology and later as the German member of the Executive Committee. He was Speaker for the chemistry division of the AG Solar in North Rhine–Westphalia and joined the European Union Renewable Energy Centers (EUREC) agency at Brussels. For five years up to 2001, he was member of the EUREC Executive Bureau. Professor Becker had numerous publications including scientific studies and reports both as author and editor. At the University of Wuppertal he gave lectures on Heat Transfer. In 2001, Becker retired from DLR and began consulting for KfW, VDI, GIZ and some solar companies in Germany. He received the Ernst Mach Award 1972 for his doctoral thesis and was honored by the SolarPACES Life Time Award.

Estefanía Caamaño-Martín was born in Madrid (Spain). She received her Ph.D. in Telecommunications Engineering — Photovoltaic Solar Energy Program from the Technical University of Madrid (1998). Since 2000, she has been an Assistant Professor and Researcher at the Solar Energy Institute of the same university, where she has developed her research career in the field of Photovoltaic Engineering. In particular, her research interests include PV systems design, characterization, performance evaluation, integration in urban environments and intelligent management of PV electricity generation. Dr. Caamaño-Martín has participated in more than 50 national and international Research & Development projects, in half of them as principal researcher. She is the author/co-author of 12 book chapters, 20 articles in international journals of high impact index, and more than 50 contributions to national and international scientific conferences related with Photovoltaics and Distributed generation. She is also a member of the Spanish standardization committees related to Electricity Supply systems and Photovoltaic Solar Energy systems. From a training perspective, she regularly teaches engineering and architecture students in degree courses and post-graduate specialization courses at the Universidad Politécnica de Madrid, in the field of Solar Energy systems.

Emilio Muñoz Cerón is a Researcher of the Centre for Advanced Studies in Energy and Environment at the University of Jaén. He graduated in Industrial Engineering in 2008, obtaining the Best Student Award of his promotion. During his Engineering studies, he spent a semester at the University of Texas at Austin and afterwards, in his Ph.D. research period, he was awarded scientific fellowships at the Research Centre for Smart Buildings and Energy Efficiency (CeDint-UPM, Madrid) and in Black-Photon Instruments GmbH (Fraunhofer ISE spin-off) in Freiburg, Germany. His international experience has been also focused on cooperation projects in developing countries, where he travelled to do field work in Bolivia, Peru and in the Saharan

Refugee Camps in Algeria. In 2014, he completed his Ph.D. at the University of Jaén, with Professors Juan Carlos Miñano and Gabino Almonacid as doctoral advisors. He obtained the International Doctorate mention with the thesis titled: "New trends in photovoltaic economy and development of a different technological spectral approach in concentrator photovoltaics". At the Centre for Advanced Studies in Energy and Environment, he is in charge of coordinating the scientific activity and the promotion of the Centre through International funding and projects. He is also in charge of the use and maintenance of the laboratories and equipment. Dr. Muñoz Cerón has published around 20 papers in peer-reviewed scientific journals, proceedings of international conferences and chapters of books.

Richard Corkish graduated from the Royal Melbourne Institute of Technology in 1986 with B.Eng. in Communications Engineering and researched satellite ground station antennas with the Commonwealth Scientific and Industrial Organisation for four years before studying for the Ph.D. (Electrical Eng.) in photovoltaics at University of New South Wales and graduating in 1994. He has been working in photovoltaics research and education, mainly at UNSW, since 1990. From 2003 to 2013, he was Head of School of Photovoltaic & Renewable Energy Engineering at UNSW. Dr. Corkish is currently Chief Operating officer of the Australian Centre for Advanced Photovoltaics, an Advisor to the Indian National Centre for Photovoltaics Research and Education, External Examiner for the Clean Energy Program at Singapore Polytechnic, and an Editor for the *IEEE Journal of Photovoltaics*.

Miguel A. Egido is a senior scientist with a Ph.D. in stand-alone PV systems and a minor in Telecommunication Engineering. All his research activity has been carried out at the Instituto de Energía Solar in Madrid and has focused on PV systems: solar radiation, sizing, installation, standards normalization and rural electrification. He has more than 20 years of experience in PV rural electrification projects and programs (in Senegal,

Cabo Verde, Brazil, Bolivia, Cuba, Nicaragua, Perú, Panamá, etc.). In the last few years, he has been working on Quality Control Assurance procedures for PV rural electrification. Dr. Egido has also taught about PV systems at the Universidad Politécnica de Madrid since 1988.

Hans-Martin Henning is Professor and Deputy Director of the Fraunhofer Institute for Solar Energy Systems (ISE) in Freiburg, Germany and Director of the division of Thermal Systems and Building. He is also Professor for "Technical energy systems" in the department of mechanical engineering at the Karlsruhe Institute of Technology. Professor Dr. Henning finished his Ph.D. in 1993 at the University of Oldenburg and started his work at Fraunhofer ISI in Freiburg in 1994. Key areas of his research are in the following topics: solar cooling and refrigeration, adsorption technology, heat storage based on sorption technology and phase change materials, heat pump technology and in general, technical energy systems for buildings. He has been responsible for many national and international research projects and in the years 1999 to 2011 he was the responsible Operating Agent for two large Tasks whicxh were carried out in the framework of the Solar Heating and Cooling programs of the International Energy Agency. Recently, Professor Henning has been actively involved in the development of computer models for the simulation of complex energy systems and their application to investigate future German energy systems including all conversion technologies and end-use sectors.

Wojciech Lipiński received his Magister Inżynier degree from the Warsaw University of Technology (2000), and doctorate (2004) and habilitation (2009) from ETH Zurich. He is an Associate Professor and the Leader of the Solar Thermal Group at the Australian National University, and a Privatdozent at ETH Zurich. He previously held Senior Research Associate and University Lecturer positions in the Department of Mechanical and Process Engineering at ETH Zurich, and an

Assistant Professor position in the Department of Mechanical Engineering at the University of Minnesota. His research interests are in thermal and thermochemical sciences, optics, and energy, environmental and space technology applications. Lipiński has published more than 100 articles in peer-reviewed journals and conference proceedings, and contributed to several books. He was awarded the 2006 Hilti Award for Innovative Research from ETH Zurich, the College of Science and Engineering 2010–2011 Outstanding Professor Award from the University of Minnesota, and the 2013 Elsevier/JQSRT Raymond Viskanta Award in Radiative Transfer. He has served as the Associate Editor for the *ASME Journal of Solar Energy Engineering*, and the Elected Member of the Scientific Council of the International Centre for Heat and Mass Transfer.

Daniel Mugnier has professional experience in engineering solar thermal systems for large DHW applications and above all solar heating and cooling systems. Managing the Solar Cooling Department of TECSOL — one of the French leading solar engineering company — Daniel Mugnier is involved as well in numerous R&D projects on solar cooling at the national, European and international level. He is also author of several publications and presentations at international conferences on solar cooling. TECSOL has achieved more than 50 feasibility studies on solar cooling and designed 10 working installations at the moment since 1990 on solar heating and cooling. He is currently Vice Chairman of the European Solar Thermal Technology Platform and Operating Agent of the IEA Solar Heating and Cooling Programme.

Professor Brian Norton is President of Dublin Institute of Technology (DIT) in Ireland. He received a B.Sc. in Physics from the University of Nottingham in the UK in Physics and MSc and Ph.D. in Energy Engineering from Cranfield University also in the UK. He has also received a D.Sc. from University of Nottingham. He is a Chartered Engineer both in the UK and Ireland. Professor Norton is the author or co-author of nine books, over 420 papers principally in solar energy research including 170 in major international learned journals. His work has ranged from solar water, air and space heating, optical and luminescent solar energy concentration, vacuum glazing, photovoltaic systems design, energy storage and solar drying. Currently Professor Norton is elected to the Council of Dublin Chamber of Commerce. He is a Director of The Greenway, Ireland's first green technology corridor, chairs "Action Renewables" responsible for facilitating renewable energy development in Northern Ireland and is Vice-President of the European Sustainable Energy innovation Alliance. He is a Fellow of the Energy Institute, Engineers Ireland and the Higher Education Academy. Among his awards are the Napier Shaw Medal of the Chartered Institute of Building Services Engineers and the Roscoe Award of the Institute of Energy. He is an Honorary Fellow of the Chartered Institute of Building Services Engineers and is a Fellow of the Irish Academy of Engineering. He is an Honorary Professor of the University of Ulster, Harbin Institute of Technology, China and University of Houston, USA.

Robert John Patterson graduated from the Engineering Physics program at Queen's University in Ontario, Canada in 2001. He then worked for a company fabricating TEM samples by ion beam micromachining for failure analysis of semiconductor devices and corrosion of superalloys. He went on to do an M.Sc. in Physics at the University of Alberta in Canada studying colossal magnetoresistance and magnetic phase changes in doped transition metal oxides, graduating in 2005. After two years in the department of physics at the University of British Columbia in Vancouver Canada, Dr. Patterson decided to move to renewable energy. He joined the School of Photovoltaic and Renewable Energy Engineering (SPREE) at the University of New South Wales (UNSW) in Sydney, Australia where he studied thermalization energy loss via numerical modelling of the vibrational properties of nanostructured materials for application to advanced photovoltaic cells. He graduated with a Ph.D. in 2012 and is currently an ARENA postdoctoral fellow at the Australian Centre for Advanced Photovoltaics in SPREE at the UNSW.

Univ.-Prof. Dr.-Ing. Robert Pitz-Paal has been one of two directors heading the Institute of Solar Research with about 120 members of staff since June 2011. His main research areas are the technical analysis and optimization of concentrating solar power systems for generating electricity and producing fuel. A challenge he finds particularly exciting is that of coupled heat transfer problems involving radiation, convection, conduction and chemical reactions. He has published more than 100 papers, book chapters and conference contributions out of which more than 60 are published in ISI journals. Born in Kleve in 1963, Pitz-Paal read Physics at the Ludwig–Maximilian University in Munich from 1983 until 1988. In 1993, he obtained a Ph.D. from the Department of Mechanical Engineering at Ruhr Universität Bochum with a doctoral thesis on concentrating solar technology. In the same year he was recruited by the German Aerospace Center in Cologne where he spent the following years working in various positions at the Energy Technology Division. In 2002, Pitz-Paal was appointed as head of Solar Research at the DLR Institute of Technical Thermodynamics in Stuttgart, and held the post of Deputy Director of the Institute from 2009 until 2010. Since 2003 Dr. Pitz-Paal has been a professor for solar technology at RWTH Aachen. He also spent the summer semester of 2008 as a guest professor at the Institute of Energy Technology at ETH Zürich. His scientific publications won several awards from the American Society of Mechanical Engineers (ASME). His department was awarded Centre of Excellence status by the DLR Executive Board in 2006 and 2011 and 2014 in recognition of its work. Professor Pitz-Paal is vice chairman of the executive committee of the international network "Solar Power and Chemical Energy Systems (SolarPACES)". He is also chairman of the CSP (=Concentrated Solar Power) working group of the European Academies of Science Advisory Council as well as heading the Board of Trustees of the Jülich Solar Institute of the Aachen University of Applied Sciences and is member of the board of the German association of the CSP Industry (DCSP). He served as associate editor for the *ASME Journal on Solar Energy Engineering* and is actually associated editor of *Journal of Solar Energy.*

Gabino Almonacid Puche is Professor in the Department of Electronic and Automatic Engineering and Director of the Centre for Advanced Studies in Energy and Environment at the University of Jaén. He graduated in Telecommunications Engineering at the Polytechnic University of Madrid (UPM) in 1978. In 1987, he completed his Ph.D. at UPM with Professor Antonio Luque as doctoral advisor. At the University of Granada he has been Director of the Polytechnic High School (1986–1990), and at the University of Jaén he has been Head of the Technology Transfer Office (1993–1997) and Head of the Department of Electronic and Automatic Engineering (1999–2005). He is currently member of the Management Team of ELAND Private Equity, member of the *Instituto de Estudios Giennenses* and member of the Renewable Energy Institute of Andalusia. He established, in 1989, the Solar Energy Research Group of the University of Jaén. This research group comprises more than 50 researchers who have made relevant contributions over the past 25 years to engineering and to technological development of PV systems. Two of his most outstanding R&D projects are (a) UniVer Project (a 200 kW PV generator integrated into the campus of the University of Jaén) and (b) REBUILD Project (EU pilot project for renewable energies integrated into the historical city centers of Amsterdam, Coimbra, Corfú, Jaén, Perugia, Prato and Rhodes). He has published around 50 articles in peer-reviewed scientific journals and almost a hundred papers in the proceedings of international conferences. In 2001, he was awarded the *Distinción San Alberto Magno* of the Polytechnic High School of Jaén, a special honor awarded to scholars who have made outstanding contributions to sciences.

Peter K.J. Robertson is Professor of Chemical Engineering at Queen's University Belfast. He originally studied Chemistry and Economics at the University of Ulster, where he also obtained his D.Phil (1989) and D.Sc. (2013). On completing his D.Phil in 1989, Professor Robertson joined the Faraday Centre a spinout company from Carlow Institute of Technology in Ireland. In March 1991, he moved to the Northern Ireland Civil Service as a Higher Scientific Officer attached to the Industrial Research and Technology Unit. Professor Robertson joined the staff of the School of Applied Sciences at Robert Gordon University in September 1995 where he was involved in teaching and research in environmental science and technology, analytical science and photocatalysis. In July 2000 Professor Robertson was appointed to the Chair of Energy and Environmental Engineering in the School of Mechanical and Offshore Engineering at RGU where he established the multidisciplinary Centre for Research in Energy and the Environment. Professor Robertson subsequently was appointed as Vice-Principal and Pro Vice-Chancellor (Research and Commercialization) at Robert Gordon University in September 2006. In this role he had responsible for setting and implementing the Research, Knowledge Transfer and Commercialisation Strategy for the University. In August 2012 his role at RGU was redefined as Vice-Principal and Pro Vice-Chancellor (Research and Academic Support Services). In January 2015 Prof. Robertson joined the School of Chemistry and Chemical Engineering at Queen's University Belfast as Professor of Chemical Engineering. His main research interests are in semiconductor photochemistry for energy and environmental applications. He has co-edited two books and authored over 90 peer reviewed journal articles and book chapters. Professor Robertson was a non-executive director of Gas2 Ltd (2012–2014) and the Aberdeen Renewable Energy Group (2006–2009). In 2013 he was appointed to the Detao Masters Academy in Beijing as a Master in Photocatalytic Technology. Professor Robertson is a member of the editorial board of the *International Journal of Photoenergy*, the *Central European Journal of Engineering* and Invited Editor of *The Handbook of Environmental Chemistry* Series Published by Springer Verlag, specifically co-editor of the Volumes on Environmental Photochemistry. Professor Robertson is a Chartered Engineer, Chartered Energy Engineer, Chartered Scientist and Chartered Chemist, a Fellow of the Royal Society of Chemistry, the Institute of Chemistry in Ireland, the Energy Institute and an Associate Fellow of the Institution of Chemical Engineers.

Ignacio Rey-Stolle is a Professor at the Solar Energy Institute of the Technical University of Madrid (Madrid, Spain) since 2003. He got his Ph.D. in 2001 from the Technical University of Madrid working on high concentrator GaAs solar cells. From his Ph.D. and on, Prof. Rey-Stolle has accumulated more than 18 years working in the field of Photovoltaic Solar Energy. In particular, the research efforts of Prof. Rey-Stolle are focused on III–V multi-junction solar cells doing material growth by MOVPE, device design and simulation and characterization and reliability studies. This work has resulted in several world-record efficiency solar cell devices that have been produced at Prof. Rey-Stolle's research group. He has co-authored more than 100 scientific papers, three book chapters, has co-edited the *Handbook on Concentrator Photovoltaic Technology*, holds one patent and has given numerous talks at Research Institutes and Photovoltaic Conferences worldwide. He has been the principal investigator of eleven national and international research projects, has participated as an associate researcher in more than twenty projects, and has coordinated three technology transfer projects to photovoltaic industries. In 2007, Prof. Rey-Stolle received the *Best Young Professor Award* of the Technical University of Madrid for outstanding scientific research and innovation. He teaches Electronics and Photovoltaic Solar Energy at the Technical University of Madrid, where he also supervises graduate student research. Since 2011, Prof. Rey-Stolle is the director of the *Master of Science on Photovoltaic Solar Energy* of the Technical University of Madrid, a leading international Master program designed to meet the intense demand for highly qualified professionals in the area of photovoltaic engineering.

Nathan Skillen is a Research Fellow in the School of Chemistry and Chemical Engineering at Queen's University Belfast (QUB). He obtained his B.Sc. in Biology with Molecular Biosciences from the Robert Gordon University in Aberdeen (2009). In December 2009, he joined the Centre for Research in Energy and the Environment (CRE+E) at RGU as a Ph.D. student under the supervision of Professor Robertson. His Ph.D.

project was part of an Engineering and Physical Science Research Council (EPSRC) funded collaborative project entitled "Enhancing Direct Photoelectrochemical Conversion of Carbon Dioxide". During this period, Dr. Skillen collaborated with the University of St. Andrews and the California Institute of Technology while developing a photocatalytic system that was capable of generating alternative fuel using novel catalysts and solar illumination. After completing his Ph.D. (2013), he continued to work with Professor Robertson and CRE+E as a Research Fellow on the EPSRC funded project entitled "Photocatalytic Bioethanol Production". In this project, he is responsible for the design and construction of multi-stage photocatalytic systems which incorporate membrane technology. In February 2015, Dr. Skillen joined the School of Chemistry and Chemical Engineering at QUB, continuing his work on the EPSRC bioethanol project. Under his current role, Dr. Skillen's main responsibilities are research based along with the supervision of bachelor and masters students during their projects. His main research interestes are in the application of photocatalytic technology for use in energy production and environmental processes inlcuding the investigation of reactor engineering, semiconductor materials, catalyst platform and solar energy conversion. He has authored and co-authored peer reviewed journal articles and a book chapter in the fields of photocatalytic water splitting, photo-reactor engineering and photocatalytic applications. He is also an Associate Member of the Royal Society of Chemistry.

Wes Stein is the Program Leader for Solar Technology at CSIRO (Commonwealth Scientific and Industrial Research Organisation), Australia. He leads a team of 45 engineers and scientists with the majority of the work focusing on high temperature concentrating solar power (CSP). This includes optics, thermodynamics, materials, control systems, power cycles and modelling in order to optimize the overall performance. He has also established CSIRO's National Solar Energy Centre which is a pilot scale research facility comprising two high flux solar towers, thermal storage and balance of plant. Wes has a strong background in power station technology and a particular interest to merge targeted advances in CSP with state-of-the-art power cycles to provide competitive, market-scale electricity and solar fuels. To this end, he has grown and manages an extensive portfolio of CSP research projects, including optimization of central receivers and optics, high temperature steam and thermal storage, solar air Brayton systems, advanced supercritical CO_2 cycles and solarized fuels. Technology developed by his team has been deployed in the form of a solar heliostat field for the Cyprus Institute, and in other international projects. Wes has championed the development and implementation of CSP technology in Australia and continues to provide leadership, strategic R&D, demonstration projects and innovative energy strategies that foster the development of a national and global sustainable energy industry. He is a member of the Australian Renewable Energy Technology Advisory Committee, international member of EU-SOLARIS Advisory Board for Technical and International Cooperation, Australian member of the IEA's SolarPACES Executive Committee (Solar Power and Chemical Systems), member of the Australian Solar Thermal Institute Project Management Committee and member of UNIDO International Solar Energy Centre for Technology Promotion and Transfer. He was a Lead Author for the IPCC Special Report into Renewable Energy, 2011, co-authored a study for the World Bank on the global Status and Opportunities for Concentrating Solar Power and co-editor for Concentrating Solar Power Technology (Woodhead Publishing 2012). He has been awarded an Australian Institution of Engineers scholarship to develop high temperature solar cycles, and a Confederation of British Industry scholarship to investigate solar thermal hybridization of coal and gas-fired power cycles, and won the UK Rendel Award for his paper on "Global Energy Issues for the next 30 years".

Jorge Solórzano is a young researcher with a Ph.D. and Master's degree in photovoltaic solar energy, specializing in Photovoltaic Systems. Both of his degrees are from the Polytechnical University of Madrid. He is an engineer with more than 8 years of experience in photovoltaics. He has published various articles in JCR indexed journals and presented papers at international congresses. His areas of research are mainly: optimization of PV generation through DMPPT techniques (microinverters and power optimizers), quality control of PV systems, PV generation forecasting and optimization of mini-grids. He is also a lecturer in three different Masters programs related to renewable energy.

Anthony Turhollow is a staff economist in the Environmental Sciences Division of Oak Ridge National Laboratory (1982–1993, 1998–present). He received a Ph.D. in Agricultural Economics from Iowa State University in 1982 and has been working in the field of biomass for energy since 1980. His particular expertise is in the economics (costs), logistics, energy and greenhouse balances, and resource assessments of biomass energy crops. From 1983 to 1993, he managed herbaceous energy crop screening and selection research and rapeseed research. He was one of the main authors on the 2005 Billion Ton Report and 2011 Billion Ton Update. Economics analyses include estimating the costs of producing herbaceous and short rotation woody energy crops. He has estimated costs associated with the harvest, handling, transport and storage of energy crops and crop residues. He has served on a number of US Department of Energy and US Department of Agriculture panels to review and select biomass-related research projects for funding.

Pedro Gómez Vidal is currently Professor and Director of the Department of Electric Engineering at the University of Jaén, Spain. He was Director of the Polytechnic High School of the University of Jaén from 1999 to 2006. Professor Gómez Vidal graduated in Industrial Engineering at the Engineering School of the University of Seville in 1982. In 2000, he completed his Ph.D. at the University of Jaén with Professor Gabino Almonacid as doctoral advisor. He is member of the Solar Energy Research Group of the University of Jaén. His research focuses on the areas of electrical hazards, safety and protection as well as on grid-connection of PV systems. In 1999, he first introduced floating DC systems along with new concepts and measures for improving safety and protection in PV systems. Professor Gomez Vidal has published over 25 articles in peer-reviewed scientific journals and over 60 papers in the proceedings of international conferences.